3-5 main arguments
clearly state each w/ supporting
explanation, evidence, and example
in 2-5 complete grammatical sentences

scoring
5 - excellent work, demonstrates deep
understanding of the readings
4 Very good work, demonstrates
understanding of the readings
3.5 acceptable work, demonstrates some
understanding of the readings.
3.0 Unacceptable work, demonstrates
lack of understanding of the
readings.
0 Works is missing OR demonstrates
little or no understanding of the
readings

INVENTING FOR THE ENVIRONMENT

LEMELSON CENTER STUDIES IN INVENTION AND INNOVATION

Arthur Molella and Joyce Bedi, editors, *Inventing for the Environment*

INVENTING FOR THE ENVIRONMENT

EDITED BY ARTHUR MOLELLA AND JOYCE BEDI

THE MIT PRESS
CAMBRIDGE, MASSACHUSETTS
LONDON, ENGLAND

IN ASSOCIATION WITH

THE LEMELSON CENTER
SMITHSONIAN INSTITUTION
WASHINGTON, D.C.

Set in Engravers Gothic and Bembo by Achorn Graphic Services. Printed and bound in the United States of America.

Library of Congress Cataloging-in-Publication Data

Inventing for the environment / edited by Arthur Molella and Joyce Bedi.
p. cm. — Lemelson Center studies in invention and innovation)
Includes bibliographical references and index.
ISBN 0-262-13427-6 (hc : alk paper)
1. Technological innovations—Environmental aspects. 2. Technology—Environmental aspects. 3. Ecology. I. Molella, Arthur P., 1944–. II. Bedi, Joyce. III. Series.

T173.8 .I6185   2003
363.7—dc21

2002035775

10 9 8 7 6 5 4 3 2 1

CONTENTS

FOREWORD

ERIC LEMELSON

The genesis of *Inventing for the Environment* was a conversation I had in 1997 with Arthur Molella, Director of the Lemelson Center for the Study of Invention and Innovation. Art and I were discussing alternative energy technologies and the effect of new technologies on sustainable development. I suggested that the Lemelson Center consider exploring the complex relationship among invention, innovation, and the environment.

To many people, the word "environment" brings to mind images of untouched, wild nature, of remote landscapes separated from daily life that are (hopefully) protected and enjoyed for their recreational and spiritual values. Other people might think of pollution and related issues such as human health and government regulation. The words "technology" and "innovation," on the other hand, suggest to many people the gadgets that permeate modern life—the personal computer, the cell phone, the fax machine, and thousands of other devices. When I consider the words "technology" and "environment," my background as an environmental lawyer often leads me to think of how technological innovation might help solve many of the myriad threats to the global environment that face us in this new century.

For several decades, we have been taught by the environmental movement and by scientists that we're not separate from our environment. In an increasingly urbanized world, the practical implication of this revelation is that for many people "the environment" means the city and areas surrounding it. Similarly, limiting our thinking to the ubiquitous products of the electronic and information revolutions ignores the importance of other technologies, such as architecture, that shape the built environment of the urban landscape.

The essays in this volume demonstrate the importance of viewing this complex subject from a broad perspective. Human beings have used technology since the dawn of history to shape and craft the environment we live in. We are entering an era in human history when technology has the potential to supply us with plentiful, clean energy from the sun and from other sources, such as hydrogen. The scale of the environmental problems we face challenges us to apply human ingenuity (the basis of technological innovation), to use resources more efficiently and equitably, to reconsider our relationship to the natural world, and to reduce the impact of our species on the biosphere. Readers of this volume will find new and unusual perspectives that will affect the way they think about technology and the environment.

The Jerome and Dorothy Lemelson Center for the Study of Invention and Innovation, based at the Smithsonian's National Museum of American History, was founded on the simple belief that history matters. It is our mission to enhance public understanding of the creative processes involving invention and innovation and to examine these processes in a broad historical context. Guiding all our activities, from symposia and museum exhibits to school programs and book projects like this, is the conviction that historians and innovators have much to learn from each other. We believe that if we bring these two groups together in an informed dialogue on issues of common interest, significant and unexpected findings will emerge.

This strategy seemed particularly appropriate in the case of environment-related inventions, where so much of current practice is based on assessments of past conditions and patterns of change. When we began to explore environmental topics, we were struck by the increasing role, since the nineteenth century at least, of innovative technologies and methods, including the invention of whole new fields such as public health and industrial ecology. In addition, environmental activities are complex and inherently collaborative, involving contributors from many different fields, unified in a common goal of improving the human condition. Coming to grips with such a complex set of activities and approaches requires a broad interdisciplinary perspective. Hence, this volume draws upon the expertise of a wide variety of specialists, including environmental, science, technology, and business historians as well as engineers, scientists, public health experts, architects, and town planners. The subject of invention provides the unifying theme, highlighting contributions from the creative fronts of disparate disciplines.

The main questions raised in this volume grew out of a year-long interdisciplinary program series sponsored by the Lemelson Center in 1998 with generous support from the Lemelson Foundation and AT&T, which also collaborated with us in the organization of these events. Throughout that year, we addressed questions about how invention may help—or sometimes unintentionally harm—the environment, recognizing from the outset that inventions are not socially neutral. In addition to issues of benefit and detriment, we considered the implied social arrangements of environmental inventions that promote the status quo or seek to forge a new order. Advocating the use of solar energy or alternative building materials like straw bales, for example, carries with it a call to restructure society as presently conceived. Siting photovoltaics on rooftops and making each home its own power plant eliminates the need for centralized generation and distribution of electricity, thereby providing enormous flexibility in housing patterns. The New Town phenomenon, both today and in the past, predicates a new community paradigm on environmental innovation. Even technologies that improve existing ones, like catalytic converters that make cars more fuel efficient and less polluting, have a broad range of consequences, from the larger economic effects of retooling factories down to the transportation choices made by individual commuters. Inventing for the environment, therefore, includes changing not only technology but also the day-to-day way of life of millions of people. We defined "the environment" in the broadest sense—in terms of the interaction of humans and "nature"—arguing that it is impossible to separate human from natural systems. It is this synthetic approach that guides this volume.

"nature"

The authors were drawn from the lecture series, the symposium, and the historical tours that were offered by the Lemelson Center. Each part of the book focuses on a question about applying invention to environmental issues. In an attempt to answer these questions, each part features two essays, one by a historian and the other by a practitioner, designed to present a balanced dialogue between history and current practice. The "Portraits of Innovation" highlight individuals whose inventive energies have made significant improvements in the environment.

historian
practitioner
innovator

The essays represent what we believe to be among the most innovative areas of current environmental practice. They explore topics in environmental history, issues of public policy, and examples of technological innovation to question how inventions have affected and can affect the environment. Each aims to take an innovative approach to understanding the interconnections of human and natural systems. Thus, the contents of the book lead from discussions of nature itself, through the built environment, to more specific technologies in such areas as public health and energy. To bring these ideas together, we conclude with an examination of applications of the principles of industrial ecology.

Mixing practitioners and historians is the key, since, as has already been noted, the very concept of the environment is deeply embedded in time and change. Statements about the environment are inevitably teleological and relative, measuring present and future conditions against the past. Advocacy positions, both for and against specific environmental policies and reforms, typically invoke the authority of historical precedent. Defenders of the automobile, for example, point out that the internal-combustion engine, for all its harmful effects on the air we breathe, actually helped to improve urban environments, previously befouled by horses. Champions of alternative energy sources call our attention to neglected stories of roads not taken in such technologies as wind, solar, and tidal energy. Historical examples can provide significant lessons for the present, sometimes even leading to the rediscovery of an old technology, as in the revival of straw-bale construction described in one of our essays. As it explores the history of "inventing for [the benefit of] the environment," it is hoped that this book will indeed put the past to use for the common good.

For some time now, the specialty of environmental history has been a growth industry within the field of history, but the role of invention in that story is still relatively unexplored. When technological invention is introduced, it is often with reference to technogenic problems, such as those associated with nuclear energy or the internal-combustion engine; if offered as a solution, it is usually in the simplistic terms of the technological fix. Rarely has it been examined critically or from multiple perspectives. Perhaps one reason for such one-dimensional interpretations is that invention itself has been viewed too restrictively as a gadget-based approach to technological improvement. When the definition of invention is broadened to include not only mechanical devices but also complex innovative processes of all sorts, social as well as technological, the possibilities expand dramatically.

Despite the interdisciplinary nature of the subject, studies of the environment that cut across disciplinary lines are still relatively rare. The majority of books in both environmental studies and environmental history deal with a single facet of technological or social studies. For example, books on alternative energy sources or the conservation movement abound. In contrast, *Inventing for the Environment* fills a need to cross specialists' boundaries, bring together a range of expertise, and assess the relationships among technologies and philosophies.

The diverse perspectives represented in this book suggest a sense of integration and unification that forms an ecological mindset once popularly known as holism. Not only must the relationship between technology and the natural world be looked at holistically; as Richard White and Steven Pyne argue, this must be done with a recognition that there may be no distinction between the natural and the artificial, between nature and human culture, in the first place. Simply put, technology is not separate from nature. History shows that the distinctions that humans and societies draw between the two are themselves cultural artifacts that have often been politically and ideologically based. But, as a number of papers in this volume argue, once the porosity of the boundary is admitted, all kinds of inventive possibilities open up. Rejecting bipolar concepts of nature and culture can allow for more interesting, seemingly paradoxical strategies, such as those offered by the new field of industrial ecology. Most of all, the integration of nature and technology widens the field of play for the creative imagination, encouraging inventive solutions that view technological society in the broadest ecological terms—and that is what *Inventing for the Environment* is about.

The Lemelson Center gratefully acknowledges the generous support of the Lemelson Foundation and AT&T in the production of *Inventing for the Environment*.

THOMAS LOVEJOY

Most of us in the environmental field would not like to admit it, but at least part of the time we are perceived to have a negative agenda. "Don't do this." "Stop that." "You can't do that."

In reality there is much more to it, for the environmental agenda is fundamentally a positive one about conserving opportunity. It also is, at heart, about stimulating human creativity and finding better ways to fulfill aspirations. It is much too simplistic to see the environmental challenge as the negative effects of what E. E. Cummings called the "world of the made" on the "world of the born." It is also about how the "world of the made," which includes ideas as well as the tangible, can contribute to a better environmental future. It is about inventing for the environment.

Sometimes it involves looking at a problem in an new way. When I was young, New York City was famous for the quality of its water emanating from the Catskill Mountains. It even won in blind testings against elegant bottled water from Europe. In subsequent decades, the watershed deteriorated in quality until the US Environmental Protection Agency was about to require the city to invest $6 billion in a water treatment plant. Someone was bright enough to figure out that it might be cheaper, and a permanent solution (not requiring infrastructure replacement 30 or 40 years later), to restore the watershed to the point where its biological diversity could deliver the high-quality water of my childhood. That was the ultimate solution, paid for by a $600 million bond issue, which bought easements, critical pieces of the watershed. The "green" solution cost an order of magnitude less than the technological one. There is renewed interest in watersheds, which the former mayor of Quito and current president of Ecuador terms "water factories."

Sometimes the trick is rediscovering old ways of looking at a problem. In Bolivia's altiplano, agriculture has always been difficult because of little rain and nighttime temperatures frequently dropping below 0°C. As part of the archeological investigations of pre-Incan Tiahuanaco, near Lake Titicaca, Alan Kolata and his colleagues re-created the raised bed agriculture system of that ancient time complete with surrounding water-filled trenches. The water modifies the microclimate and blunts the effects of frost: an ancient system as effective today as then.

In other instances, the solution involves very modern technology. The need to reduce greenhouse gas emissions poses a major challenge to the energy industry. Some leaders of that industry, including John Brown of BP, have the prescience to recognize that their companies are in the energy business, not just the fossil-fuel business. Energy solutions encompass energy conservation and energy efficiency; however, they also have to include alternate energy such as hydrogen fuel cells, as well as solar and wind sources.

One major element in inventing for the environment will be an increased use of biological resources and processes, which are inherently better because they are, by definition, biodegradable. One example of this is bioremediation, the use of enzymes and organisms from nature to clean up wastes—toxic substances, oil spills, and the like. This takes advantage of the early history of life on Earth, which experienced widespread harsh environments physically different from most of those found on the planet today. The consequence is a variety of contemporary microorganisms with weird metabolisms and appetites which science has been discovering in extreme environments over the last 20 years. One example is a bacterium, first found in the sediments of the Potomac River, that can reduce iron compounds. It also turns out to be able to break down chlorofluorocarbons, the man-made molecules that are so destructive of the ozone layer. Other microbes break down hydrocarbons in oil spills or remove heavy metals in the environment.

Some assert that bioremediation has a modest future at best because industry will soon operate in ways that will not release toxic waste into the environment. This view ignores the reality that, in part, that goal will be achieved by bringing bioremediation into the factory. This also ignores the subset of bioremediation that works as bioconcentration, in which biological processes remove valuable items such as gold from effluent and concentrate them so they can be reused. Each of these remedial strategies will play a part in achieving the dream of industrial ecology in which the waste stream of one industry becomes the feedstock for another.

Technology is obviously a central element in inventing for the environment, but it is also essential to realize that no technology is necessarily an unalloyed blessing, that technology is essentially neutral, and it can be used for good or for ill.

The current state of biotechnology is a case in point. Increasing pest resistance in agricultural crops through genetic engineering would seem an exciting new achievement, but the recent evidence that corn pollen with Bt genes (from *Bacillus thuringiensis,* a biocontrol agent) can produce substantial mortality in monarch butterfly caterpillars certainly came as a surprise to most. Little work has been done so far to understand the impact on soil biodiversity (a basic contributor to soil fertility) of treating plants possessing herbicide resistance genes with commercially available herbicides such as Roundup. Perhaps the potential exists to inadvertently create new super weeds if certain genes from genetically engineered plants get into wild populations. At the same time, I believe the challenge of feeding the human population while minimizing further destruction of natural habitat and biodiversity will require important contributions from biotechnology.

Another form of inventing for the environment recognizes that a substantial portion of environmental travails comes from activities which are managed or decided as if in isolation from one another but which, when added together, constitute a serious problem. An example is the degradation of the South Florida ecosystem through individual decisions affecting water. The combined effect led to a reduction of 25–50 percent in the annual sheet flow of water.

The solution for or the prevention of such problems comes from inventing a form of integrated decision making. The habitat conservation plans promulgated by the US Department of the Interior and the State of California are cases in point. They involve wide participation in land-use planning and decision making. Another case in point is the current InterAmerican Development Bank project for the Darien region of Panama. The latter project recognizes that road building in the tropics has almost always been followed by spontaneous colonization and deforestation and that the interest in the Darien was for a better connection with Panama City rather than in a road through the Darien Gap to Colombia. The IADB project is a package of elements, including land titling for present residents, improvements in health services, and education, all intended to stabilize the existing population and to enable it to resist illegal and spontaneous colonization before any road improvement takes place. Roads should not be

isolated projects; they should be elements of sustainable-development pack-
ages. This is essentially a new way of thinking.

Essential to inventing for the environment will be a continuing capacity
for evaluation and assessment. In the management of natural resources, this
is thought of as adaptive management—that is, as designing natural-
resource projects as if they are scientific experiments so that one can critically evaluate their successes and failures and, if need be, make mid-course
adjustments.

Such a capacity for evaluation should not be confined to natural-
resource projects. In the late 1950s, the Connecticut city of New Haven
was famous as a pioneer in urban renewal, and its mayor, Richard Lee, was
asked to explain this new element in urban life all over the United States.
New Haven's first effort included a "connector" superhighway that replaced
a poor neighborhood. It overlooked the social impact of population dis-
placement. Today social assessment would be a part of the design process.

We should not be afraid of learning by doing. That, of course, is not an
excuse to do just anything or to be unthoughtful in the design stage, but it
does recognize that the challenges are complex and that it is smart to keep
an eye out for surprises.

Central to inventing for the environment will be linking economics in
new ways, much as in the example of the New York watershed. For example, the former mayor of Quito, Roque Sevilla, was confronted with the
effect of heavy rains on storm sewers, exacerbated by residential creep up
the slopes of the surrounding mountainsides. He calculated that it would be
dramatically cheaper to buy out those properties and create a green zone
above a certain altitude than to build a new sewer system. Quito is also confronted with severe air pollution, especially from diesel buses, and the most
economic public-transportation solution for this city (44 kilometers long
but only 3–6 km wide) was not a subway or a rail system but electric buses
with their own right of way. (The latter idea was borrowed from the great
urban innovator Jaime Lerner of Curitiba in Brazil's Panama state.)

One of the great innovations using economics was the initiation of emissions trading under the Clean Air Act in the United States. At first regarded
with some suspicion that it conferred a "right to pollute," emissions trading
allows the owner of an old-style plant that exceeds its emissions limit to offset the excesses by buying unused emission rights from an ultra-modern
plant. Not only did this lead to dramatic reductions in sulfur dioxide emis-

sions and acid rain; by using market forces, it did so at approximately 5 percent of the initial estimated cost of the solution.

Today, emissions trading relating to carbon and greenhouse gases is being considered on a worldwide scale under the Kyoto Protocol of the Climate Convention. While there is considerable complexity and (as yet) lack of agreement about this "Clean Development Mechanism," it is essential to making rapid progress in reduction of greenhouse gases (which emanate both from the burning of fossil fuels and from deforestation).

Inventing for the environment as outlined in this volume is absolutely central to meeting the environmental challenge. It must include the best of the old, the best of the new, and the ability to learn from what we do. It must range in scale from the particular to the global. It should transcend borders: the "peace park" between Ecuador and Peru is an exciting example. It must succeed: the well-being of future generations depends on it. And it draws on one of the best attributes of our species, namely responding with creativity to challenges. There can be no greater challenge. Are we up to it? In one sense, it is just a matter of wanting to be.

ON NATURE AND TECHNOLOGY

TEMPERED DREAMS

RICHARD WHITE

I come from a discipline, history, whose whole orientation is retrospective. Except on the grandest of scales, neither I nor most of my colleagues can offer much in the way of predictions that anyone who cares to examine our track record will care to trust.

History's strengths are more humble; they are context and contingency, and they are an odd pair. For historians context equals historicizing. It is their basic enterprise. It is about what particular pieces are, and are not, on the table at any given time, how they got there, and, just as critically, the relationships between them. It is hard to play the game if you don't know the rules and you don't know the score. Context is about the rules and the score. But context is also about where we, as both pieces and players, fit in. To fail to see yourself in historical context is to badly mistake where you are and how you got there. Knowing where you are and how you got there is, at the very least, helpful in figuring out how to get where you are going and, perhaps, even in deciding where you want to go.

Unfortunately, all the context in the world doesn't explain tomorrow, which is where you always end up. Tomorrow is as much about contingency as context. Rules change. New pieces appear. The world is not, of course, totally remade, but sometimes events do take off in a dramatic new direction. Ecologists and historians are the academics whose disciplines most appreciate contingency. Things are connected; thus, after something happens, most things that follow are different. The film *It's a Wonderful Life* is about contingency. It became Stephen Gould's metaphor for evolution. Contingency is the stock of the historical trade.[1]

Contingency depends on scale. Historically, the smaller the scale, the greater the amount of contingency. Our individual lives are more contingent than the national life, but even the national life can shift because of relatively small events. On the grandest scale, large meteors crashing into the Earth, with the extinctions that followed, are examples of single events that changed everything, but they are rare. Because contingency varies with scale, and because larger structures tend to endure longer, attempts to plan for and structure the future, while always difficult, do have efficacy.

History, with its emphasis on context and contingency, is not a crucial component for those disciplines or political movements that believe they know a set of universal rules that always apply. Contingency hardly matters in a world of inviolable rules. For modern secular fundamentalists who believe that the market knows best or that nature knows best, history matters largely as a set of just-so stories.

The problem with history for fundamentalists of any stripe is that it historicizes the very entities that are supposedly generating the universal rules. Because certain strains of environmentalism embrace a fundamentalist assumption that nature knows best, environmental history has a very edgy relationship with environmentalism. It historicizes not just environmentalism, but nature itself. For my generation of environmental historians, at least, environmental history has been pretty much a one-trick horse. Although we may have only one trick, it is a good trick and we will keep on using it: Where others see culture we see nature, and where others see nature, we see culture.

Many environmentalists want a purity in nature, but environmental history constantly portrays human beings and the natural world (at least on the scale at which we operate) as so entangled, so inseparable, that we do not produce the kind of purity that nature/culture divisions demand. Much of what we value as nature is partially the result of our own prior manipulations.

The landscapes that environmentalists see as natural, for example, many of us see as historical: that is, as blends of the cultural and natural. For Stephen Pyne, North America at contact was already shaped by human beings who wielded fire.[2] This means that wilderness is not so much a thing or even a place as an idea. But the opposite is also true. Where environmentalists, or for that matter most academics, see only culture (as in the city), environmental historians insist on nature. Joel Tarr and Martin Melosi have been doing this for a long time.[3] A whole shelfload of environmental histories of cities are now in the dissertation stage. They point out the obvious: Cities cannot function unless the natural systems that they have shaped and modified continue to function. The built environment remains entwined with the natural environment.

In all of this, many modern environmental historians are intellectually closer to Thomas Jefferson and Ralph Waldo Emerson—to Enlightenment and even early Romantic America—than to the later Romantic America of John Muir and the modern cult of wilderness.

There are real limits to Jefferson and Emerson, but they did not divorce the human from the natural. They retained a sense of the human body as natural and a sense of human work as providing a connection between culture and nature.

There springs from Jefferson and Emerson a line of thinking about technology, culture, and nature that is more interesting, and I think more influential, than cruder views of a separation between technology and nature and a process in which the cultural, the technological, subordinates the natural. For Jefferson, the farmer and his plow represented not a conquest of nature but a finishing of nature. For Emerson, "Nature, in the common sense, refers to the essences unchanged by man: space, the air, the river, the leaf. Art is applied to the mixture of his will with the same things, as in a house, a canal, a statue, a picture. But his operations taken together are so insignificant, a little chipping, baking, patching, and washing, that in an impression so grand as that of the world on the human mind, they do not vary the result."[4]

Emerson obviously had little conception of how far a little baking could take us in modifying the world. But the general conception of the entwining of the human and the natural, the inability to separate the two even in human works and art, did not die with a recognition of the extent of human alterations of the planet.

One response to the power of humans and their technology to transform the planet has consisted of an arrogant joy and a desire for more of the wealth produced. A second set of responses have been anti-technological and anti-industrial; these romantic responses have shown up in a variety of back-to-nature movements and nostalgia for pre-industrial societies. But it is the third set of responses that I think is more critical. The Jeffersonian-Emersonian responses have tried to distinguish between environmentally beneficial technologies and environmentally malignant technologies. Believers think the right technology will transform us and our world for the better.

Belief in environmentally transformative technologies is probably most easily discerned in the work of Lewis Mumford, perhaps the most influential and certainly the longest-running public intellectual of the middle of the twentieth century. He made the relationship between new technologies and the environment a centerpiece of his social theory.

Mumford thought that he and his contemporaries lived in a moment when society was passing from the Paleotechnic Age—the age of coal and steam. The Paleotechnic Age had relied on steam engines that produced mechanical energy through shafts and belts. It had represented an "upthrust into barbarism." It had produced a world of steady, unremitting, repetitive, monotonous toil. It had lived off the accumulated energy capital of the past instead of "current income," and then had wasted 90 percent of that energy. The defining mark of the Paleotechnic was pollution; its by-products yielded a "befouled and disorderly environment; the end product an exhausted one." The Paleotechnic had treated the "environment itself . . . as an abstraction. Air and sunlight because of their deplorable lack of value in exchange had no reality at all." The Paleotechnic had made abstractions into realities "whereas the realities of existence were treated . . . as abstractions, as sentimental fancies, even aberrations."[5]

Electricity represented what Mumford called the Neotechnic. Electricity and alloys, particularly aluminum, were replacing coal and iron. Hydroelectric power would purify polluted industrial cities, and they would also purify human society. Electricity would free workers from factories centered on large steam engines and disperse them to the countryside, where small factories depending on electric motors could thrive. Workers would take joy in their work. The steam engine tended toward an economy of monopoly and concentration; electricity would promote independence and decentralization. Through electricity, Mumford envisioned greater individual independence, more cohesive communities, and an end to crowding, pollution, and waste.

Mumford's proclamations about the Neotechnic seem naive now. His Neotechnic may have doomed the steam engine, but it was not the end of coal or pollution. The workers that the electric motor was supposed to liberate now await their liberation by the computer. It will, I think, be a long wait. The hydroelectric plants that were supposed to ensure environmental purity did offer environmental benefits; however, they brought environmental problems of their own, and now plans to restore a more pristine natural world in the Pacific Northwest center on tearing down Mumford's monuments to modernity. His solution has become a contemporary problem. This is ironic, but irony is cheap and not very valuable. What is most interesting to me is not that Mumford misjudged the future, but that today's critiques of the dams that Mumford praised repeat the premises that built the dams without even knowing it. There is a piece of folklore in the Pacific Northwest which, like any good piece of folklore, cannot be killed by a historian like me. It is the folklore of Bonneville Dam, which supposedly was originally designed without any provision for fish to pass upstream. The origins of this story, as far as I can tell, are in an artist's rendition of the dam that did not include fish ladders. But the plans for the original dam not only included fish ladders, they included fish elevators. The dam had redundant systems to move fish upstream. What the planners failed to consider adequately was how to get them back downstream.

The reason the "no fish ladder story" is so persistent is that it is so comforting. Why are we in such a mess? It is because engineers, technicians, politicians, and industrialists ignored and defied nature. We, on the other hand, know that we must mimic nature and follow nature if we are to succeed. But this is precisely what many of the engineers who built the dams thought. Carl Magnusson, an engineer who taught at the University of Washington and was one of the main promoters of Grand Coulee Dam, put it this way:

The dam will accomplish intentionally the result achieved capriciously, by all the rampant natural forces of the Pleistocene period. It will raise a portion of the Columbia back into the Coulee, and again the floor of the ancient gorge will be inundated. After being dry and arid throughout almost all the history of mankind, the Grand Coulee once more will be a waterway.

Magnusson was right, once an ice dam had blocked the Columbia and forced water into the Grand Coulee. All kinds of things happen in nature. Mimicking nature is not necessarily always a safe course. Distinguishing between good and bad technologies on the basis of their resemblance to natural processes is not a reliable guide. It is a logic that has sometimes caused the problems we seek to solve.[6]

The issue is, however, more complicated than this. The designs were flawed in that they produced unintended consequences, but some of those consequences turned out to be desirable. Many of the environments that people in the United States now treasure as natural are actually contingent environments that are the results of human manipulation. A hundred years of intense environmental manipulation in the Pacific Northwest has created, for example, irrigation systems whose very inefficiencies have created pothole lakes and wetlands which host abundant migrating waterfowl and irrigation ditches with trout and kokanee and beaver and muskrat. These contingent and accidental environments complicate our world and blur even further the lines between the cultural and the natural.[7]

There is a truth in the line of thinking (descended from Jefferson and Emerson and maintained by Mumford) that has refused to draw sharp lines between the human, the technological, and the natural, but it has not, at least recently, proved to be a particularly useful truth. It needs rethinking. Its problems are threefold. The first two are practical. First, in its oldest forms it sees human technologies as simply natural processes in a new form. Much has been gained by this refusal to separate the human and the natural, but it blinded Jefferson and Emerson to environmental damage. Second, in its more modern statements, this view recognizes environmental damage from bad human technologies, but it still identifies good human technologies with nature. They mimic natural processes. But mimicking nature has, on the Columbia, often produced significant environmental problems. The last problem is more conceptual. Although this line of thinking refuses to sever the connections between the human, the technological, and the natural, it treats technology as a capstone that sits atop the natural world. But in the modern world, at least at the scales on which historians operate, this is a difficult thing to do. Good technology is not icing on nature's cake; the technical has melted and blended down into the cake itself. Separating the two is not only difficult, it might be undesirable.

I am talking here about an idea that, I must admit, I am getting tired of— hybridity—but it still has some miles left in it. It is an idea, but not a word, that I used in my own work on the Columbia River, where the dams, the irrigation system, and the power networks have become what I call an organic machine. Machines don't just sit atop natural systems. Machines at once modify, become part of, and depend upon natural systems. Lines blur.

Similar ideas appear in the work of Bruno Latour, but perhaps the clearest expression of what I am after appears in the work of the French geographer Henri Lefebvre. Lefebvre, for example, describes a modern house and its street as a seemingly localized expression of human work and technical skill. Both have an air of stability and immovability. They would seem to personify the human and the local. But, Lefebvre writes, "critical analysis would doubtless destroy the appearance of solidity of this house, stripping it, as it were, of its concrete slabs, and its thin non-load bearing walls, which are really glorified screens, and uncovering a very different picture. In the light of this imaginary analysis, our house would emerge as permeated from every direction by streams of energy which run in and out of it by every imaginable route: water, gas, electricity, telephone lines, radio and television signals, and so on. Its image of immobility would then be replaced by an image of complex mobilities. . . ."[8] The house becomes analogous to a machine manipulated by its inhabitants.

The house is connected to large systems that are themselves both natural and technical. The local is not simpler than the national or global. Organizational complexity is not simply a function of greater size as one goes from a house to a street to a neighborhood to a city; different spatial scales—some far greater than the city—have already interpenetrated the house itself and are necessary to explain it. Where do the water, the gas, and the electricity come from? To understand the nexus of spatial relations produced by the house, it is necessary to understand regional, national, and even global relationships to the natural world, because each interpenetrates the house.

I think about Lefebvre's house a lot. It is, on everyday terms, an expression of the hybridity, the mixture of the natural and the technical, that human beings create and inhabit. If you separate out the natural and the technical, you don't improve the house; you destroy it.

The bad part about history, the essentially conservative part of the discipline that I often struggle against, is quite simple: You can't ever begin from scratch. You can only work on the basis of what has gone before. We live in a world of hybrids. To ask how technology affects nature may be to ask the wrong question. Look at the technology and you will see it interpenetrated with nature; look at nature and you will find the imprint of human technology. This is admittedly an issue of scale. Go to the smallest or the largest scale and there is nature, but this is not where we live. We live on the scales historians examine.

If we begin with hybrids, if we see a landscape always already transformed, then our concern becomes less one of maintaining an existing purity (nature or "the" environment) on which we depend than one of imagining new and better worlds. These worlds will themselves be hybrids, and we will always be constrained by processes never quite under our control. But these are the only choices that history seems to have left us.

NOTES

1. Stephen J. Gould, *Wonderful Life: The Burgess Shale and the Nature of History* (Norton, 1989).

2. Stephen Pyne, *Fire in America: A Cultural History of Wildland and Rural Fire* (Princeton University Press, 1982).

3. Joel Tarr, *The Search for the Ultimate Sink: Urban Pollution in Historical Perspective* (University of Akron Press, 1996); Martin V. Melosi, ed., *Pollution and Reform in American Cities, 1870–1930* (University of Texas Press, 1980).

4. Ralph Waldo Emerson, "Nature," in Ralph Waldo Emerson, *Essays and Lectures* (Library of America, 1983), 8.

5. Lewis Mumford, *Technics and Civilization* (Harcourt, Brace, 1934), 157, 168, 169.

6. I treat this in much more detail in *The Organic Machine: The Remaking of the Columbia River* (Hill & Wang, 1995).

7. For a good example of such unintended consequences see Mark Fiege, *Irrigated Eden: The Making of an Agricultural Landscape in the American West* (University of Washington Press, 1999).

8. Henri Lefebvre, *The Production of Space* (Blackwell, 1991), 93.

THE TOOL THAT IS MORE: AN INQUIRY INTO FIRE, THE ORIGINAL PROMETHEAN INVENTION

STEPHEN J. PYNE

FLICKERING FLAME

Gaston Bachelard once declared that flames had a hypnotic effect that rendered them unsuitable for rational analysis. Instead one sank into reverie. He then demonstrated his thesis indirectly by writing another 200 pages of flickering, discursive text.[1] Yet he did grasp one core fact: that the more one stares into the flames, the less obvious their identity. Fire is the electron of the human-scaled world, at times a wave, at other times a particle; in some circumstances, a process, in others a seeming element; in some contexts, a natural phenomenon, and in others a cultural creation. Or better still, it simply exists beyond our dichotomizing categories and common-sense imagery. What is undeniable is its unremitting bonding with humanity.

THE CURIOUS CHARACTER OF FIRE

In nearly all myths, when people get fire, they move beyond the rest of creation; they become distinctively human. Aeschylus had Prometheus proclaim that, in bestowing fire on humanity, he had invented "all the arts of man." That's a claim as reckless as it is bold. But it is certainly the case that humans are tool users, that fire is among the oldest of human technologies, probably the most pervasive, and likely the most enduring. Since they first met, people and fire have rarely parted. Together they have crossed deserts and glaciers, passed into rainforest and oak grove, sailed over oceans and flown through clouds, landed on Mars and the moon. Everything humans have touched, fire has touched as well.

Yet it remains as curious a technology as it was for the ancients, an odd "element." In one form, it is a tool that behaves like other tools. It can apply heat the way an ax can apply impact. A candle holds flame the way a handle holds an axhead. Yet in other forms, it more resembles a domesticated species. It must be birthed, tended, trained; it compels people to change their habits to accommodate its own. Field fires have more in common with dairy cows than with shovels. The hearth fire cannot be put on a shelf as a hammer can. It has more akin to a draft horse that needs a barn, feed, currying, and a bridle. Fire is a captured ecological process that people can, broadly, harness. We can tap into the power of air and water to turn gears and millstones, but we cannot call forth floods or gusts in the way we can flame. In brief, fire roams across a wide spectrum of human technologies. Moreover, fire is perhaps the ultimate interactive technology because it makes possible other tools. Even where fire does not dominate—where in fact it might seem absent—somewhere along the technological chain it almost certainly serves as a catalyst or enabling device that allows events to proceed, without which a link or two would break.

Its variants do matter, however. To the extent that fire is a simple tool, it is possible for another tool to replace it. An acetylene torch can replace a forge, an incandescent wire an oil lamp. This process has so progressed that the industrial world has hardly any use for open flame at all, which it regards as unacceptably dangerous. Much as early life incorporated oxygen into the molecular machinery of the cell, constraining it to single, well-controlled acts, so modern technology has absorbed fire, until combustion has replaced fire altogether, and concentrated heat combustion. It is harder to substitute for fire as a kind of domesticated creature because burning was essential to the task. Fire did a variety of things, not easily replaced one by one. But to the extent that it burns in a built setting (even one "built" of natural materials), it is possible to reconstruct that setting, piece by piece, with surrogates for fire at each point. This, for example, is the logic of industrial farming. When fire serves its purposes as a loosely controlled ecological process, however, no substitution is truly possible. What is needed is fire, and fire as it burns freely in a roughly natural context. The ability to start and stop this process is surely a technology, but it is not a "tool" as commonly understood. One can break a campfire down into its constituent parts to find alternative sources of heat, light, and social attraction. One cannot so break down a fire sweeping through a pine forest. The range of its interactions with its surroundings is too complex and interactive. To speak of such fires as "tools," as though they were equivalent to chain saws, tractors, and ammonia fertilizers, is to miss the point of their presence.

Its titles are thus important. Treating domesticated fire as though it were a mechanical device can cause troubles. It is a truism that how people perceive fire will influence how they respond to its powers and problems. Such perceptions are also complex, for fire's symbolic power has always matched its practical power. The care of fire became the paradigm for domestication; the application of fire became equally the paradigm of technics, of the innumerable crafts that require fire or rely on the tools that fire renders and assists. Fire remains, above all, the great transmuter. It is, for poets and philosophers as much as for engineers, the essence and model of change, not solely for the things it personally combusts but more widely for the infinity of things its applied heat softens, melts, molds, speeds up, and powers.

Over millennia fire has itself been transmuted. No Paleolithic hunter would likely recognize the fire in a pump-action shotgun; no Neolithic swiddener the flames buried in a tractor or the nitrogenous fertilizer sprayed by a portable power pump; no priest the theophanous fire behind a fluorescent lamp; no natural philosopher the fiery prime mover, fed on fossil fuel, beyond the sublunary world, turning the geared wheels of industry; no poet the quintessential combustion that makes software possible. In truth, as the third millennium dawns, one can improve little on the observation of Pliny the Elder, the great Roman naturalist of the first century A.D., as he pondered the role of controlled fire on remaking rock.

At the conclusion of our survey of the ways in which human intelligence calls art to its aid in counterfeiting nature, we cannot but marvel at the fact that fire is necessary for almost every operation. It takes the sands of the Earth and melts them, now into glass, now into silver or various forms of lead, or some substance useful to the painter or the physician. By fire minerals are disintegrated and copper produced: in fire is iron born and by fire is it subdued: by fire gold is purified: by fire stones are burned for the binding together of the walls of houses. Fire is the immeasurable, uncontrollable element, concerning which it is hard to say whether it consumes more or produces more.[2]

PROMETHEUS UNCHAINED

Call them, collectively, pyrotechnologies. Begin, however, with the technology of fire itself, because the power fire promised could happen only if one could create and control fire at will. Fire had to be present when needed and had to exist in a form that was usable. This required devices to start fire, special fuels to stoke it, and appliances to store and regulate it. They are among the most ancient of technologies and the most familiar, or were until industrialization rendered them alien, almost magical.

FIRE STARTERS, FIRE PRESERVERS

Nature has not been an easy source for fire, however. Some places have little flame; others have it only as the whim and seasonality of lightning or volcanic eruption allow. Nor, for early hominids, was fire easy to make. They had to hold on to it once they had it. If they lost it, they could get more only by begging, borrowing, or stealing from others. Yet it was rare for groups to give fire away: it was too precious. They shared only within a clan, from a common source, and shared with outsiders only during core ceremonies like marriage or treaty-signings where the commingling of their fires symbolized the merging of their interests. To lose fire could be disastrous, the very symbol of catastrophe.

So they strove to preserve fire. Slow matches, banked coals, embers insulated with banana leaves or birch bark, and perpetually maintained communal hearths kept fire constantly alive. With suitable kindling and coaxing, new fires could be ignited from this source. The effort to preserve the hearth fire or the sacred fire of the larger community had thus an immensely practical purpose, eventually coded in elaborate ceremony and symbolism. Many peoples, moreover, carried their glowing fires with them when they traveled. It was once believed that Australian Aborigines, Tasmanians, and Andaman Islanders did not know how to start fire because for decades they were not seen to kindle one. Instead they carried their firesticks with them. They were right that fire was usable only if it was portable. Most groups, however, substituted fire-starters for fire itself. Three kinds of devices prevailed—the fire drill, the fire piston, and the fire striker. The first includes fire plows and saws, as well as drills proper, and works by vigorous rubbing to the point that the heat of friction can kindle tinder. The second, more restricted, works like a diesel engine by quickly plunging a tinder-draped piston into a small chamber and then pulling it out. The rapid buildup of heat and sudden release into oxygen results in ignition. The third embraces a wide variety of instruments that showers sparks onto tinder. Drills and strikers closely mimic the stone and bone tools of *Homo sapiens*, and almost certainly date from the same Paleolithic epoch; their geography tracks a map of human migrations. To coax fire from wood or flint must have seemed like the deepest conjuring. Certainly, the ability to call fire forth on command signaled a revolution in fire history.

Over time, certain fire-starters triumphed, almost to the point of becoming universal. Conquerors and colonizers imposed their own devices; trade bolstered others; Europeans, in particular, promoted the strike-a-light, favored since Neolithic times. (Even the 5,000-year-old "ice man" recovered from a glacier in the Ötztaler Alps had one, along with a pouch for tinder.) Eventually pyrite and flint gave way to steel and flint and joined European traders, missionaries, soldiers, and colonists as they tramped around the world. The technics were, after all, the same as that exploited for flintlock rifles. Then a chemical revolution replaced the awkward strike of steel with the smooth friction of the match. The first (the sulfur-reeking "lucifer") appeared in 1827, succeeded by a phosphorus version in 1830, and the safety match in 1852. No longer did fire-starting require either cost or skill. Anyone could carry it, anyone could call it forth. The ancient bonds of fire-tending and codes of fire-related behavior disappeared into pants pockets. But by then, other than smoking tobacco, there was little reason to haul it on one's person.

FUELS: THE GREAT CHAIN OF FIRE'S BEING

Spark was only a start, as easy to carry as an idea. What mattered was not ignition but preservation. What mattered were the fuels necessary to keep a fire aglow: a spark was only as robust as its tinder. One solution was to store kindling in pouches to ensure it was ready. Another was to combine fuel and flame in a slow match or a firestick. When one torch burned out, another would be kindled from it. The firestick could then transfer flame to a campfire, perhaps sheltered, from which another firestick could be wrested when the time came to move on. The flame became constant. The role of fire keeper was essentially that of fuel provider.

Whether closed or open, a tended fire was really a fire well fed. The fire-keeper might equally be called a fuel-keeper. The search for combustibles was endless and often time consuming. It frequently extended broadly over the countryside, and was a consideration in the periodic relocation of villages. Most settled, agricultural places had to grow their fuels, which they did by coppice or the use of stubble or by reliance on the dried dung of their livestock. Regardless of where they got it, they had to stockpile it, keep it dry, split it into suitable forms. It was hard to say which most controlled the other—the fire or the fire tender.

The need for fuel prompted its own technologies. Not surprisingly, most relied on fire—fire-killed forests, fire-pruned coppice, fire-distilled wood such that fire created the fuel for more fire. Perhaps the best-known practice involves charcoal, a twice-cooked substance (once without oxygen, once with it). The slow heating of wood in a sealed dome leaches out by pyrolysis the volatiles that encourage flaming. The solids that remain will then burn steadily through conduction, glowing with a steady heat, rather than flame wildly.

Still, fire could burn everything people brought to it: it could quickly exhaust, if people chose, whole countrysides. The lust for more fire—checked only by the ability of surrounding landscapes to grow biomass and people to convert them into combustibles—eventually led to an unbounded fuel source: fossil biomass. Fossil fuels existed as coals, lignites, oil shales, natural gas, and petroleum. Petroleum, in particular, inspired its own pyrotechnology for chemical distillation, which made it also immensely portable and vastly more potent. But refined fuels required refined combustion chambers. Automobiles could not run on wood or coal; refrigerators and heat pumps could not function easily with furnaces; power lawnmowers could not survive on steam. The creation of new fuels, in brief, not only made possible but demanded new tinder pouches and new hearths. The fusion of fossil fuels with fire engines, each rapidly redesigning the other, traces the fast spiral of industrial fire.

FIRE APPLIANCES: CREATING SPECIALTY HABITATS FOR FIRE

The place where spark and fuel met decided the traits of the domesticated fire. Fire proved enormously malleable—flame had no fixed form, firelight no necessary brilliance, the heat of combustion no inevitable flow. All could be molded, and over time each property was selected much as dogs and horses were bred for size, speed, coloration, and sense of smell. The chosen means was the combustion chamber, which controlled not only the movement of heat and fuel but also that of air. And more: refining the fire required that air be refined into oxygen and rough biomass into its chemically active parts. What oxygen was to air, this distilled combustion was to fire.

Until recently, however, these contrived keepers of specialty flames still put fire before its human tenders in a very direct way. Fire's presence, as fire, was undeniable, however encased in brick or metal or sited above tallow or pipe. But industrialization has changed that. Flame no longer appears before people or, for that matter, before nature in a visible way. Rather, technology has progressively separated combustion from flame and segregated the chambers where burning occurs from the sites where its energy is felt. No one cooks over a dynamo, as they might a hearth or a forge; electricity has erected a firewall between source and sink greater than any masonry bulkhead. Instead, fire exists covertly in its products rather than overtly by its active presence. It flourishes subliminally in the cement, brick, tile, glass, silicon wafers, metal, incandescent lights, refrigerators, heat pumps, and gas-propelled vehicles that populate the modern world. Industrial appliances have done for the evolution of natural fire what genetic engineering promises to do for the evolution of life.

So, too, industrial fire rarely meets directly with the biological Earth. Combustion occurs outside the biosphere and within mechanical casings that have so broken burning into its constituent reactions that the outcome qualifies only minimally as fire. Controlled flame rarely strikes trees, soils, or scrub, or the creatures that live amid them. It encounters fire through its servant machines; yet this is sufficient for industrial combustion to fundamentally restructure the ecology of fire on Earth. The modern world's flow of matter, energy, and organisms increasingly follows the stream of industrial combustion. Even the climate teeters on a geologic tightrope as long-buried biomass, passed through the pyrotechnic flames, bursts forth into its atmosphere, layers its continents, and sinks into its oceans. No true flame could do more.

HOW FIRE FIGHTS FIRE

Controlled fire has come full circle. Its first seizure led to a program of captive breeding that ended with fire crumbling into chemical shards. The once-visible fire is becoming a virtual one. Pre-industrial fire could always, if it escaped, revert to type, leave the hearth or the forge, and become feral; industrial fire cannot. Pyrotechnologies have refined the hearth fire to the point of extinction. Still, the process of replacement does not stop at the hearth—is not content to merely displace open burning—but has pursued flame wherever it appears. It has sought to remove all free-burning fire, indirectly by substituting for it, directly by suppressing it.

From the beginning, controlled fire has been humanity's primary means to contain wildfire. People protectively burned fuelbreaks and patches to retard fire's spread; they countered wildfire with backfire. Industrial fire has changed that, however, and has removed fire even from firefighting, just as it has removed flame from houses. For industrial countries firefighting has ceased to mean the clash of one flame against another, and means instead the suppression of free-burning fire by the engines and pre-burned bricks and cement of industrial combustion. Two fires cannot, it seems, both claim the same niche. If a new species of burning arrives, it somehow means the old ones must depart.

CYCLES OF PYROTECHNOLOGY

HOW FIRE HAS COOKED THE EARTH

Just as fire turns the gears of ecological cycles, so it has cranked the cycles of many of the things people do to make that ecosystem habitable. Consider three examples, all of them variants of cooking: the cooking of food, the "cooking" of rough biomass, and the "cooking" of rock. For each, fire is the great enabler. Remove it and the cycle collapses.

COOKING AS PYROTECHNIC PARADIGM

Let us begin with cooking, which is where the technology originates. In many fire-origin myths, a proto-humanity laments as a cruel hardship that it must eat food cold or raw and has no means to preserve food other than by drying it in the sun. The capture of fire changed all this. Cooking became the very emblem of the domesticated fire. Out of the campfire and the hearth arose the kiln, the furnace, the forge, the crucible, the oven, and the metal-encasing combustion chamber. From cooking food it was a short step to cooking other matter—stone, wood, clay, ore, metal, the air, even sea water, whatever fire could transmute into forms more usable to people. In effect, humanity began to cook the Earth.

Cooking was, in fact, only one phase in a long-wave cycle of food preparation for which people might resort to fire at nearly every stage. Fire helped pluck or massage the food out of the larger biota; fire cooking followed fire hunting, fire foraging, fire-based farming and herding. Fire helped ready meat, grain, or tubers for eating, improving taste, leaching away toxins, and killing parasites. Fire—its heat, its smoke—then helped preserve for the future what was not instantly eaten.

It is difficult today to comprehend how pervasively fire could affect this process, but as an interesting illustration consider the ways by which pyrotechnology shaped the economy of food for sixteenth-century American Indians as recorded in Thomas Harriot's *Briefe and True Report of the New Found Land of Virginia*. In paraphrase: The axis of the village passes through a great fire, around which the tribe stages its "solemn feasts." The hunting grounds for deer they keep open by regular burning, and the deer themselves may also be fire-driven into streams or coastal tidewaters during a fall hunt. The crops of maize are swiddened. The houses have hearth fires. But, unexpectedly, the cycle extends even to fishing. With fire the Indians felled trees and hollowed them into boats. They carried fire in the craft while they speared for their prey and at night the torch would draw fish toward them. They broiled their catch over flames, or cooked it in an earthen pot along with maize and other foodstuffs. They could dry and preserve any surplus fish, also with fire and its trapped smoke. After the meal they could celebrate or offer prayers around a "great fyer." Like their village, their lives and their economy centered around fire.[3]

COOKING WOODS

If fish, venison, maize, and cassava could be cooked, why could fire not "cook" the landscape for other goods? Or, indeed, the land itself? Ancient chemistry was largely cooking applied to assorted substances. Whether the change sought was physical (a change of state) or chemical (a change of substance), fire wrought it. Fire could break apart, distill, soften, stiffen, encrust, melt, or transmute a landscape.

Outside of Nile-like flood plains, where water could do the work, agriculture looked to fire to purge a site of bad features and promote good ones. A good burn did what fire ceremonies claimed it could do: it cleansed a site of weeds, pathogens, and competing species, and it fertilized with combustion-freed nutrients even as it opened a site to the sun. A good fire, however, required good fuel. Farmers could get it by plunging into new lands or by growing it; the agronomic term for such fuel is "fallow."

Thus fields, whether farmed or grazed, moved. Either the field cycled through the land, or the landscape (as a succession of plants) cycled through a given plot. The first is classic swidden, or slash-and-burn cultivation; the second, field rotation. But both relied on fire. The cycle of fallowing obeyed the logic of fire ecology. Returns were excellent the first year, less each subsequent year. In the absence of weeding and manuring, by the third year, a site was overwhelmed with indigenous growth and abandoned. Remove fire completely and at some point the fields will fail.

Such fire practices shaped whole landscapes. Other pyrotechnologies, however, could chisel features on a more delicate scale. Consider how people could cook the boreal forest of northern Europe to feed their general economy. The range of things heated, steamed, boiled, or roasted is huge. Of course, there was widespread swidden farming, without which cultivation was impossible, and broadcast burning for pasturage, essential to livestock and especially dairy products. Beyond that, however, it was possible to chop up and cook the remaining forest to human purposes. One could collect and open-burn the unfarmed woods (aspen was particularly desirable) to get potash, a valued source of potassium used as fertilizer in farming and in the manufacture of goods from soap to gunpowder. One could anaerobically burn hardwoods to get charcoal. One could slow-cook pine to siphon off tar, pitch, turpentine, and other fugitive distillates that made up the "naval stores" industry (so called because the products were vital to wooden ships). Scoring patches of pines—a kind of raw orchard—assured a good supply of pitch as the trees poured forth sap, which then hardened, to cover the injuries.

Through such means people could colonize an otherwise uninhabitable forest, one often sited on morainic soil resistant to the plow and in a climate hostile to winter grazing. What foods people could not cultivate locally, they could trade for. That traffic, of course, relied on wooden ships, which got their masts from the Scots pines that sprouted in dense throngs in the aftermath of fires, their caulking from the tars and pitch distilled from lesser pines, and their ropes from the hemp that flourished on burned plots. Little of the landscape escaped: its human residents bent such places to their will with a kind of second-order firestick farming, sometimes on the scale of individual trees slashed for pitch or tapped for resin, often of swidden-sized patches cultivating charcoal or potash. Without fire to rework the woods, however, their labor meant nothing. Without their fires they were little better off than moose or voles.

COOKING STONES

The firing of rock is perhaps more spectacular because it has no obvious natural origins, save perhaps volcanoes. (The Roman philosopher Lucretius thought that a forest fire had led to the discovery of metallurgy by melting outcrops which then dripped copper and iron, but most readers parse those passages as poetic license.) The more likely inspiration was cooking. Miners roasted ore as they might pork, boiled down liquids as they did syrups, poured molten glass and iron as they might jelly. A mining complex resembled nothing so much as a vast industrial kitchen.

Pre-industrial mining exploited fire at every turn. Prospectors burned over hillsides to expose rock. Miners relied on fire to tunnel, to smelt, to forge. Only the very richest and nearest mines could afford to haul raw ore very far. Rather, they had to crush and process as much as possible on site, and nearly every stage demanded fire. Accordingly, mines were only as good as their fuel supply, which until recently meant wood or charcoal. The great copper mines of Cyprus, for example, grew, cut, and regrew the surrounding pine forests a score of times over the centuries; the Rio Tinto mines in Spain engorged 42 tons of wood a day, amounting to 3.2 million hectares of woodlands over its lifetime. The origins of forestry in Sweden and Russia lay in the state's desire to promote the growth of fuel-laden woods around great iron mines.

Within the mining cycle, fire figures repeatedly. Georgius Agricola's great treatise *De Re Metallica* (1556) is a grand introduction, cataloguing practices that date from ancient times to the onset of the industrial era. Where the veins resisted their iron picks, hardrock miners lit fires to shatter the stone sufficiently to pry one out. This was dangerous work, requiring that mines consider ventilation, but miners already relied on fire to illuminate the shafts, and it was only a matter of degree to put their torches to the stone directly. Eventually gunpowder replaced wood and steam. Yet "fire in the hole" endured.

With fire, assayers tested the ore to determine its character and value. With larger furnaces or pyres, some open, some enclosed, they roasted and cooked crushed ore. The actual process varied with the properties of the metals involved, the abundance of fuels, and local traditions. But at some point all metallic ores would be heated either to separate them from the country rock or to liquefy them so they could be poured and shaped; most often both. Hotter fires required a special chamber, proper fuels (at a minimum, charcoal), and control over air, preferably by means of a bellows. Eventually the furnace becomes a forge to further refine and mold.

But non-metallic stones often demanded firing as well. Turn to Vannoccio Biringuccio's *Pirotechnia,* published in 1540, for a splendid survey of fire's pervasive presence in every metallic (and any other) mining that involved chemical changes. Limestone could be roasted into calcinated lime suitable for cement, sand melted into glass, clay baked into ceramics. Sulfur, mercury, and alum all depended on chemical fire to pluck them loose from gangue and then to purify them into their elemental core. ("Purify," in fact, derives from the English *purifien* which is cognate to the Greek word *pyr,* meaning fire.) Then there are the distillates: salt from sea water, nitric acid from *aqua fortis,* alcohol, oils, and "sublimates" in general. Almost any chemical reaction—the "art of alchemy," whether true in its larger claims or not, thought Biringuccio—relied "on the actions and virtues of fires." Fire was the chemical fulcrum by which humanity could leverage even its mechanical power, by which it could make and move the hard tools that together reshaped first-world nature into a second world of humanity. (More ominously, he concludes his treatise with fire weaponry, cataloguing devices that rely on fire to hurl projectiles or on the projectiles to kindle fire.) In the end, the lithic cycle feeds itself: the iron burned out of the Earth becomes the picks and shovels by which miners can dig more ore and the axes by which to cut the timber they require for shoring and—most ardently—the fuel they need for smelting and forging.[4]

The cycle turns back on itself. While Biringuccio concludes with an extended metaphor on "the fire that consumes without leaving ashes, that is more powerful than all other fires, and that has as its smith the great son of Venus," the fires of *Pirotechnia* needed something real to burn. Here biomass had an advantage: it could be more easily cooked because it could itself burn. Stone could not, until industry found ways to burn fossil biomass. What had once seemed an absurdity, the self-combustion of rock, has in fact become the basis for our modern pyro-civilization.

FIREPOWERS

CONTROLLED—AND NOT-SO-CONTROLLED—FIRE AS A FORCE OF CHANGE

Burning trees for ash and pitch could appeal to nature for its inspiration; burning stone less so; but in both cases fire set by human hands met natural objects. There is no intrinsic reason, however, why humanity had to restrict its torch to the things nature presented to it. Nor did they: pyrotechnology could go where people pleased and could just as readily obey a logic other than that proposed by nature.

Consider warfare and engines, whose dynamics derived from politics and economics rather than wet-dry cycles and the pyric chemistry of living biomass. Their ecological impact was sometimes overt, as when battles set fires that roamed across fields and woods. More often their ecological clout was disguised, an iron fist hidden in a velvet glove of economics. Fire weapons and fire engines restructured the flow of goods and peoples, they influenced how people used the land, they quickened the tempo of technological change. They rearranged fuels, they invented new fire devices. They plunged whole landscapes into a forge of human fury and ambition.

WAR AS FIRE ECOLOGY

War has been associated with fire for so long that the image of one often equates with the other. "Fire and sword" very nearly says it all: open fire, as a tactic of battle, as the scorched earth of retreating armies, as the laying waste by victors; closed fire, as the means of forging weapons, of casting cannon, of powering ordnance. "Firepower" remains yet today the code word for military strength.

Few battlefields have lacked fire. Fires have burned on prairies and in woods, amid ships and cities, flung over ramparts and scattered with artillery shells. Fire weapons have traveled on land, sea, ice, and air. Yet open fire could be problematic, and nowhere more than amid the havoc of battle. Clausewitz's "fog of battle" was most often a cloud of smoke. A broadcast burn could, with a change of wind, turn on those who set it; smoke screens obscured the field for both sides. Even in naval battles, the ideal was to hurl enough controlled fire to disable a wooden ship, not enough to destroy it as a prize. Sieges sought to burn out defenders, while soldiers on the battlements poured down flame on assault troops. In the ancient world, Greek fire (a sulfurous liquid) was a weapon to dread. For gardened societies, especially, the chaos of war invited the chaos of wildfire, since the breakdown in social order exposes niches for fire and strews the landscape with fuel.

In the second millennium, two revolutions in firepower shook the conduct of war. One was gunpowder (which gave new meaning to the expression "to fire"); the other was industrialization, which mechanized war and expanded its range. While each fabricated a host of new fire weapons, it is often easy to miss the flames for the roar. The worst casualties of World War II resulted from blasting cities with a mix of "conventional" blockbuster bombs and incendiaries; even the atomic bombs on Hiroshima and

Nagasaki wreaked their greatest damage through the fires they kindled. The US Strategic Bombing Survey concluded that four-fifths of the destruction wrought on British and German cities by aerial bombardment was "fire damage," that "incendiaries, ton for ton as compared to high explosive bombs, were approximately five times as effective in causing damage," and that the aerial assaults on Japan were "frankly fire attacks." If fire seems increasingly invisible on modern battlefields, it is because the flames have vanished into tank engines, cartridges, and rockets. But even the Gulf War, fought on incombustible sands, ended with burning oil fields. It was, after all, another fire war, fought over the fuels of modern industry. Perhaps not so oddly as it seems at first, those flames will likely endure as the unquenchable symbol of that conflict.[5]

As that black pall, spreading over the sky like an oil slick, shows, waging war with fire has ecological effects. For some landscapes—temperate shade forests, mangrove swamps, cities—war-hurled fire is a major disturbance. Battlefields are shaken landscapes; fire ordnance is a great slasher and burner of towns and forests. A little weirdly, this is not always ecologically evil; training fields in East Germany churned by tanks and shells led, after unification, to nature reserves of exceptional biodiversity. Mostly, though, the biological impacts of military fire are muted and hidden, as with other forms of industrial combustion. War quickens the pace of technological development, redefines and sometimes replaces societies and their economies, and realigns politics, all of which can break and burn landscapes as thoroughly as any outright conflagration.

THE POWER WITHIN: HOW FIRE ENGINES BECAME PRIME MOVERS

Still, the more revolutionary fire is that encased in metal and used to power pistons. With the steam engine, the stationary fire became more than a hearth-evolved furnace: it apotheosized into a prime mover. The fast combustion of fire engines could compete directly with the push and pull of slow-combustion muscle. As Matthew Boulton, James Watt's partner in combustion, succinctly told a visitor, "I sell here, Sir, what all the world desires to have—power."[6]

The problem, as so often, was fuel. The steam engine could not by itself break down the ancient ecology that bonded burning to biomass. The early engines were furnaces, not unlike distillation systems, except that the boiled-off steam could drive a piston. They burned cordwood (or charcoal), which left combustion ultimately at the mercy of what the countryside could grow and operators could glean from it. Those engines could consume staggering quantities of wood; they could rapidly burn up whole landscapes. That set in motion the search for a more robust fuel, a quest that ended with coal.

Fossil fuels had long been burned, but locally and specifically. They suffered from the lack of a place in which to combust usefully. They could not be spread over fields like branches, or be rolled like smoldering logs, or be loosed as flame could over once-living fallow. The steam engine thus gave coal what it most lacked: a combustion context. In return, coal granted to the new fire engines what they most hungered for: abundant fuel. They soon worked on one another, coal encouraging better designs and engines seeking more refined fossil fuels. Together they revolutionized power machinery and transport, and through transport, all the landscapes internal fire could touch. The steam engine soon spawned other combustion-driven prime movers that could burn more portable fossil fuels like petroleum and natural gas. Each innovation bred others. Eventually this swarm of fire-breathing machines forced fire ecology into another order of being. They made possible industrial fire.

Even oblique means can sometimes yield awesome ends. That is what steam did to fire. Combustion no longer flowed from living source to living sinks: it burned biomass from the geologic past and released its outflow to a future Earth. The ancient chain of combustion no longer resembled anything in its past. Industrial combustion added fires in the form of its prime movers and the machines they in turn goaded into being. It subtracted fires by substituting its closed fires for open ones and by attacking free-burning fires seemingly wherever it found them. And it relocated fire ecologically by breaking down and rearranging landscapes, helping decide what might burn and when it should burn and by what means. Although a robust ecological understanding of industrial combustion still eludes us, its effects everywhere surround us. Through its engines, industrial fire has become the prime mover of Earth's fire regimes.

THE FUTURE OF FIRE

The future promises both more and less. It looks different for fire as tool, as domesticated technology, and as captured ecology.

As technology quickens and industrialization sprawls, the scope of controlled combustion will explode. More fire appliances and more fossil-fuel combustion may mean more fire broadcast over the planet than at any time in history. Yet the counterpressures will also build. Advanced technology will further separate the desired properties of fire from its free-burning forms. Future prime movers may no more resemble open flame than does the energy chain of the cellular Krebs cycle. In mature economies, energy is becoming "decarbonized": we are extracting more energy with fewer hydrocarbons. If some forms of solar and nuclear energy develop, combustion may have a powerful rival. In any event, the Big Burn that has characterized the early phases of industrialization has led to a Big Dump that threatens to overwhelm ecological sinks. It may even alter the overall climate. The trend will continue to disaggregate fire's effects from fire itself.

The big loss will be fire as domesticated technology. Open burning for farming, for pastoralism, for rural life—these are being replaced by fire appliances, or the chemicals distilled from fossil biomass. The pastoral landscape of agriculture, in all its varied forms, is vanishing before urban encroachment and nature reserves. The one has no liking for open flame, the other little desire for fire kindled by human artifice. Fire as a catalyst for rural economies will slip away with those lapsing economies. The largest domain of anthropogenic fire technologies will disappear along with the family cow, the woods pasture, the fallowed wheat field, and the pruned orchard. Biocentric philosophies challenge the very legitimacy of domestication as a model for the human relationship with the natural world; such fires will share in that disdain, and that loss.

Fire promises to thrive, however, within the nature reserves that the industrial world is amassing. Not all such sites crave fire; some shun it, and will wither under the flame. But many have adapted to fire regimes and will suffer from fire's withdrawal as surely as they would a shift in rainfall. The fact is, fire can be as ecologically powerful when removed as when applied. A fire drought can be as serious as a fire flood. Yet if fire is not directed in some way, the outcome will be fire eruptions, fire in patterns different from those to which the biota accommodated, wildfire. The alternative is some kind of controlled but still freely burning flame. The domain of fire as a captured ecological process will likely expand.

What may be most curious will be the future's perception of fire and of humanity's monopoly over it, the sense that we remain the keepers of the planetary flame, whose ecological power and responsibility descend from our mastery over that inconstant flame. The extinction of fire from daily life has encouraged an erasure of fire from social memory. Whether the future will continue to honor fire as the founding art, as Prometheus proclaimed, is unclear. It may choose to dismiss it as a mythic curiosity from an era that sucked marrow out of charred bones, herded cattle to fresh-burned forage, and sat contentedly around a flickering campfire instead of a big-speaker, big-screen home entertainment center. It may prefer the virtual fire of computers to the robust, and dangerous, half-tamed fire of nature.

NOTES

1. Gaston Bachelard, *The Psychoanalysis of Fire* (reprinted translation: Beacon, 1964).

2. Pliny, quoted in Cyril Stanley Smith and Martha Teach Gnudi, eds., *The Pirotechnia of Vannoccio Biringuccio* (reprint: MIT Press, 1966), xxvii.

3. Thomas Harriot, *A Briefe and True Report of the New Found Land of Virginia* (reprint: Dover, 1972), 69, 55–57, 60, 63, 66.

4. Smith and Gnudi, eds., *Pirotechnia,* 336.

5. Percy Bugbee, "Foreword," in *Fire and the Air War,* ed. H. Bond (National Fire Protection Association, 1946).

6. James Boswell, *Life of Samuel Johnson,* vol. 2 (London, 1934 edition), 459.

WHAT ROLE DOES INNOVATION PLAY IN URBAN LANDSCAPES?

Most city dwellers appreciate their public green spaces as a respite from the hustle and bustle of urban life. What they may not think about, though, is that many of these idyllic locations are as much a product of design and construction as skyscrapers. The historian Timothy Davis reflects on "the nature of Nature" in the city, specifically addressing the planning of Washington, D.C. From the formal landscape of the Mall to the more romantic tumble of greenery in Rock Creek Park, he exposes the civil engineering and aesthetic principles that have guided the city's landscape plans.

Michael Robinson, former director of the Smithsonian's National Zoological Park, follows Davis's lead and discusses the manipulation of landscape for the purposes of education. Robinson points out that zoos teach about exotic species through the flora, as well as the fauna, on display. This includes not only creating accurate copies of animals' native habitats, but also increasing visitors' awareness of the local species that have made the zoo their home.

Complementing Robinson's essay is a portrait of Jon Coe, an award-winning designer of zoo exhibits. One of the earliest developers of immersion exhibits, Coe is a proponent of decreasing the barriers between animals and humans. His goal is to transform visitors from passive watchers to active participants in the animals' landscape.

The various forms that "Nature" takes within city limits may indeed owe as much (if not more) to construction as to biology. However, a common belief in the importance of combining innovations in building techniques and materials with aesthetics and pedagogical goals connects these essays.

INVENTING NATURE IN WASHINGTON, D.C.

TIMOTHY DAVIS

Washington has long been known as a city of beautiful parks and boulevards, tree-lined watercourses, and stately monuments set in stunning natural surroundings. Few capital cities are favored with such an attractive array of soothing parks, shady streets, and verdant riverbanks. Washington, it seems, is a city where the works of man and nature coexist in natural harmony.

What many observers fail to realize, however, is the degree to which the "natural" environment of Washington is a human invention. Washington's "natural landscape" is a cultural construction, both literally and figuratively. Not only has the physical environment of the nation's capital undergone significant transformations over the past 200 years; the very idea of what nature is and how the city's residents should use, shape, and interpret it has changed dramatically. Since the first Native American occupation, humans have been <u>actively involved in reconceptualizing, transforming, and constructing ("inventing") the nature of nature</u> in Washington. Throughout this process, successive generations have redefined Washington's natural environments both physically and imaginatively in the pursuit of a variety of social, practical, and personal goals.

In fact, <u>changing cultural perceptions about what nature is, was, or should be</u> have been responsible for most of the major transformations in Washington's physical environment. The city's basic plan is a testament to Enlightenment ideals of man's rational mastery of the natural world. Mid-nineteenth-century plans for the Mall reflect the period's penchant for an idealized nature rooted in picturesque aesthetics and romantic philosophies. The city's basic infrastructure of paved streets, public water systems, and artificial illumination exemplifies the late-nineteenth-century determination to discipline and control chaotic and unhealthy urban environments.

Its parks, parkways, and playgrounds reflect evolving ideas about the need to counterbalance the regimentation of urban life with outdoor recreation and contact with nature. Recent improvements to air and water quality, along with plans to restore species and habitat, underscore the growing influence of environmentalism and ecological sciences.

Not every new idea about urban nature was transmitted into physical form, however. Nor was every change to the local environment desired or anticipated. Some reconceptualizations of Washington's nature were too vague, too impractical, or too controversial to produce significant physical results. Bureaucratic constraints, financial limitations, and political differences delayed the implementation of many proposals for decades or more. Some significant alterations to Washington's ecology and topography were both unplanned and unwelcome—most notably the siltation of the Potomac that transformed the formerly robust river into a torpid, foul-smelling mud flat by the middle of the nineteenth century. Dramatic feats of civil engineering were required to restore the river to a semblance of its original configuration, and landscape architects labored for decades to create the waterfront parks and parkways that strike most modern observers as fortuitous survivals of an earlier, more natural and authentic era in the history of the nation's capital.

Washington is by no means unique in this regard. The seemingly natural elements of most American cities are products of prolonged and extensive human interventions. The cultural construction of urban nature has ranged from the creation of impressive showpieces such as New York's Central Park and Boston's Emerald Necklace, with their elaborate displays of artistry, botany, civil engineering, and sociological experimentation, to the more mundane and often forgotten reconfigurations of topography and hydrology that transformed heterogeneous amalgams of swamps, streams, hills, and valleys into orderly and disciplined metropolises. The cultural construction of urban nature reflects shifting intellectual fashions, changing recreational patterns, base economic concerns, and broad-based technical and legislative developments such as the evolution of metropolitan water and sewer authorities, scientific research and legal statutes aimed at promoting clean air and water, and appropriations to acquire, develop, and maintain playgrounds and park systems. Zoos, botanical gardens, museums, research and educational institutions, media productions, and outdoor-oriented businesses also contribute to the totality of urban nature, as do private landowners, associations, and political groups. Everyone who ventures into a city's parks or its less overtly natural spaces engages in a conscious or subconscious process of "inventing nature" that changes from day to day and season to season. Each person continually reconceptualizes his or her own surroundings and relationships to nature in response to a variety of influences that range from the immediate physical environment to the vagaries of the weather and the actions of other individuals sharing the same social space. This individual invention of nature is further tempered by personal predisposition, shifting moods, and the broader social climate created by a multitude of complex and often competing cultural influences.[1]

Planners, landscape architects, engineers, and ecologists obviously engage in a process of mentally inventing Washington's nature before they commit their ideas to paper, but so do families planning Sunday outings, individual strollers, residents or former visitors ruminating on their real or imagined experiences, or even people who never plan to set foot in the city but mentally formulate their impressions of what Washington's nature is, was, or should be. Though intensely personal, most of these individual perceptions can be related to broader social experiences and cultural trends.

Despite certain peculiarities stemming from its iconic status as the nation's capital, Washington is fairly typical of the broader American experience and can serve as an illustration of evolving cultural conceptions of nature in general and urban nature in particular. From a biological perspective, the complex of flora, fauna, and other physical phenomena we conventionally categorize as nature may have remained relatively unchanged throughout human history, but the meaning and idealized forms of nature have been anything but constant. Nature has been seen as a source of fear and danger and as a fount of inspiration and enlightenment. It has been presented as a positive moral, cultural, and political influence and as an invitation to anarchy and indolence. The natural environment has been viewed as an economic resource destined for economic exploitation and as a biological entity deserving of protection for scientific reasons, for spiritual purposes, or in its own non-anthropocentric right. Tracing the changing nature of nature in Washington offers a fresh perspective on the development of the nation's capital while expanding the boundaries of the term "inventing for the environment" to encompass a broad range of cultural processes and physical activities.[2] The cultural invention of nature in the national capital region began long before 1790, when the federal city was established. The fertile soils, mild climate, and abundance of fish, game, and waterfowl made the area attractive to a succession of Native American groups, who actively shaped the environment long before the arrival of European explorers. Archeological evidence suggests that a highly mobile population created temporary campsites as far back as 6000 B.C. At the time of initial European contact the region was dotted with small semi-permanent villages, where Native Americans cultivated corn, pumpkins, and other crops in addition to engaging in hunting and gathering activities. Native agricultural patterns and the practice of burning undergrowth to maintain a favorable environment for deer and other game produced a more open, seemingly natural park-like landscape than the dense second-growth forest that prevails in undeveloped areas today. John White's 1585 depiction of an Indian village along the Potomac (figure 1), while undoubtedly idealized, suggested the degree to which the indigenous inhabitants transformed their "natural" surroundings for a variety a purposes ranging from sustenance and habitation to ritual and recreation.[3]

The first Europeans to reach the Washington area viewed its natural attributes largely in terms of their potential for economic exploitation. In 1631 Henry Fleet touted the region's salubrious climate and abundant natural resources. John Smith similarly noted the ease with which the native inhabitants supplied themselves with the necessities of life and suggested that European settlers could easily turn a profit from the region's rich natural advantages. The fertile lowlands below the future site of Washington were ideal for agricultural development, and tobacco plantations dotted the banks of the Potomac by the middle of the eighteenth century. Natural harbors such as Georgetown and Alexandria rapidly grew into bustling communities. The low-lying land between Rock Creek and the Anacostia River, which would eventually become central Washington, remained sparsely settled, with a few proprietors maintaining large but mostly undeveloped holdings.[4]

Geography combined with politics and philosophy to establish the nation's capital in its current location. The result was a brand new city situated by congressional fiat and surveyed as a highly rational geometric abstraction yet strongly influenced by local topography and broader geographic concerns. While the 10-mile-square District of Columbia epitomized the era's Enlightenment-based concern for order and balance, this intellectual abstraction was intimately tied to the physical properties of the site. The starting point of the original survey was the tip of Jones Point, a natural promontory extending into the Potomac just south of Alexandria. The ring of hills surrounding the federal city presented natural boundaries

FIGURE I
A 1590 engraving, based on drawing by John White, depicting the Indian Town of Secota. (Library of Congress)

that would also provide excellent defenses in the event of foreign attack or domestic hostilities. Located midway between Maine and Georgia astride a river that penetrated deep into the interior of the eastern seaboard, the proposed capital would theoretically help unite the North and the South while forging a vital link to future inland states. While some objected to placing the nation's capital in an unpopulated "wilderness," others argued that, because it was free of the moral vices, public health disorders, and political corruption associated with large cities, the sparsely settled region offered a stable and virtuous location for the capital of a democratic nation. Located near the head of navigation on the Potomac River, with a vast source of potential water power near at hand at Great Falls, the proposed city seemed ideally situated to fulfill its destiny as a political, commercial, and industrial center.[5]

Once a site had been chosen, the next step was to develop a plan for transforming the fields, woods, marshes, and streams that skeptics derided as a swamp in the midst of a wilderness into a capital city that would be—in an oft-repeated phrase—"worthy of the nation." The man assigned this task was Pierre Charles L'Enfant, a French-trained military engineer who had gained a modicum of renown for his role in remodeling New York's city hall as the temporary headquarters of the federal government. L'Enfant's plan for the nation's capital was a bold invention. Rather than develop a modest agrarian-scale town as Jefferson and other skeptics advised, L'Enfant called for an impressive urban edifice expressing the young nation's grand aspirations through ostentatious formal and symbolic gestures. Drawing heavily on his boyhood familiarity with the Baroque splendor of grand European designs, L'Enfant proposed a highly stylized urban geometry of sweeping diagonals superimposed on an underlying grid. This composition was anchored by strong axial relationships formed by grand avenues extending from the president's house and the seat of Congress, which intersected at the site of what would eventually become the Washington Monument. Lesser monuments and public buildings would occupy smaller squares and circles at the intersections of the diagonal avenues and major streets. The visual grandeur afforded by these sweeping vistas and harmonious geometric arrangements was heightened by the political symbolism underlying the design. L'Enfant's bold reinterpretation of the site was not just an exercise in formal composition; it was an attempt to inscribe the American political system into the raw landscape of the Potomac shoreline. The distant but equitable relationship between the president's house and

Congress was meant to symbolize the separation and balance of powers, while the avenues were named and located to enshrine the historical and political significance of the individual states.[6]

The rigid geometry, complex symbolism, and grandiose scale of L'Enfant's plan gives the impression that it was a pure abstraction imposed on the capital of an undeveloped and (to the human eye) undifferentiated natural landscape. This interpretation has some merit, but it fails to do justice to the subtlety and complexity of L'Enfant's design. While L'Enfant's proposal epitomized the Enlightenment determination to reshape the natural world according to rational principles, it also reflected the innate physical characteristics of the site. L'Enfant carefully reconnoitered the landscape of the proposed city and designed his plan so that its explicitly artificial aspects worked in concert with existing natural features to display both qualities to best advantage. Like most contemporary designers and philosophers, L'Enfant believed that "raw nature" was a starting point for artistic expression. The artist's task was to recognize the promising attributes of a site and develop them into more refined and idealized forms based on complex aesthetic and philosophical criteria. After surveying the site of the proposed capital, L'Enfant declared: "Nature has done much for it, and with the aid of art it will become the wonder of the world."[7] He expressly rejected a more mundane gridded plan favored by Jefferson on the grounds that its formulaic repetition ignored local land features that could be developed to produce imposing visual and symbolic effects. Not only would a gridded capital appear "tiresome and insipid," L'Enfant insisted, but its mechanical geometry would lack "a sense of the real grand and truly beautiful only to be met with where nature contributes with art."[8] By embracing and enhancing the latent potential of the site's natural advantages, L'Enfant proclaimed, his design would "turn a savage wilderness into a Garden of Eden."[9]

The striking degree to which "natural" conditions shaped the basic form of L'Enfant's manifestly "artificial" design is most apparent in Andrew Ellicott's 1794 topographical survey, which shows how the city nestled into the point of land formed by the junction of the Potomac and the Anacostia, with its inland borders defined by the valley of Rock Creek and the rugged ground that marked the beginnings of the Piedmont Plateau. L'Enfant's second major bow to local topography was to locate the Capitol atop a conspicuous promontory known as Jenkins Hill. L'Enfant observed that the hilltop would serve as a natural "pedestal" providing impressive views to and from the structure and underscoring the institution's symbolic importance. Many of the lesser circles and junctions were also placed to take advantage of subtle differences in elevation and prospect.

While L'Enfant looked for design cues in the local landscape, he viewed nature as subservient to man's needs and showed no compunction in radically transforming existing natural features to enhance the splendor of his grand design. L'Enfant's most radical idea was to reinvent the stream known as Tiber Creek as a spectacular and emphatically artificial water feature that would rival the hydraulic extravagances of European gardens. He proposed to divert the stream into an underground conduit at the city boundary and route it to the base of the Capitol, where it would emerge as a man-made cascade 100 feet wide and 40 feet high. Between Capitol Hill and the Potomac, the Tiber would be transformed into an ornamental canal defining the north edge of a grand central promenade. A continuation of this canal would extend through the southern part of the city and empty into the Anacostia. Several lesser streams would be enclosed in culverts and buried without a trace to accommodate the geometric street plan and increase the amount of usable land. The water from these streams would be diverted to create impressive fountains, cooling the air in summertime and providing festive adornments throughout the year.

The striking geometrical quality of L'Enfant's plan together with the broad and barren expanse of the current Mall creates the impression that the French designer was rigidly opposed to the picturesque landscapes that later theorists often cast as more natural and organic. Close examination of L'Enfant's plans and writings reveals that the original design called for significant expanses of picturesque landscaping, however. This mixture of formal and informal elements was entirely in keeping with contemporary practice; it seems unusual only because of the twentieth-century emphasis on the Mall's broader formal aspects. The picturesque style in landscape gardening had come to the fore in England by the time L'Enfant composed his design and was beginning to influence developments elsewhere. L'Enfant also knew that even the grandest Baroque compositions, such as Versailles and Vaux-le-Vicomte, contained densely wooded and naturalistic segments to help delineate and complement their dominating formal elements.

L'Enfant's desire to balance formal and picturesque elements was evident in the terminology he used to describe features of the plan. At 400 feet wide and close to a mile long, the "Grand Avenue" extending from the

Capitol to the monument grounds would be as imposing as the heroic vistas of French chateaux, but it would be "bordered with gardens" and flanked by tree-sheltered public walks. The monument grounds would be embellished with "artfully planted trees" and the Capitol would be set within a "garden park." The chief executive's mansion would reside in an area designated as "President's Park."[10] Combining traditional antipathies toward unimproved nature with democratic opposition to the political symbolism of royal European parks, one critic complained that the proposed woodlands around the president's house would create an "immense and gloomy wilderness" that would contradict the image of openness and accessibility an elected leader should strive to project.[11] The dominant view of the political symbolism of naturalistic landscape design, however, was that its informality, implied freedom of movement, English associations, and absence of rigid hierarchies or a single dominant perspective represented a democratic and egalitarian alternative to the centralized authority implicit in the formal continental style. By juxtaposing the "natural" and formal styles and positing a harmonious relationship between the two, L'Enfant's proposal embodied the political tension between Jeffersonian democracy and the more centrist Federalist interpretations of American government.[12]

Inventing Washington as a formal and theoretical construct was an imposing intellectual feat; translating the paper plan into an actual, functioning city proved to be an even greater challenge. L'Enfant's proposal provided the basic blueprint, but the transformation of Washington from an undeveloped countryside to an imposing modern metropolis was destined to be a long, contentious, continually evolving, and inevitably incomplete process. L'Enfant himself engendered a variety of personal and political conflicts and was soon forced to resign. Heavily encumbered by war debts, the young nation had little to expend on improvements to its capital city. In 1800, when the federal government officially moved from Philadelphia to Washington, only one wing of the Capitol had been completed and the president's house was still under construction. Pennsylvania Avenue, the grand connection between these two edifices, was derided as "a deep morass, covered with alder bushes."[13] A few groups of houses were scattered about the unimproved streets, but contemporary visitors observed that most of the city remained an unkempt wilderness.

During the first few decades of the nineteenth century, Washington began to take on the appearance of an extended village, if not a capital city to rival Paris, London, or even Boston. Clusters of development formed around major government buildings and streets, while small farms, gardens, and pastures gradually supplanted much of the thickets and woodlands that once dominated the landscape. When Jefferson became president, he had Pennsylvania Avenue ornamented with rows of fast-growing poplars, an explicitly artificial but visually effective practice he may have admired while serving as ambassador to France. While many observers ridiculed the aspiring capital as a "city of magnificent intentions," some found charm in its slow pace and bucolic scenery, which appealed to Jeffersonian beliefs in the virtues of agrarian democracy as well as to the contemporary penchant for pastoral imagery. Gazing over the city from the heights of the Capitol, Washington confidante Margaret Bayard Smith exclaimed: "Indeed the whole plain was diversified with groves and clumps of forest trees which gave it the appearance of a fine park."[14]

Traveling artists such as George Parkyns and George Beck encapsulated this romantic vision of early Washington. Their turn-of-the-century views presented the nascent capital as an Arcadian idyll and reassuring emblem of Jeffersonian democracy. In contrast to ostentatious European capitals, Washington was a diminutive "city on a hill" embowered in nature and innocent of the oppressive environment and moral corruption that theoretically afflicted larger metropolises (figure 2). While such images provide a general sense of the city's development, their depiction of local scenery was strongly influenced by contemporary pictorial conventions that added imaginative accents to emulate traditional landscape painting. A few wealthy landowners took this process a step further, creating attractive country seats by embellishing existing landscapes according to the tenets of contemporary aesthetic theories. One of the most impressive of these estates was Kalorama, developed in the early years of the nineteenth century in the picturesque highlands bordering Rock Creek. Kalorama garnered widespread praise for its artfully constructed mixture of natural and man-made scenery. A contemporary painting by Charles Codman (figure 3) presented a highly romanticized view that epitomized contemporary landscape tastes.[15]

The first successful large-scale attempts to "reinvent nature" in Washington in actual, physical terms were largely utilitarian in character. The desire to improve trade and communications, rather than lofty appeals to political symbolism, aesthetics, or science, brought about significant changes in the city's relationship with its physical environment. While the Potomac played a decisive role in siting the city, its waterfalls, shifting channels, and widely varying water flow obstructed travel and inhibited the river's viability as

bridges & canals

FIGURE 2
"George Town and Federal City, or City of Washington." Aquatinted drawing by
G. Beck, 1801. (Library of Congress)

a major commercial waterway. A number of bridges were constructed during the first half of the nineteenth century, many of which were replaced
repeatedly to repair damage caused by floods and other natural forces. Canal
construction represented an even more aggressive attempt to circumvent,
augment, and subvert the forces of nature. Four major canals were constructed in and around Washington as entrepreneurs labored to reinvent the
riverine environment of the capital region. The short-lived Patowmack
Canal at Great Falls captured considerable public attention. Begun in 1785,
this ambitious project attempted to bypass the Great Falls of the Potomac
with an elaborate system of locks located on the Virginia side of the river.
The engineering feats involved in this endeavor garnered international
acclaim and the canal became a popular tourist destination. Despite its
impressive technical achievements, the Patowmack Canal never operated
dependably and was a commercial failure. It was absorbed by the larger
Chesapeake and Ohio Canal Company in 1828.[16]

FIGURE 3

Charles Codman, "Kalorama," c. 1830. (Diplomatic Reception Rooms, US Department of State)

The Chesapeake and Ohio (C&O) Canal represented an even more ambitious attempt to transform the natural landscape for the sake of trade and commerce. The proposed canal would connect the Potomac and Ohio river basins with a 360-mile man-made waterway, paralleling the Potomac with a more navigable artificial channel. Marveling at the audacity of this plan, President John Quincy Adams declared that the project would be "a conquest over nature such as has never yet been achieved by man." Unfortunately, the Baltimore and Ohio Railroad started building its track westward on the same day that ground was broken for the C&O Canal. The canal reached Cumberland, Maryland in 1850, but could not compete with the railroad. Plans to proceed across the Alleghenies were abandoned, though the canal remained in operation until the early twentieth century. Completed in 1843, the Alexandria Canal intersected with the C&O Canal above Georgetown, allowing canal traffic to bypass the shallow and slow-moving bend of the Potomac. L'Enfant's vision of an impressive ornamental canal emanating from the base of Capitol Hill was not realized, but the architect Benjamin Latrobe oversaw the construction of a more utilitarian waterway in the same approximate location. The Washington City Canal was narrow, shallow, and poorly constructed. As commercial traffic declined, it functioned primarily as an open sewer and a breeding ground for mosquitoes, collecting noxious waste from nearby buildings and markets. Widely reviled as a public health hazard, the canal was covered over during the 1870s, taking with it the last trace of Tiber Creek, one of the fundamental elements of Washington's original "natural" topography.[17]

L'Enfant's lofty ambitions for the Mall remained unrealized at the middle of the nineteenth century. Jefferson's poplars had grown up to flank Pennsylvania Avenue with stately rows of greenery, and the grounds of the Capitol were embellished with an assortment of ornamental plantings, but the Mall itself was little more than an ill-defined cow pasture. The Mall's undeveloped status was a continued source of embarrassment and concern. Aside from the unsightliness and practical problems posed by the unkempt assemblage of pastures, hillocks, dirt tracks, and marshes, the undeveloped expanse of the Mall presented a troubling symbol for a nation whose self-proclaimed destiny was to subdue the American continent and transform its wastes and wildernesses into a blooming garden, an agrarian paradise, and an exemplar of moral, social, and economic progress.[18]

Despite the slow pace of improvements, there was no shortage of proposals for the Mall's embellishment during the first half of the nineteenth century. Most of these plans shared a common faith that nature could be improved to serve a variety of practical, political, educational, and inspirational ends. Rather than simply cordon off a parcel of untrammeled nature, or produce scenes of purely aesthetic value, the proper public park should combine pictorial beauty and opportunities for leisurely strolling with instructional horticultural displays, facilities for botanical research, and less explicitly didactic demonstrations of artistic, moral, and social principles. In addition to the immediate practical benefits of serving as models for the embellishment of home grounds, encouraging horticultural experimentation, and promoting scientific and aesthetic progress, such public gardens were thought to have more subtle and far-reaching effects. Contemporary landscape designers and moral philosophers believed that properly designed public gardens would elevate the taste of the citizenry and improve the behavior of those who came in contact with beautiful and well-ordered scenery. While many of these ideas had their basis in English landscape theories, they had a particular resonance in the United States at a time when many believed that the country was beginning to stray from its mythic roots in agrarian-based republican virtue. Public parks were also thought to promote a sense of egalitarianism and democratic solidarity, as the various classes of society interacted in safe, soothing, and suitably refined public environments. By combining Enlightenment rationality with romantic sentimentality, promoting a domesticated hybrid of aesthetic nature, democratic idealism, and Jeffersonian agrarianism, and inculcating a shared value system rooted in genteel notions of taste and decorum, public gardens might help mediate the threats posed by Jacksonian mobocracy and the disconcerting growth of commerce, manufacturing, and urban culture.[19]

The first major addition to the Mall to reflect this view of urban nature was the botanical garden. Botanical gardening, experimentation, and collecting had long been regarded as gentlemanly pursuits. Both Washington and Jefferson engaged in horticultural research and Washington had advocated that a portion of the Mall be set aside for a botanical garden as early as 1796. Additional proposals ensued over the next 20 years. In 1820 Congress finally authorized the creation of a five-acre garden at the base of Capitol Hill. Minor improvements were made over the next few years including the laying out of graveled walks and the installation of floral borders. A major addition to the Botanical Garden was the construction of a greenhouse in 1842 to house exotic plant specimens collected by Charles Wilkes's federally sponsored South Seas expedition.[20]

Individual Washingtonians also transformed their surroundings for the sake of horticultural experimentation and botanical display. One of the most prominent of these gentleman gardeners was Joshua Peirce, who propagated a wide variety of ornamental trees, shrubs, and plants at his country place in what is now Rock Creek Park. This estate, known as Linnaean Hill in honor of the inventor of the modern system of biological nomenclature, exemplified the confluence of scientific, commercial, and aesthetic impulses that characterized contemporary attitudes toward nature. The attractive grounds drew visitors from far and wide, while Peirce's nursery business supplied a wide range of plants that were used to embellish home grounds and public spaces in Washington and surrounding cities. Peirce's experiments with camellias played an important role in the plant's transition to American soil.[21]

The first major plan for transforming the Mall was produced by architect Robert Mills in 1841. Mills addressed the problem at various scales. His most conservative scheme called for a combination arboretum/pleasure garden laid out with curvilinear walks, irregularly shaped plantings, and a pair of fountains. This modest garden was limited in extent to the area in

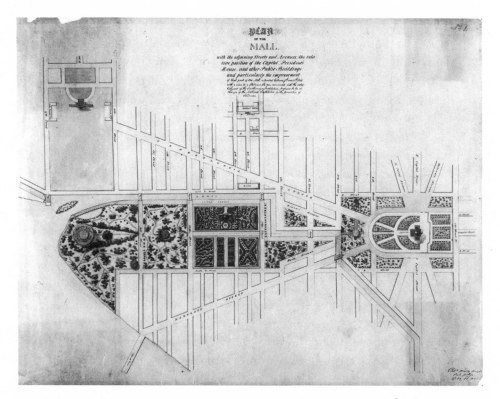

FIGURE 4
Robert Mills, "Plan of the Mall," 1841. (National Archives)

front of the proposed national museum, for which Mills developed a
romantic pseudo-medieval design. In a more ambitious proposal, Mills
expanded the scope of his commission to encompass the entire Mall (figure 4). The area between Seventh Street and Twelfth Street was devoted to
an eclectic array of horticultural displays that mixed informal paths and
plantings with more utilitarian linear beds ostensibly intended to evoke
medieval medicinal gardens. Together with the architectural treatment proposed for the national museum, these archaic references reflected the influence of the Gothic Revival that was beginning to permeate American art,
architecture, and popular culture. Mills's proposal for the western portion of
the Mall was more open and spacious, with specimen trees and informal
plant groupings punctuating rolling lawns interlaced with winding paths
in the informal "natural" or "gardenesque" manner popularized by the
English landscape designer John C. Loudon and his American acolyte

Andrew Jackson Downing. Mills's proposals for the Mall and for a national museum were not adopted, but James Renwick's design for the Smithsonian Institution evinced the same spirit of gothic revivalism, and Downing's subsequent plan for the Mall displayed a similar gardenesque sensibility.[22]

Downing's 1851 plan for the Mall epitomized mid-nineteenth-century perceptions of idealized urban nature. This was hardly surprising, as the New York designer, nurseryman, and writer played a preeminent role in articulating the reigning fashions in landscape design, rural architecture, and public park development. Downing adapted English theories to elite and middle-class American audiences, designing estate grounds along the Hudson River, publishing influential books on architecture and landscape design, and editing *The Horticulturist,* a popular journal that set the pace in such matters. Downing viewed the Mall commission as an ideal opportunity to produce a national object lesson in the aesthetics and ideology of the American public park. The biggest problem, Downing noted, was that no such thing existed. Several older cities had transformed obsolete commons and fortifications into public promenades, and a few were graced with ornamented squares and garden-like cemeteries, but the rapid growth of American cities had left urban populations without easy access to the edifying influences of nature and attractive scenery. Downing and his associates lobbied energetically for the establishment of bona fide parks in major American cities. Transforming the Mall from an undeveloped morass into a model public pleasure ground would provide an ideal opportunity to demonstrate the social value and the design characteristics of a truly American approach to the long-cherished ideal of "rus in urb."[23]

Downing outlined his vision in a brief essay submitted with the 1851 plan. First and foremost, he proposed to create America's first "national Park." Downing's vision of the ideal national park was not an enclosure of pristine wilderness, which could be found in abundance throughout the frontier regions at the time, but an explicitly designed landscape in the center of the nation's capital. Second, by inventing a prototype for the American public park, he hoped that his scheme would serve as a model for similar developments throughout the country. Third, he emphasized the proposal's educational value as "a public museum of living trees and shrubs," where visitors to Washington could steep themselves in horticultural knowledge. Downing declared that this combination of artistic, scientific, and democratic principles "would serve, more than anything else that could be devised, to embellish and give interest to the Capital."[24]

Downing's design stretched from the Capitol to the monument grounds and the White House (figure 5). Though it had six distinct segments, it displayed greater unity than Mills's eclectic efforts, the entire composition being held together by a simpler and more forceful and consistent series of paths and picturesque landscape effects. The area south of the White House, now known as the Ellipse, would consist of a circular field or "parade ground" bordered by a ring of trees. An ornate suspension bridge over the moribund Washington Canal provided access to the Mall. The monument grounds consisted of broad open lawns adorned with looping paths and artfully placed plantings of native trees. The next section, between 12th Street and 14th Street, was the most anomalous to modern eyes. Downing proposed to create an evergreen garden, with an encyclopedic array of native and foreign examples clustered around concentric paths. In addition to its educational value, this garden would provide greenery in the winter months, when Congress was in session and the tourist season was at its pre-air-conditioning height. Where the Mall narrowed between Pennsylvania Avenue and Maryland Avenue, Downing proposed a "Fountain Park" that reinterpreted L'Enfant's grandiose formal cascade as a modest circular fountain balanced by an irregularly shaped man-made lake flanked by picturesque plantings. At the base of the Capitol, Downing called for the construction of additional greenhouses and a few minor alterations to the existing Botanical Garden.

If Downing's proposal for the Mall had been enacted, Washington would undoubtedly have played a pivotal part in reinventing the form and function of urban nature in mid-nineteenth-century America. Unfortunately, Downing died in a steamboat fire a year after submitting his plan. The Smithsonian's grounds were developed more or less along the lines he had proposed, but the broader outlines of his plan were never realized. Washington's potential influence as a model was soon eclipsed by that of New York, where Central Park played the leading role in sparking the trend toward naturalistic park development that transformed the American urban landscape in the second half of the nineteenth century.

While park advocates experienced continued setbacks in their efforts to reinvent Washington's natural landscape, the District of Columbia government and the US Army Corps of Engineers radically transformed the topography, ecology, and hydrology of the nation's capital during the second half of the nineteenth century. One of the fundamental steps in this reinvention of Washington's urban environment was the provision of a safe

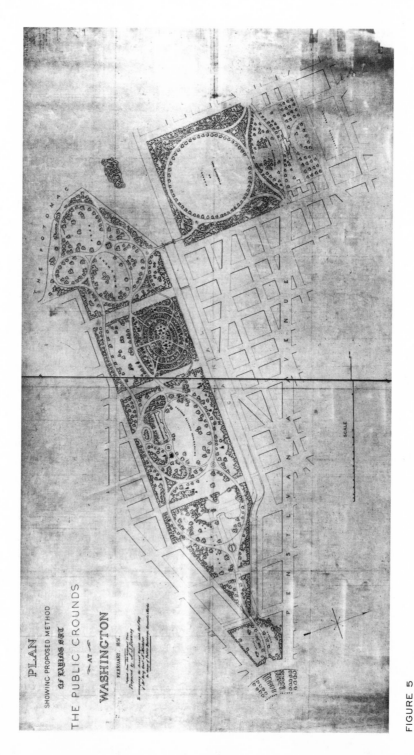

FIGURE 5

A. J. Downing, "Plan Showing Proposed Method of Laying Out the Public Grounds at Washington, 1851." (National Archives)

water supply

and dependable water system. As Washington outgrew the dimensions of a simple country town, local water sources proved inadequate and, more often than not, unsafe. To remedy this situation, the Army Corps of Engineers developed an elaborate supply system that drew water from the Potomac above Great Falls, diverted it to a purifying reservoir near Little Falls, then conveyed it through a large conduit to Georgetown, where it was piped to subsidiary reservoirs located in various sections of the city. The system was improved and expanded repeatedly over the ensuing decades to keep pace with the city's growing population and geographic expansion.[25]

The Army Corps of Engineers' most striking environmental intervention was the radical transformation of the city's relationship with the Potomac, which involved reconfiguring the depth and breadth of the stream, constructing the self-scouring Tidal Basin, and creating extensive expanses of filled land that are now taken for granted as natural elements of Washington's topography. By the middle of the nineteenth century, the siltation problems that had plagued the bend of the Potomac near Washington since the city's founding had grown intolerable. The decreased water flow and increased siltation brought on by deforestation and development had reduced the once-imposing stream into a shallow sheet of water that expanded and retracted with the tide, exposing hundreds of acres of noxious and intractable mud flats. In addition to drastically curtailing the utility of local harbors, the waterfront morass of swamps and mud flats was unsightly and unwholesome. The combination of stagnant waters and raw sewage discharged from the city's rudimentary sanitation system posed significant public health hazards. It was frequently pointed out that the executive mansion was located scarcely more than a stone's throw from this pestilent slough, posing a constant threat to the president and his family. In 1881 a particularly severe spring flood inundated low-lying reaches of the city, backing sewage-laden water as far inland as the Botanical Gardens and convincing Congress that something had to be done to discipline the unruly river.

re-shaping Potomac

During the 1880s the Army Corps of Engineers embarked on an ambitious construction campaign aimed at confining the main flow of the river to a single clearly defined and navigable channel, draining and filling in the marshes and tidal flats, and establishing a solid and permanent shoreline. Steam-operated dredges excavated thousands of tons of sedimentary material from the river bottom, transforming the Potomac into a much deeper but narrower river (figure 6). The dredged material was deposited on the

FIGURE 6

"Improvement of the Potomac Flats, Washington, D.C." Illustration from *Scientific American,* September 19, 1891.

surrounding swamps and mud flats to create hundreds of acres of new park land that was carefully drained and shaped to produce broad lawns, well-defined water features, attractive paths, and a mixture of formal tree-lined boulevards, curving carriage drives, and naturalistic plantings. So seamlessly were these new lands blended with the surrounding environment that few visitors realize that the western half of the Mall from 17th Street to the Lincoln Memorial, West Potomac Park, and East Potomac Park is entirely man-made.[26]

While the Army Corps of Engineers transformed Washington's waterfront, the District government busied itself with equally ambitious alterations to the city's physical fabric. Determined to modernize the sleepy southern town, Board of Public Works leader Alexander "Boss" Sheperd initiated an aggressive civic improvement campaign. Sheperd's men laid down an extensive system of gas and sewer pipes, greatly improving public health conditions and literally turning darkness into light with widely popular street illumination programs. More controversially, Sheperd set about redressing disparities between L'Enfant's rigidly geometric street plan and the anomalies of Washington's natural topography. Hills were leveled and declivities filled to produce uniform street grades and a more orderly and urbanized appearance. The street improvement campaign was a boon to speculators building subdivisions outside the original urban core, but in previously developed areas the heavy-handed imposition of uniform street grades left houses and businesses perched on hillocks or overshadowed by embankments carrying the new roadways. On a more positive note, city forces initiated a street tree planting campaign of monumental proportions. By 1881 Washington could boast of 120 miles of verdant avenues shaded by over 53,000 newly planted trees. Combined with federal efforts to embellish the city's squares and circles, this campaign helped transform Washington from a raw and unsophisticated country town into an increasingly tidy and mature, if still somewhat sleepy, metropolis. The legacy of shady tree-lined streets and neatly tended small parks lasted well into the twentieth century, when the ravages of Dutch elm disease, automobile-induced street widening, and shrinking maintenance budgets diminished the sense of leafy repose that characterized large portions of the city for the better part of a century.[27]

Despite radical transformations to the city's physical environment, the nation's capital entered the last quarter of the nineteenth century without a proper public park along the lines envisioned by Downing and developed with great success by his successor Frederick Law Olmsted, whose under-takings in New York's Central Park set off a mania for park building throughout the country. These expansive reservations contained hundreds of acres of rolling lawns and carefully composed naturalistic plantings. While they were still steeped in the ideology of romantic landscape appreciation, the new generation of park advocates cast nature in a more intellectually accessible and essentially therapeutic role as an anodyne to the forces of industrialization and urbanization that were rapidly transforming American society. Scientifically organized horticultural displays and the intricate, highly stylized "gardenesque" planting arrangements favored by Downing gave way to broad vistas and simpler, more naturalistic groupings of trees and shrubs that were meant to offer soothing relaxation rather than taxing mental stimulation. Though less obviously manipulated, the topography and vegetation of these urban and suburban parks were meticulously trans-formed to present an idealized version of the pastoral scenery and wood-lands that was increasingly beyond reach for most urban Americans. According to Olmsted, parks developed in this vein would act "in a directly remedial way to enable men to better resist the harmful influences of town life and to recover what they lose from them."[28]

As urban development began to spread into the rolling hills beyond the boundary of the original L'Enfant plan, park advocates insisted that some-thing had to be done to provide Washington with the sort of extensive naturalistic playground that had come to be regarded as an essential component of the metropolitan landscape. A consensus emerged by the end of the Civil War that the largely undeveloped Rock Creek Valley afforded the most advantageous location, though lengthy debate ensued over the extent and boundaries of the proposed park and the question of who should bear the responsibility of paying for its acquisition and development. Advocates for the proposed Rock Creek park combined the flowery rhet-oric of romantic landscape appreciation with democratic idealism and Olmsted's concerns for the psychic needs of an increasingly congested urban population. In 1866, Army Corps of Engineers Major Nathaniel Michler asserted that the desirability of turning the upper reaches of Rock Creek into a park was so self-evident that the basic merits of the proposal scarcely needed to be addressed. Summarizing prevailing beliefs in the moral, therapeutic, and democratic virtues of park development, Michler advised that the proposed reservation would afford opportunities for "all classes of society" to engage in "healthful recreation and exercise" while enjoying a pleasing and uplifting landscape that would "cultivate an appreciative and refined taste in those who seek its shade for the purpose of breathing the free air of Heaven and admiring nature."[29]

Michler articulated the prevailing vision of an idealized "natural" environment in his report to Congress. "There should be a variety of scenery," Michler maintained, "a happy combination of the beautiful and the picturesque—the smooth plateau and the gently undulating glade vying with the ruggedness of the rocky ravine and the fertile valley, the thickly mantled primeval forest contrasting with the green lawn, grand old trees with flowering shrubs. Wild, bold, rapid streams, coursing their way along the entire length and breadth of such a scene, would not only lend enchantment to the view but add to the capabilities of adornment."[30]

All of these qualities could be found along the valley formed by Rock Creek, Michler insisted. To nineteenth-century eyes, however, even this intrinsically attractive landscape required significant human intervention to fulfill its potential as an exemplar of nature's beauty and beneficent influence. Trees, rocks, and hillsides were not ecological entities to be preserved for their own sake, but raw material from which to develop pleasing compositions and edifying experiences. While the proposed park site was richly endowed with diversified natural amenities, Michler advised that it still required "the taste of the artist and the skill of the engineer to enhance its beauty and usefulness." Michler proposed "gentle pruning and removing what may be distasteful, improving the roads and paths and the construction of new ones, and increasing the already large growth of trees and shrubs, deciduous and evergreen, by adding to them those of other climes and countries." He also advocated the construction of dams along Rock Creek to produce ornamental lakes and ponds. Though the existing woodlands had many attractions, the "natural" forest was also found wanting. Michler advocated a carefully considered campaign of selective vista cutting to provide more varied and expansive views.[31]

While this sort of aggressive reshaping of nature might strike today's readers as inappropriate and even "unnatural," by the time Rock Creek Park was finally developed at the turn of the century, the culture of nature had changed significantly. Traditional landscape design and appreciation were becoming lost arts. In fact, the notion that park-making was an art and not simply a matter of legislative protection was gradually disappearing, first among the general public and then within the profession itself. Rather than attempting to reshape existing conditions to produce classically ordained visual compositions, <u>park-makers increasingly devoted their energies to protecting surviving tracts of woodland from adverse development.</u> Existing forests were often improved through scientific management practices, but development was generally limited to the provision of circulation systems and simple shelters. Selective vista clearing was still commonly practiced, but the elaborate pictorial effects favored by mid-nineteenth-century designers were increasingly viewed as fussy, antiquated, and "unnatural." Parks were cast not so much as picturesque set pieces meant to inspire traditional aesthetic or religious contemplation, but as relics of primeval America where soft and over-urbanized modern Americans could associate with the same environments that had theoretically ennobled their pioneering forefathers or experience the more abstract spiritual virtues and physical pleasures that wilderness advocates such as John Muir ascribed to encounters with untrammeled nature. The rapid expansion of the national park system around the turn of the century reflected this shifting emphasis. On the metropolitan level, Massachusetts led the way with the creation of a system of reservations designed to preserve scenic areas in the greater Boston area. When Rock Creek Park was officially authorized in 1890, it united these two trends, joining Yosemite, Sequoia, and General Grant (now King's Canyon) as the first generation of new national parks since Yellowstone.[32]

Between 1890 and 1918, the Army Corps of Engineers constructed a series of paths and carriage drives to promote public access to Rock Creek Park. When it came to the management of vegetation and other aspects of the park's "natural" environment, however, biological processes were more or less allowed to take their "natural" course. The result of <u>this "hands-off" management strategy, which stemmed at least as much from budgetary constraints as from conscious intent,</u> was that the grassy meadows, open woodlands, and varied understory that had endowed much of Rock Creek Valley with a picturesque, park-like character as former estates and recently abandoned farmlands gradually gave way to an overgrown environment of dense second-growth forest and riotous underbrush. Landscape architects influenced by traditional park design practices called for modest vista-clearing efforts and other aesthetic improvements throughout the first third of the ~~nineteenth-century~~ but to little avail; "nature" continued to reclaim the park. The resulting landscape may have lacked the refinement and picturesque variation envisioned by Michler and his contemporaries, but the public clearly enjoyed the park's sylvan paths and shady drives, along with

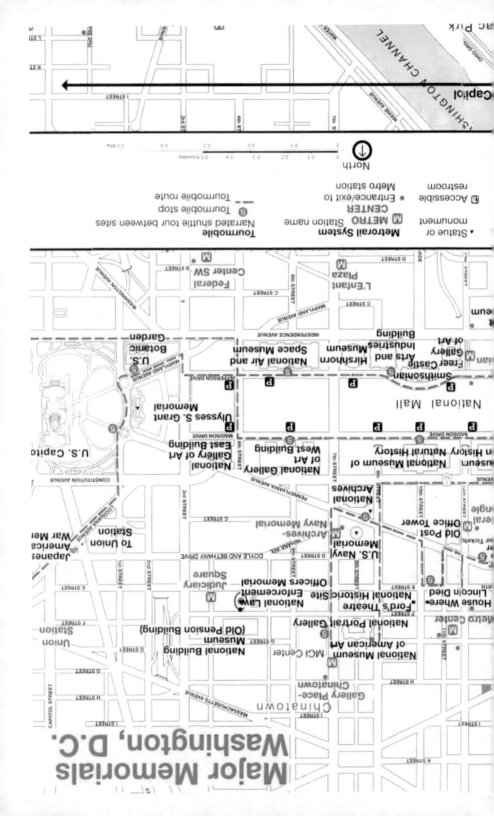

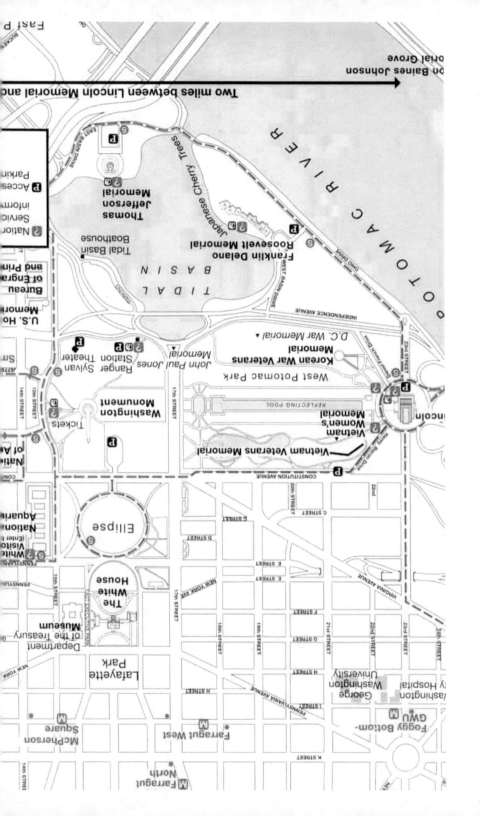

the numerous picnic areas and the picturesque artificial pond constructed alongside the historic Pierce Mill.[33]

The next major attempt to reinvent the nature of nature in Washington was crystallized by the Senate Park Commission plan of 1901–02. This influential compendium of proposals epitomized the goals and methods of the City Beautiful movement, which exercised a powerful influence on American urban development from the 1890s through the 1930s. The basic tenet of the City Beautiful movement was that American cities were ugly, chaotic, and inefficient, and that coordinated planning improvements based on European models could transform them into models of civic achievement worthy of a great and powerful nation. City Beautiful spokesmen urged architects, landscape architects, sculptors, artists, and businessmen to work together to improve the appearance of American cities.[34]

The City Beautiful movement is usually identified with ostentatious neoclassicism and associated with grandiose architectural projects and civic center developments rather than with attempts to reconfigure urban nature. City Beautiful reformers were strongly influenced by the nineteenth-century park movement, however, and their proposals generally included plans for comprehensive systems of informal parks and parkways that were intended to spread the benefits of nature throughout the city and its suburbs. These new parks were cast not simply as Arcadian retreats, but as multi-faceted contributions to urban improvement. Parks systems were proposed as means of transforming under-utilized and often menacing urban wastelands into sanitary, attractive, and profitable civic amenities. They were also seen as benefiting their surroundings, as improving the character of adjoining districts, and, according to the Darwinian notions of environmental determinism that informed the growing science of urban sociology, as elevating the habits of neighboring residents, as well.

The City Beautiful movement had an enormous effect on Washington. It was bad enough that the average American city was ugly and amorphous, but politicians, architects, and reformers of many stripes expressed outrage that the nation's capital was disorganized and with a few notable exceptions, unattractive. There were a few grand mansions and impressive public buildings, along with the smattering of minor parks embellished during the preceding quarter century, but Rock Creek Park was all but inaccessible and the Mall remained a major disappointment. The Smithsonian area and the monument grounds were attractively landscaped in the gardenesque style advocated by Downing, but railroad tracks and a train station marred the east end and much of the rest remained haphazardly developed and poorly maintained. Rock Creek Park provided an attractive sylvan retreat, but it could only be reached through busy city streets or by laboriously negotiating the lower valley, which had degenerated into a heavily polluted industrial zone and public dumping ground.

In 1900, the centennial of the government's removal to Washington provided a prime opportunity for critics to call for a major effort to restore the order and dignity envisioned by Washington, Jefferson, and L'Enfant a century before. The American Institute of Architects held its convention in Washington that year and focused on the issue of improving the city's appearance. Together with local park advocates, the AIA convinced Senator James McMillan of Michigan, chairman of the Senate Committee on the District of Columbia, to authorize a special commission charged with developing guidelines for the coordinated improvement of public grounds in the nation's capital. The Senate Park Commission (the McMillan Commission, as it became popularly known) reconnoitered the situation in Washington and conducted a thorough study of existing plans. They even sailed to Europe for inspiration, examining the parks and public spaces of London, Paris, Rome, and other capitals before submitting their report, which is recognized as one of the classic documents of American city planning.[35]

Eloquently written and lavishly illustrated, the Senate Park Commission report called for "the treatment of the city as a work of civic art" through informed, tasteful, and carefully coordinated planning.[36] Criticizing the previous century of piecemeal and seemingly haphazard development, the commission praised L'Enfant's clear and well-ordered plan and proposed to resurrect its emphasis on geometric precision and monumental civic grandeur. Downing's "naturalistic" landscaping was condemned as an egregious misinterpretation of the Mall's intended form and function. In keeping with the contemporary affinity for grand beaux arts gestures, the commission insisted that the Mall should be redeveloped to achieve the Baroque splendor envisioned by L'Enfant. The nineteenth century's eclectic romanticism would be swept aside in favor of a vast "green carpet," 300 feet wide and $2\frac{1}{2}$ miles long that would be bordered by quadruple rows of elm trees to enhance its imposing formal effect. Both the Botanical Garden and the picturesque plantings fronting the Smithsonian were to be removed along with the railroad tracks and other affronts to the monumental dignity envisioned by the commission. The eclectic romanticism of

Renwick's Smithsonian building was also sacrificed on the altar of order, clarity, and geometric precision. The ungainly hill at the base of the Washington Monument—one of the last relics of the Mall's original topography—would be replaced by an architectonic extravaganza of terraces, steps, pools, and fountains. On the newly filled land at the Potomac end of the Mall, a grand neoclassical temple surrounded by rigidly geometric plantings would honor the memory of Abraham Lincoln and provide a dramatic punctuation to the monumentally formal space envisioned by the commission (figure 7).[37]

The Senate Park Commission plan would eventually play a prominent role in guiding the development of Washington's public spaces. Yet it was neither universally endorsed nor comprehensively adopted. Many politicians (including Speaker of the House Joseph Cannon) objected to its extravagant price tag. Funding shortages plagued every aspect of the plan's development and many elements of the original vision were cast aside to cut costs or placate politicians. Local opposition arose to the commission's proposals to eliminate the Botanical Gardens and strip the Mall of its existing plantings, which were prized by many people for their picturesque charm, cooling shade, and seemingly natural appearance. The Washington

FIGURE 7
"General View of the Monument Garden and Mall, Looking Toward the Capital." Illustration from William Howard Taft and James Bryce, *Washington: The Nation's Capital,* 1913 and 1915.

Evening Star editorialized against "the destruction of all the noble shade trees in the People's Park for the sake of a sixteen-hundred-feet-wide track of desolation as arid and as hot as the Desert of the Sahara." The sentiment that Washington's "natural" landscape was under assault by auto-cratic planners bent on imposing a rigidly formal artificial environment was epitomized in a contemporary cartoon depicting axe-wielding commis-sioners marching on Capitol Hill, while a frightened bear peered out of the trees at a phalanx of gardeners lugging cube-shaped boxwoods. The elabo-rate monument gardens were rejected as technically infeasible due to the unstable nature of the underlying soils, but the broader outlines of the com-mission's plans for a formal open Mall were eventually realized.[38]

The treatment of Washington's monumental core has always generated the most attention, but the commission also prepared a detailed plan call-ing for a comprehensive system of parks, parkways, and playgrounds spread-ing throughout the nation's capital. Acknowledging that the elaborate park system it proposed was not part of L'Enfant's original plan, the commission insisted that it reflected "the need, not recognized a hundred years ago, for large parks to preserve artificially in our cities passages of rural or sylvan scenery and for various spaces adapted to various forms of recreation."[39]

Warning that the rapid pace of development called for quick action, the commission recommended the improvement of existing parks, the creation of tree-lined parkways linking "spots of exceptional beauty," the conversion of the banks of the Potomac into attractive recreational areas, and the estab-lishment of a continuous ring of parkland connecting the abandoned Civil War forts that encircled the capital city. This extensive park system would not only be beautiful and uplifting, it would provide demonstrable public-health benefits. The new parks would restore polluted beaches and streams, reclaim debris-choked valleys and industrial zones, and eliminate the unsightly habitations and lower-class populations that tended to materialize on these marginal lands. In the days before air-conditioning, a well-developed park system was also seen as a palliative to Washington's oppres-sive summer heat. Forested parks would provide soothing shade and breezy hilltops. In addition, the cool air that settled in the preserved stream valleys would offer a respite from hot summer temperatures.[40]

One element of this broader park plan was the development of a park-way along Rock Creek from the Potomac waterfront to Rock Creek Park. By this time parkways were regarded as essential components of any well-considered city plan. By providing attractive links between civic centers and suburban parks or residential neighborhoods, they exemplified the City Beautiful goal of combining traditional social and aesthetic motivations with more modern concerns about the health, efficiency, and economic vitality of rapidly growing metropolises. Parkways were often located along polluted urban watercourses, where they were used to eradicate public health menaces, rejuvenate degraded environments, and replace threatening lower-class neighborhoods with soothing naturalistic landscapes devoted to middle-class leisure and recreation.[41]

The lower portion of Rock Creek Valley seemed ideally poised for this type of development. While the region from Dupont Circle north to the southern boundary of Rock Creek Park remained attractively wooded, the lower valley and the Potomac waterfront had become a combination industrial zone and public dumping ground. Ramshackle tenements and huge banks of refuse lined the creek. The C&O Canal and a number of small industrial concerns robbed the creek mouth of any semblance of natural beauty. Small factories, a brewery, and the towering facilities of the Washington Gas Light Company loomed over the Potomac Waterfront in the area now occupied by the Watergate apartment complex and the Kennedy Center. Conditions along the creek were so bad that a number of citizens' groups lobbied to enclose the stream in a culvert and fill in the valley to create a level connection between Washington and Georgetown. A grand formal boulevard would trace the route of the covered creek and, it was hoped, serve as the spine of an attractive residential district along the lines of Boston's Back Bay, which had also been constructed on land reclaimed from an urban "wasteland." Another group of local reformers argued that the stream should be cleaned up and the valley restored to a semblance of its original conditions so that it would afford an attractive and seemingly "natural" link between the Mall and Rock Creek Park.[42]

The Senate Park Commission weighed both options and ruled in favor of the restored valley plan, which it deemed preferable on aesthetic, economic, and practical grounds. Restoring the valley would require an enormous investment of time and money, but the commission insisted that a winding creekside drive surrounded by naturalistic scenery would afford a more appropriate and attractive addition to the city's park system. A major attraction of the streamside location, it was noted, was that there would be no cross-streets to interfere with recreational traffic on the parkway drive. There were fewer than 8,000 automobiles in the entire country, and the commission made no reference to the needs of motorists, but carriage owners were equally enamored of long, uninterrupted pleasure drives. A mixture of formal and informal parkways would perform similar functions in other parts of the city. Longer parkways would extend along the Potomac to Great Falls and Mount Vernon.[43]

Despite the Senate Park Commission's prestige and its success in initiating the wholesale transformation of the Mall, little progress was made on broader park and parkway development initiatives during the first quarter of the twentieth century. Rock Creek and Potomac Parkway was not authorized until 1913, and land acquisition for the project dragged on into the 1920s. Citing the loss of natural areas to development, park advocates pressed the case for additional protection and improvements. Local and national lobbying efforts led to the creation in 1924 of the National Capital Park Commission, an advisory federal agency intended to promote and coordinate park development throughout the region. Reflecting the increasingly broad-based agendas of such bodies, the agency was renamed the National Capital Park and Planning Commission 2 years later. The removal of "park" from the title 25 years later bore witness to the waning influence of traditional park-making concerns on the evolution of the modern urban fabric.[44]

By the 1920s, the traditional vision of the urban park as a place of pastoral repose for the contemplation of picturesque scenery was fast disappearing. The traditional landscape park ideology was giving way to the notion that urban nature should provide opportunities for active sports and games. Aesthetic resistance to the rigid geometry of playing fields and class prejudices against the probable participants in publicly subsidized sports had led Olmsted and other nineteenth-century park designers to banish these elements from the genteel confines of their edifying compositions. By the 1920s, however, groups such as the National Recreation Association and the National Conference on Outdoor Recreation were vigorously promoting the creation of parks designed for active sports and structured play. Playgrounds and athletic fields would theoretically improve the physical and moral fiber of urban Americans, who were thought to be not only less robust than their predecessors but also inclined to juvenile delinquency, Bolshevism, and just plain laziness. While the Senate Park Commission plan had included modest provisions for playground development, the district and federal governments increased their activities in this area during the 1920s. Depression-era relief programs helped provide funds and manpower to build new playgrounds and transform sections of existing parks into places for neighborhood residents to engage in sports and games. Many of Washington's playgrounds and recreation facilities date from this period.[45]

The rapid rise in automobile ownership after World War I confronted park managers with new challenges. On the one hand, the proliferation of automobiles overwhelmed existing park roads, which had been designed to accommodate the carriages of the urban elite. Questions were quickly raised about the impacts of automobile traffic on parks, which ranged from noise, congestion, and noxious fumes to the rapid deterioration of road surfaces and the introduction of an ill-behaved class of park users, who raced about in their motorcars with little concern for the finer points of landscape appreciation. On the other hand, the spread of middle-class and even working-class automobile ownership enabled many urbanites to range far and wide in search of nature and rural scenery. This unprecedented mobility helped ease the burden on urban parks while providing a strong constituency for the development of wide-ranging county, regional, and statewide park systems. Park planners in Washington and throughout the country expanded their horizons, developing ambitious proposals for elaborate networks of parks, beaches, and recreation areas that would be linked with each other and tied to city centers with lengthy and lavishly developed automobile parkways. While parkways had been integral to urban park schemes since the 1870s, the automobile elevated them to even greater prominence as both city planners and the general public rapidly realized they could serve double duty as attractive recreational corridors and efficient commuter arteries.[46]

The evolution of Rock Creek and Potomac Parkway exemplified national trends. The long delay between the parkway's conception and its completion coincided with America's transition from horse and buggy to automobile. Changing perceptions of the parkway's intended form and function were evident in both the evolution of the project's design and the shifting language employed to explain its attractions and significance. Early plans called for an elaborate network of carriage drives, pedestrian promenades, and bridle paths surrounded by elaborate picturesque plantings. Subsequent development, together with the wider, straighter roads required by automobiles, resulted in the elimination of many picturesque design features, a significant reduction in the number of exits and entrances, and a general simplification of the parkway landscape. Extensive excavations, grading, and plantings were still required to transform the polluted valley into a seemingly "natural" landscape, but topographic manipulations and

FIGURE 8
Constructing Rock Creek and Potomac Parkway: excavations between M and P
Streets, 1934. (Washingtoniana Division, D.C. Public Library)

plant massings were broader and less complex. Not only were motorists
likely to be more concerned about getting to work on time than with comparing the views through their windshields with traditional landscape
paintings, but higher speeds made it impossible to linger over the complex
pictorial compositions favored by earlier designers. Limiting access from the
surrounding neighborhoods improved efficiency, but it made the parkway
function less as a local park and more as a green tunnel linking northwest
Washington and the central city. No longer called upon to provide illustrations of moral, political, or aesthetic theories, "nature" was merely asked to
screen motorists from unsightly development and create a comforting
impression of encompassing greenery. Still, when the parkway was completed, it contained a bridle path, foot trails, neighborhood parks, and picnic areas and proved popular as a multi-use linear park.[47]

FIGURE 9
Inventing nature along George Washington Memorial Parkway. (Library of Congress)

Newspaper coverage of the parkway's completion reflected these changes. Referring to the largely man-made landscape as a "great natural garden," the *Washington Post* underscored its original role as a component of Washington's park system. The *Evening Star,* however, emphasized its role as a traffic artery: "The special value of the parkway is that it will afford an uninterrupted passage to the downtown area, or to Virginia, by avoiding the many intersections and traffic congestion that plague motorists on the regular street routes." Conflating nineteenth-century nature worship with the concerns of modern-day commuters, the paper rejoiced that, with the project's completion in 1936, "motorists from the Chevy Chase–Bethesda area will have the privilege of riding downtown through a veritable fairyland, a natural setting for nature's own worship, and not so much as a traffic light to impede progress. There is, perhaps, no city in the world offering so much beauty for those going to work."[48]

During the 1920s and the 1930s, parkway development seemed like an ideal means of combining recreation, natural resource protection, and transportation. Buoyed by the examples of Rock Creek Parkway, Potomac Parkway, and Mount Vernon Memorial Highway, which afforded an attractive tree-lined route from Washington to Mount Vernon when completed in 1932, planners proposed an ambitious parkway development program for the Washington area. George Washington Memorial Parkway would cover both sides of the Potomac between Mount Vernon and Great Falls, while the C&O Canal Parkway would follow the old canal bed all the way from Georgetown to Cumberland. Other parkways would preserve and rehabilitate threatened stream valleys, provide recreational opportunities, and service commuters and real estate development throughout the capital region.[49]

Even as these expansive plans were being prepared, however, a combination of social and technological factors made it increasingly apparent that the competing functions of "park" and "way" could no longer be accommodated in single, multi-purpose environments. Rising speeds and traffic volumes spurred demands for wider, straighter, and more efficient transportation corridors. At the same time, park advocates placed increasing emphasis on preserving natural areas as ecological reserves with a minimum of extraneous development. Park road and parkway development was increasingly viewed as incompatible with the ascendant ideology of natural resource preservation. As suburban development reduced the amount of open space surrounding the nation's capital while dramatically increasing commuter traffic, supporters of these conflicting agendas transformed Washington's parks into battle zones, waging protracted campaigns for supremacy that raged throughout the 1940s, the 1950s, and the 1960s and have not entirely disappeared.

Conflicts came to a head when highway engineers sought to accommodate rising traffic demands by routing expressways through existing parks and upgrading parkways into high-speed "freeways." From the highway engineers' perspective, parks were not "nature" but vacant space—and vacant space that was already owned by public authorities and thus exempt from the prolonged and politically charged condemnation procedures that accompanied freeway construction in residential neighborhoods. Traffic planners prepared detailed studies enumerating the ways in which Washington's traffic needs could be satisfied by converting the city's parks into bustling commuter arteries. Rock Creek would serve as a convenient funnel for traffic to the northwest suburbs. Elevated freeways would flank the Mall. New bridges would cross the Potomac, shattering the stillness of the Theodore Roosevelt Island nature preserve and obliterating the picturesque serenity of the Three Sisters Islands.[50]

Citizens' groups, park managers, and conservation associations rallied to oppose these plans. Freeway promoters countered that they were not harming the parks, but making their natural beauty accessible to thousands of motorists who would never visit them otherwise. When politically connected opponents blocked a proposed expressway through Rock Creek Park, traffic engineers attempted to shift the proposed roadway through Glover-Archbold Park instead. When this plan was also foiled, the expressway was relocated further west, where it was euphemistically labeled a parkway and routed underneath the historic Cabin John Aqueduct, eradicating the tranquil scenery that had long drawn visitors to the site. The massive Theodore Roosevelt Bridge was constructed in the 1960s, but the Three Sisters Bridge proposal was finally dropped after repeated skirmishing convinced transportation planners to retreat. Environmentalists also challenged the National Park Service's own parkway plans. The C&O Canal Parkway project was shelved after a series of high-profile protests involving figures such as Supreme Court Justice William O. Douglas. The development of the George Washington Memorial Parkway was dramatically curtailed. Broader economic and political concerns factored into these decisions, but environmentalist opposition played a prominent role in sealing the parkway's fate.[51]

Planning documents of the 1950s and the 1960s bore witness to yet another transformation in reigning conceptions of the nature of nature in the nation's capital. While highway engineers viewed parks as potential transportation corridors, and environmentalists rallied to protect scenery and ecological associations, planners presented undeveloped areas as a generic abstraction called "open space." This new terminology reflected the planning community's determination to disassociate itself from earlier aesthetically motivated ideologies and emphasize an increasingly technocratic focus on the orderly allocation of a broad array of community services ranging from parks and playgrounds to housing, employment, and transportation. The planning commission's 1950 comprehensive regional plan employed the term "open space" as a general rubric covering undeveloped private land, parks, playgrounds, parkways, and neighborhood recreation

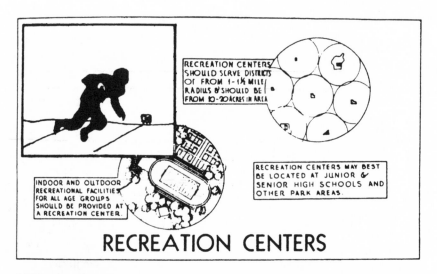

FIGURE 10

"Types of Recreation Facilities," 1950. (US National Capital Park and Planning Commission)

SUPERVISED NEIGHBORHOOD PLAYGROUNDS ON SCHOOL, PARK, OR OTHER PUBLIC PROPERTIES ARE THE BASIC UNITS OF THE RECREATION SYSTEM PLAN.

ELEMENTARY SCHOOL SITES WHEN USED AS A NEIGHBORHOOD PLAYGROUND IN THE RECREATION SYSTEM SHOULD HAVE AN AREA OF AT LEAST 5 ACRES

THERE SHOULD BE A PLAYGROUND WITHIN ¼ MILE OF EVERY CHILD.

SUPERVISED PLAYGROUNDS

EVERY NEIGHBORHOOD SHOULD HAVE ITS NEIGHBORHOOD PARK. WHERE POSSIBLE THIS SHOULD INCLUDE AREAS OF SPECIAL NATURAL INTEREST

IF POSSIBLE A NEIGHBORHOOD PARK SHOULD HAVE AN AREA OF BETWEEN 10 & 20 ACRES

SOME PARKS OF THIS TYPE MAY HAVE A PORTION OF ITS AREA USED FOR PICNICKING

NEIGHBORHOOD PARKS

areas. Elaborate tables were prepared, ratios of appropriate acreage per capita were enumerated, and the various types of spaces were described in the language of sociology and economics rather than in the traditional discourses of landscape design, environmental science, or nature worship. Despite these rhetorical changes, the 1950 open space program was not appreciably different than earlier park development proposals.[52]

A decade later, when the National Capital Planning Commission (NCPC) issued its Year 2000 Policies Plan, the conceptual framework had shifted to an even more abstract basis. Undeveloped areas were portrayed as amorphous green spaces, to be pushed and pulled into various configurations as planners pondered schematic development strategies, which they portrayed in colorful renderings that seemed to have more to do with contemporary abstract art than with actual cities or physical environments. Reacting to the monumental vagueness of this brand of open-space planning that appeared to give no heed to the physical characteristics of the landscapes it sought to manipulate, the next major attempt to articulate the nature of Washington's nature embraced the natural sciences as a model. Following the ecology-based design principles espoused by University of Pennsylvania landscape architect Ian McHarg, the 1967 NCPC monograph *Toward a Comprehensive Landscape Plan* presented the underlying geologic and biotic features of greater Washington through a series of elaborately detailed maps. Riding the rising tide of the ecology movement, this meticulously assembled and ostensibly scientific overview was intended to support an ecologically conscious approach to the design and management of Washington's physical environment. While the McHarg plan had little immediate effect, it exemplified the trend toward conceiving nature as a mosaic of ecological associations. Park management was increasingly seen as a process of identifying, restoring, and preserving biological processes in the hopes of reproducing the stable and harmonious conditions that ostensibly prevailed before European settlement. Ecology-based notions of environmental protection replaced picturesque aesthetics and social improvement as the ideological basis of natural resource management.[53]

The outdoor recreation craze of the 1960s also had a marked effect on Washington's natural areas. As in the 1920s and the 1930s, parks were promoted as places to engage in active physical exertion. The team sports and structured play activities favored by early-twentieth-century reformers remained popular—and playgrounds and athletic facilities continued to be built—but there was an increasing emphasis on individual sports. The broader natural environment was cast as a recreational space through which one could run, bike, or hike. This health consciousness, together with the desire to develop more environmentally sensitive methods of commuting, established bicyclists as rising influences on park management. During the late 1960s, Rock Creek Park officials began restricting automobile access to the main park drive on Sunday mornings. This initiative soon expanded to weekends and holidays. By 1971 the rising popularity of cycling produced a short-lived attempt to allocate one lane of Rock Creek and Potomac Parkway to bicycle commuting. The resulting traffic congestion infuriated motorists, whose complaints ended the experiment. Parkway officials paved over the old bridle path in an effort to placate bicyclists. By the 1990s, park trails had become so popular that overcrowding created significant conflicts between different types of users, forcing park managers to develop and post guidelines for "multi-use trail ethics." Calls for additional trail construction ran afoul of environmental considerations, however, underscoring the continued tension between making Washington's natural areas more accessible and protecting them from adverse development.[54]

By the 1990s, bicyclists and wilderness advocates were lobbying to close Rock Creek Park to commuter traffic on a permanent basis, proclaiming that they were protecting the park and promoting more enlightened forms of nature appreciation. Commuters and recreational motorists responded by extolling the psychic and spiritual benefits they derived from driving through the park's leafy scenery. Restricting this experience to athletic recreationalists, they insisted, reflected an elitist and exclusionary vision of the purpose of public parks and the role of nature in the urban environment.[55]

The perennial tension between preservation and access was just one of the many challenges facing Washington's natural resource managers at the turn of the twenty-first century. The "open space" issue became increasingly pressing as the ever-widening circle of suburban development consumed former farmland at an alarming rate, prompting local and regional governments to debate "anti-sprawl" policies designed to concentrate building activities and maintain significant expanses of rural scenery and protected countryside. Traditional landscape planning and design values experienced a resurgence in some circles. Growing appreciation for the history of landscape architecture led to a flurry of efforts to research, restore, and preserve public and private examples of "designed" nature throughout the city. Many neglected parks were refurbished, and the National Gallery of Art created a picturesque sculpture garden adjacent to the Mall that was explicitly cast as a modern-day successor to Downing's romantic landscape scheme. The NCPC issued a lavish prospectus for Washington's twenty-first-century development that invoked the grandiose aspirations and representational strategies of the Senate Park Commission plan. Pristine boulevards, waterfront parks, and imposing civic monuments would once again lead the way to a better—or at least more attractive and economically vibrant—future.[56]

Other turn-of-the-twentieth-first-century attempts to reinvent nature in Washington combined earlier motivations with technological innovation. One of the most ambitious elements of the NCPC's plan was a proposal for a gigantic sphere-shaped aquarium to be located along the Anacostia. While this high-tech "natural" environment remains a pipe dream, the National Zoo was already thrilling Washingtonians with artificial simulations of natural environments, which were designed to showcase its exotic collections while promoting an ecology-based nature ethic. Following the trend toward more naturalistic display environments that transformed zoological landscapes in the second half of the twentieth century, the creators of the National Zoo's "Amazonia" exhibit employed sophisticated technology to create an artificial "rain forest environment" that rapidly became one of its top attractions.

The nature of nature in Washington continues to evolve as new technologies, new cultural concerns, and new environmental factors influence the ways in which the city's physical environments are constructed and perceived. While the chronological approach employed in this essay provides a useful narrative framework, it should not be construed as an attempt to suggest that cultural conceptions of nature in Washington have exhibited any sort of teleological progression toward a more "correct" or desirable state, or to assert the primacy of one vision over another. Such claims are a common component of much writing on invention and the environment, as are implicit or explicit assertions that the most recent approach is better and more "natural" than its predecessor, so that new policies are necessary to promote it. By calling attention to the diversity of ways in which humans have attempted to experience, shape, and understand the nature of nature in Washington, this essay simply seeks to suggest that what we think of as

FIGURE 11
"Amazonia," National Zoological Park. (Jessie Cohen, National Zoological Park)

"nature" is largely a human invention and that it will continue to evolve in response to changing physical, social, and cultural conditions.

The protean quality of this idealized nature is evident in the multitude of ways in which Washingtonians have attempted to alter their environment. Washington's "natural" environment has been shaped by Native American cultural practices; by the Enlightenment rationalism of Washington, Jefferson, and L'Enfant; by the picturesque aesthetics and romantic ideologies of Downing and other nineteenth-century park advocates; by the pragmatic concerns of businessmen and public works purveyors; by the grandiose ambitions of the Senate Park Commission; the social theories of Progressive reformers and postwar urban planners; by the ecological doctrines and recreational motives underlying the late-twentieth-century environmental movement; and by the multitudes of people who have shaped, experienced, and conceptualized it in their own daily lives.

While new "natural" environments and new modes of experience will undoubtedly arise, many of the beliefs and practices outlined above have shown remarkable resilience, as have the physical landscapes they helped produce. Some invented natures and landscape ideologies have faded away, but others have taken root and flourished. So much so, in fact that their constructed natures have often been forgotten and they are assumed to be fortuitous survivals of Washington's primeval landscape. Few visitors to Washington's "natural" areas realize the degree to which the environments they encounter reflect complex legacies of social, political, and technological developments. Even overtly artificial spaces such as the Mall and the cherry-tree-bedecked Tidal Basin have become so deeply ingrained in the cultural consciousness of Washington that they now play important roles in shaping perceptions of the nature of nature in Washington. From a biological perspective, moreover, many of Washington's artificially produced environments are so thoroughly ensconced and intricately intertwined with more putatively "natural" systems that it is difficult if not impossible to say where "artifice" ends and "nature" takes over. Given the shifting contours of our cultural definitions of nature, mapping the borderline between "real nature" and "artificial nature" would appear to be a fruitless endeavor. What is clear, however, is that Washingtonians will continue to debate the appropriate form and function of the region's "natural" environment, building on past beliefs and practices to invent new landscapes and new ways of thinking about them that will inevitably be supplanted by even newer conceptions of the nature of nature in Washington.

NOTES

1. On how individuals and cultural groups mentally "invent" their own environmental experiences, see Pierre Bourdieu, *Distinction: A Social Critique of the Judgment of Taste* (Harvard University Press, 1984); Michel de Certeau, *The Practice of Everyday Life* (University of California Press, 1984); Henri Lefebvre, *The Production of Space* (Blackwell, 1991).

2. Numerous scholars have emphasized the culturally determined nature of the western concept of nature and traced its evolution over time. Artists, poets, philosophers, engineers, policy makers, and ordinary citizens have labored for centuries to define the nature of nature and promote their visions of ideal relationships between mankind and the natural environment. A good deal of the history of western thought is bound up in this discourse, which has generated a considerable body of literature stretching back several centuries. Classic mid-twentieth-century inquiries into the relationships between nature and culture with particular relevance to the American experience include the following: Hans Huth, *Nature and the American: Three Centuries of Changing Attitudes* (University of California Press, 1957); Earl Pomeroy, *In Search of the Golden West: The Tourist in Western America* (Knopf, 1957); Paul Shepard, *Man in the Landscape: A Historic View of the Esthetics of Nature* (Knopf, 1967); Roderick Nash, *Wilderness and the American Mind* (Yale University Press, 1967). A number of geographers, art historians, and scholars associated with the "cultural studies" movement have also addressed the topic, emphasizing the commingling of economic, ideological, and aesthetic impulses. Important works in this vein include the following: Raymond Williams, *The Country and the City* (Oxford University Press, 1973); Williams, "Ideas of Nature," in *Problems in Materialism and Culture,* ed. R. Williams (Verso, 1980); Ann Bermingham, *Landscape and Ideology: The English Rustic Tradition,* 1740–1860 (University of California Press, 1986); Denis Cosgrove, *Social Formation and Symbolic Landscape* (University of Wisconsin Press, 1998); Stephen Daniels and Denis Cosgrove, eds., *The Iconography of Landscape* (Cambridge University Press, 1988); Angela Miller, *The Empire of the Eye: Landscape Representation and American Cultural Politics, 1825–1875* (Cornell University Press, 1993); William Truettner, *Thomas Cole: Landscape into History* (Yale University Press, 1994). Environmental historians have also awakened to the culturally constructed nature of "nature," examining the resultant implications; see e.g. William Cronon, ed., *Uncommon Ground: Rethinking the Human Place in Nature* (Norton, 1996); Max Oelschlaeger, *The Idea of Wilderness: From Prehistory to the Age of Ecology* (Yale University Press, 1991).

3. Christian Feest, "Virginia Algonquians," in *Handbook of North American Indians,* vol. 15 (Smithsonian Institution, 1978), 253–270; Fairfax Harrison, *Landmarks of Old Prince William* (Richmond, 1924; reprint edition: Gateway, 1987), 19, 20, 143, 445; Frederick Gutheim, *The Potomac* (Holt, Rinehart and Winston, 1949; reprint: Johns Hopkins University Press, 1986), 23–26, 43; Paul Inashima, *Archeological Investigation*

of Selected Construction Locales Along the Mount Vernon Memorial Highway (US Department of the Interior, National Park Service, 1985); Herman Friis, *Geographical Reconnaissance of the Potomac River Tidewater Fringe of Virginia from Arlington Memorial Bridge to Mount Vernon* (Association of American Geographers, 1968), 6; Nan Netherton, Donald Sweig, Janice Artemal, Patricia Hickin, and Patrick Reed, *Fairfax County, Virginia: A History* (Board of Supervisors, Fairfax County, Virginia, 1978), 20.

4. Frederick Gutheim and National Capital Planning Commission, *Worthy of the Nation: The History of Planning in the National Capital* (Smithsonian Institution Press, 1977), 15–22; Junior League of Washington and Thomas Froncek, eds., *The City of Washington: An Illustrated History* (Knopf, 1977), 9–36.

5. Despite its location almost exactly midway between Maine and Georgia, the Potomac site of the nation's capital was seen as favoring Southern interests, who returned the favor by backing Secretary of the Treasury Alexander Hamilton's proposal for federal assumption of the states' Revolutionary War debts, which Southern politicians initially opposed. Washington and Jefferson were familiar with the proposed site and used their influence to ensure its selection. For a detailed account of the site of the nation's capital, see Kenneth R. Bowling, *The Creation of Washington, D.C.: The Idea and Location of the American Capital* (George Mason University Press, 1990). For brief summaries, see John Reps, *The Making of Urban America: A History of City Planning in the United States* (Princeton University Press, 1965), 240–245; Reps, *Monumental Washington: The Planning and Development of the Capital Center* (Princeton University Press, 1965), 1–4; Gutheim, *Worthy of the Nation,* 14.

6. For a detailed and sophisticated analysis of the L'Enfant plan and its precedents, see Pamela Scott, " 'This Vast Empire': The Iconography of the Mall, 1791–1848," in *The Mall In Washington, 1791–1891,* ed. R. Longstreth (National Gallery of Art, 1991), 37–58. See also Reps, *The Making of Urban America,* 240–262; Reps, *Monumental Washington,* 1–25.

7. L'Enfant, quoted in Reps, *The Making of Urban America,* 248.

8. Ibid.

9. L'Enfant, quoted in Therese O'Malley, " 'A Public Museum of Trees': Mid-Nineteenth Century Plans for the Mall," in *The Mall in Washington,* 71.

10. Scott, "This Vast Empire," passim.

11. See "President's park," " "Congress Garden," and "bordered with gardens" in the legend to L'Enfant's plan, reprinted in part in John Reps, *Washington on View: The Nation's Capital since 1790* (University of North Carolina Press, 1991), 22; "artfully planted trees," quoted in Scott, "This Vast Empire," 43.

12. David Stuart, quoted in Bowling, *The Creation of Washington,* 228. Stuart was a member of the commission appointed to supervise L'Enfant's planning efforts, which had frequent conflicts with the imperious designer.

13. Reps, *Monumental Washington,* 22–29. Connecticut Representative John Cotton Smith's description of Pennsylvania Avenue in 1800 is quoted on 29.

14. Bayard Smith, quoted in Gutheim, *Worthy of the Nation,* 40.

15. Mary Mitchell, "Kalorama: Country Estate to Washington Mayfair," *Records of the Columbia Historical Society, 1971–1972* (1973), 164–189; Frances Trollope, *The Domestic Manners of the Americans* (1827) (Vintage Books, 1960), 231.

16. Donald B. Meyer, *Bridges and the City of Washington* (Washington: Commission of Fine Arts, 1974), 3–5, 26–34; Netherton et al., Fairfax County, 201–206; Harrison, *Landmarks of Old Prince William,* 547, 565–579; Bowling, *The Creation of Washington,* 118–120; Gutheim, *The Potomac* (reprint: Johns Hopkins University Press, 1986), 252–257.

17. Gutheim, *The Potomac,* 44, 45, 256–266; Julius Rubin, "Canal or Railroad: Imitation and Innovation in the Response to the Erie Canal in Philadelphia, Baltimore, and Boston," *Transactions of the American Philosophical Society* 51 [New Series] (1961), November, 63–79; Reps, *Monumental Washington,* 29; Junior League of Washington, *The City of Washington,* 122, 123; Adams, quoted in ibid., 146.

18. The classic interpretation of America's sense of manifest destiny and its impact on the physical and cultural appropriation of western lands is Henry Nash Smith's *Virgin Land: The American West as Symbol and Myth* (Harvard University Press, 1950). William H. Goetzmann examined the roles that scientific enterprise, geopolitical concerns, art, and other contemporary cultural preconceptions played in western exploration in *Exploration and Empire: The Explorer and the Scientist in the Winning of the American West* (Knopf, 1966). Recent, more explicitly critical interpretations of the sources, strategies, and implications of the nineteenth-century concept of manifest destiny include Patricia Nelson Limerick, *The Legacy of Conquest: The Unbroken Past of the American West* (Norton, 1987); Richard White and Patricia Nelson Limerick, *The Frontier In American Culture,* ed. J. Grossman (University of California Press, 1994); Albert Boime, *The Magisterial Gaze: Manifest Destiny and American Landscape Painting, c. 1830–1865* (Smithsonian Institution Press, 1991); Peter Bacon Hales, *William H. Jackson and the Transformation of the American West* (Temple University Press, 1988).

19. Numerous scholars have traced the historical development and cultural significance of public parks in early- and mid-nineteenth-century America. David Schuyler explores Downing's career and theories in depth in *Apostle of Taste: Andrew Jackson Downing, 1815–1852* (Johns Hopkins University Press, 1996) and

surveys broader developments in *The New Urban Landscape: The Redefinition of City Form in Nineteenth-Century America* (Johns Hopkins University Press, 1986). Therese O'Malley specifically relates these issues to Washington developments in "A Public Museum of Trees" and in her Ph.D. thesis, Art and Science in American Landscape Architecture: The National Mall in Washington, D.C. 1791–1852 (University of Pennsylvania, 1989).

20. Scott, "This Vast Empire," 46–48; Gutheim, *Worthy of the Nation,* 54. Tamara Plakins Thornton discusses the cultural implications of the post-revolutionary American elite's attraction to horticulture in *Cultivating Gentlemen: The Meaning of Country Life among the Boston Elite, 1785–1860* (Yale University Press, 1989). See also Benson J. Lossing, *The Home of Washington: Or Mount Vernon and Its Associations, Historical, Biographical, and Pictorial* (New York: W. A. Townsend, 1859).

21. William Bushong, *Rock Creek Park, District of Columbia: Historic Resource Study* (US Department of the Interior, National Park Service, 1990), 29–32; Timothy Davis, "Rock Creek and Potomac Parkway, Washington, D.C.: The Evolution of a Contested Urban Landscape," *Studies in the History of Gardens and Designed Landscapes* 19 (1999), April–June, 131–133.

22. Scott, "This Vast Empire," 47–50; Gutheim, *Worthy of the Nation,* 51–53.

23. Schuyler, *Apostle of Taste,* 187–211, and *The New Urban Landscape,* 1–76.

24. Downing's 1851 plan for the Mall and selections from the accompanying notes are reproduced in Reps, *Washington on View,* 126, 127, and Reps, *Monumental Washington,* 51–53. O'Malley analyzes Downing's design at length in "Art and Science in American Landscape Architecture: The National Mall in Washington, D.C. 1791–1852" and in "A Public Museum of Trees." Schuyler summarizes Downing's Mall proposal and traces its fate in *Apostle of Taste* and in *The New Urban Landscape.*

25. Gutheim, *Worthy of the Nation,* 56, 90, 91.

26. Ibid., 57, 92; Reps, *Washington on View,* 208, 209, 228.

27. Gutheim, *Worthy of the Nation,* 84–87; Reps, *Monumental Washington,* 56–61. The street tree counts are from a 1881 Army Corps of Engineers map detail reprinted in Junior League of Washington, *The City of Washington,* 236.

28. Frederick Law Olmsted, quoted in *The Papers of Frederick Law Olmsted, Supplementary Series Vol. 1, Writings on Public Parks, Parkways, and Park Systems,* ed. C. Beveridge and C. Hoffman (Johns Hopkins University Press, 1997), 366. This volume provides an excellent introduction to Olmsted's landscape theories. Schuyler explicates Olmsted's work and writing in *The New Urban Landscape.* The definitive Olm-

sted biography is Laura Woods Roper, *FLO: A Biography of Frederick Law Olmsted* (Johns Hopkins University Press, 1983).

29. Communication of N. Michler, Major of Engineers, to the Chairman of the Committee of Public Buildings and Grounds, relative to a suitable site for a public park and presidential mansion, Sen. Doc. No. 21 to Accompany S. 549, 39th Cong., 2nd Session, 1867, 1, 2.

30. Ibid., 2.

31. Ibid., 2, 3.

32. For writings of Charles Eliot, who played a seminal role in promoting metropolitan reservations, see [Massachusetts] Trustees of Public Reservations, First Annual Report of the Trustees of Public Reservations, 1891 (Boston: George H. Ellis, 1892); [Massachusetts] Metropolitan Park Commision, Report of the Metropolitan Park Commissioners, January 1893 (Boston, 1893); Charles W. Eliot, *Charles Eliot, Landscape Architect* (Houghton Mifflin, 1903). For more on Boston's Metropolitan Park System, see Schuyler, *The New Urban Landscape,* 143–146; Norman Newton, *Design on the Land: The Development of Landscape Architecture* (Harvard University Press, 1971), 318–336; Mel Scott, *American City Planning since 1890* (University of California Press. 1969), 17–26. For broader examinations of the origins of the national park system, see Al Runte, *National Parks: The American Experience,* second edition (University of Nebraska Press, 1987); John Ise, *Our National Park Policy: A Critical History* (Johns Hopkins University Press, 1961).

33. The Olmsted Brothers' firm was engaged to produce a landscape management document for the park in 1917. While the ensuing document ("Rock Creek Park: A Report by the Olmsted Brothers, December 1918," original manuscript in Department of Interior Library, Washington) has repeatedly been cited by subsequent park managers, few of its recommendations have ever been adopted. For more detailed overviews of Rock Creek Park's early development, see Timothy Davis, "Rock Creek Park Road System," Historic American Engineering Record Report No. DC-55 (National Park Service, US Department of the Interior, 1996); Barry Mackintosh, *Rock Creek Park: An Administrative History* (Washington: Division of History, National Park Service, 1985), 1–45; Bushong, *Rock Creek Park Historic Resource Study,* 61–121.

34. Classic overviews of the City Beautiful movement include William H. Wilson, *The City Beautiful Movement* (Johns Hopkins University Press, 1989); Scott, *American City Planning,* 1–109; Newton, *Design on the Land,* 400–426.

35. Glen Brown, ed., *Papers Relating to the Improvement of the City of Washington, District of Columbia* (Government Printing Office, 1901). The Senate Park Commission's

history and its significance to the development of American city planning is recounted in numerous publications, including the following: John A. Peterson, "The City Beautiful Movement: Forgotten Origins and Lost Meanings," *Journal of Urban History* 2 (1976), August, 417–428; Peterson, "The Nation's First Comprehensive Plan: A Political Analysis of the McMillan Plan for Washington," *American Planning Association Journal* 51 (1985), April, 134–150; Peterson, "The Mall, the McMillan Plan, and the Origins of American City Planning," in Longstreth, *The Mall in Washington,* 101–113; Thomas S. Hines, "The Imperial Mall: The City Beautiful Movement and the Washington Plan of 1901–1902," in Longstreth, *The Mall in Washington,* 79–99; Reps, *Monumental Washington,* 96–154.

36. US Congress, Senate Committee on the District of Columbia, *The Improvement of the Park System of the District of Columbia* (Government Printing Office, 1902), 12.

37. Ibid., 29–52. In a curious permutation of the concept of "inventing nature," the commission asserted that the Washington Monument itself was such a primal and imposing visual force that it appeared to be "almost a work of nature" (p. 48).

38. *Washington Evening Star,* January 14, 1908. Cartoon and editorial reproduced in Reps, *Monumental Washington,* 152.

39. *The Improvement of the Park System of the District of Columbia,* 23.

40. Ibid., 75, 76. In the days before air conditioning, people were more attuned to subtle variations in local climates. Well into the 1920s, families would drive into Rock Creek Park to escape the heat. Many spent the night in their cars there to take advantage of the cooler valley temperatures. In 1922 the *Washington Herald* reported that as many as 500 cars could be found in the park on a hot summer's night, most of them containing respectable families seeking relief from the heat ("Parking in Rock Creek," *Washington Herald,* June 14, 1922; see also "Night Parking Prohibition Rule in Rock Creek Exempts Families," *Washington Evening Star,* July 18, 1922.

41. For more on American parkway development at this period, see Schuyler, *The New Urban Landscape,* 126–146; Davis, "Rock Creek and Potomac Parkway," 137–143; Davis, Mount Vernon Memorial Highway and the Evolution of the American Parkway (Ph.D. dissertation, University of Texas, Austin, 1997).

42. For a comprehensive account of the origins and evolution of Rock Creek and Potomac Parkway, see Davis, "Rock Creek and Potomac Parkway," passim.

43. *The Improvement of the Park System of the District of Columbia,* 83–122, 137–142.

44. Gutheim, *Worthy of the Nation,* 168–173; Stephen Child, "A Park Needs of Washington Awaken Interest of Public," *Parks and Recreation* 6 (1923), May-June, 403–412; Charles W. Eliot 2nd, "Planning Washington and Its Environs," *City Planning* 3 (1927), July, 177–193.

45. The Boston landscape architect Arthur Shurtleff described the changing concerns of park users and designers in *Future Parks, Playgrounds, and Parkways* (Boston Park Department, 1925). The National Conference on Outdoor Recreation published several volumes outlining the need for comprehensive recreational opportunities, including *Proceedings of the National Conference on Outdoor Recreation, May 22–24, 1924* (Government Printing Office, 1924) and *A Report Epitomizing the Results of Major Fact-Finding Surveys and Projects Which Have Been Undertaken Under the Auspices of the National Conference on Outdoor Recreation* (Government Printing Office, 1928). Among the most influential contemporary expressions of these changing concerns was *Public Recreation: A Study of Parks, Playgrounds, and Outdoor Recreation Facilities. Regional Survey Vol. 5* (New York: Committee on the Regional Plan of New York and Its Environs, 1928). Galen Cranz summarized these transformations in *The Politics of Park Design: A History of Urban Parks in America* (MIT Press, 1989). For more on "The New Deal and The New Play," see Phoebe Cutler, *The Public Landscape of the New Deal* (Yale University Press, 1985), 8–28. For details of recreational developments in Washington during this period, see Office of Public Buildings and Public Parks of the National Capital, *Annual Report of the Director of Public Buildings and Public Parks of the National Capital, 1926–1932* (Government Printing Office, 1926–1933).

46. For broader accounts of the automobile's impact on urban parks and the evolution of the motor parkway, see Gilmore Clarke, "The Parkway Idea," in *The Highway and the Landscape,* ed. W. Snow (Rutgers University Press, 1959), 32–55; Newton, *Design on the Land,* 596–619; Christopher Tunnard and Boris Pushkarev, *Man-Made America: Chaos or Control?* (Yale University Press, 1963), 159–167; Clay McShane, *Down the Asphalt Path: The Automobile and the American City* (Columbia University Press, 1994), 31–40, 203–228; Davis, "Mount Vernon Memorial Highway and the Evolution of the American Parkway."

47. Davis, "Rock Creek and Potomac Parkway," passim.

48. "Linking the Parks," *Washington Post,* July 23, 1934; "Road to Link The Potomac, Rock Creek Parks," *Washington Post,* June 11, 1935; "New Rock Creek Road Link to Open Artery," *Washington Evening Star,* October 24, 1935; "New Parkway Here to Rank with Finest," *Washington Evening Star,* April 17, 1936.

49. Isabelle Gates, "What Could Be a Finer Tribute to Washington's Memory?" *American Motorist—District of Columbia Edition* (June 1930), 67, 107. On the development

of the Mount Vernon Memorial Highway and its successor, the George Washington Memorial Parkway, see "Mount Vernon Memorial Highway and the Evolution of the American Parkway" and "Mount Vernon Memorial Highway: Changing Conceptions of an American Commemorative Landscape," in *Places of Commemoration, Search for Identity and Landscape Design,* ed. J. Wolschke-Bulmahn (Washington: Dumbarton Oaks, 2000), 123–176.

50. For a summary of Washington's postwar transportation plans, see Gutheim, *Worthy of the Nation,* 229–248, 271–343. For one of the most ambitious attempts to convert parkland into freeways, see J. E. Greiner Company and De Leuw, Cather & Company, *Transportation Plans for Washington* (prepared for Board of Commissioners, District of Columbia, 1946), which included the proposal for elevated freeways flanking the Mall.

51. For a chronicle of Washington's postwar parkway and freeway battles, see Davis, "Rock Creek and Potomac Parkway," 192–209; Davis, "Mount Vernon Memorial Highway and the Evolution of the American Parkway," 787–885; Barry Mackintosh, "Shootout at the Old C. & O. Canal: The Great Parkway Controversy," *Maryland Historical Magazine* 90 (1995), summer, 141–163.

52. *Washington Present and Future: A General Summary of the Comprehensive Plan for the National Capital and Its Environs. Monograph No. 1* (NCP&PC, 1950); *Open Spaces and Community Services: A Portion of the Comprehensive Plan for the National Capital and Its Environs. Monograph No. 4* (NCP&PC, 1950).

53. NCPC, *Year 2000 Policies Plan* (NCPC, 1961); Ian McHarg, *Toward a Comprehensive Landscape Plan for Washington, D.C.* (NCPC, 1967). Runte summarized the influence of the ecology and environmentalism movements on the national parks in *National Parks: The American Experience,* 181–208. Stanford Demars perceptively critiqued the growing dominance of ecology-based park management in *The Tourist in Yosemite, 1855–1985* (University of Utah Press, 1991), 122–158. Allston Chase called attention to the scientific shortcomings of purportedly ecological park planning policies in *Playing God in Yellowstone: The Destruction of America's First National Park* (Harcourt Brace, 1987).

54. National Capital Open Space Program, *Technical Report No. 3: Outdoor Recreation in the National Capital Region* (Washington: National Capital Regional Planning Council, 1965); Davis, "Rock Creek and Potomac Parkway, Washington, D.C.: The Evolution of a Contested Urban Landscape," 215; idem, "Mount Vernon Memorial Highway and the Evolution of the American Parkway," 900–902; Mackintosh, "Rock Creek Park: An Administrative History," 89–94; Roger Moore, *Conflicts on Multiple-Use Trails: Synthesis of the Literature and State of the Practice* (Federal Highway Administration, 1994).

55. Linda Wheeler, "Weighing the Future of Rock Creek," *Washington Post,* July 27, 1996; National Park Service, *Rock Creek Park General Management Plan/Environmental Impact Statement Newsletter: Preliminary Alternative Scenarios, No. 3* (June 1997); "Rock Creek Park: The Citizens Have Their Say," letter from Rock Creek Park Superintendent Adrienne Coleman to the editor, *Washington Post,* August 2, 1997; National Park Service, *Rock Creek Park General Management Plan/Environmental Impact Statement Newsletter: Preliminary Alternative Scenarios, Number 4* (January 1998); Roger K. Lewis, "Rock Creek Park: Deciding Its Role and Its Future," *Washington Post,* July 19, 1997. The NPS's proposed management plan for Rock Creek Park and public reactions to the various management schemes are discussed at length in Davis, "Rock Creek Park Road System," Historic American Engineering Record Report No. DC-55 (National Park Service, US Department of the Interior, 1996), 244–255.

56. Charles Birnbaum and Heather Barrett, eds., *Making Educated Decisions 2: A Landscape Preservation Bibliography* (National Park Service, 2000); NCPC, *Extending the Legacy: Planning America's Capital for the 21st Century* (NCPC, n.d.).

BIOLITERACY, BIOPARKS, URBAN NATURAL HISTORY, AND ENHANCING URBAN ENVIRONMENTS

MICHAEL H. ROBINSON

In the course of their development museums have clearly undergone an evolution that on the one hand converges with universities and on the other with zoological and botanical gardens. . . . What a splendid institution it would be that combined under one well-coordinated management and on one tract of land, university, museum and gardens—botanical and real zoological gardens—not merely gardens of nothing but vertebrates.

—Frank Lutz, 1930[1]

Urban environments, in the sense of areas occupied by extra-human living components—what as a biologist I mean by the word "environment" as opposed to an architect's definition—are greatly enriched by the presence of parks of all kinds, whether they are simply open areas of unmodified countryside, conventional, eccentric, zoological, or biological. (I exclude from this definition of "environment" extra-human inhabitants of human dwellings or buildings built for human activities. Thus, pets and household pests are excluded.) To provide a name for biological parks, described below, I have coined the word "biopark." Bioparks can provide a greater augmentation of species richness and exciting natural history than the other kinds of parks.

What is a biopark, and why should zoos transform themselves into a new entity with a new name? The need for a new entity stems from the burgeoning world environmental crisis. I have argued elsewhere[2] that the major threats to the living world accelerated mightily in the second half of the twentieth century, and that we cannot hope to sustain the levels of biodiversity necessary for ensuring a habitable planet unless we are bioliterate. Bioliteracy will be as essential to an all-round education in the twenty-first century—indeed in the third millennium—as Latin, Greek, and theology were thought to be in medieval times. Bioexhibits (broadly defined to include natural history museums, anthropology museums, zoos, "safari" parks, botanic gardens, arboretums, aquariums, and oceanariums) are potentially, and sometimes actually, major forces for informal biological education. Institutions exhibiting living organisms (mainly restricted to plants and animals, but potentially capable of exhibiting microorganisms and fungi) have great powers of attracting human curiosity and interest.[3] Despite their affective attractiveness, as well as that arising from rational and aesthetic impulses, these institutions are blighted by a Victorian compartmentalization of knowledge. This division, once productive in science, is now a barrier to environmental holism. To state it simply, it is an obsolete approach to separate the exhibition of living animals from that of living plants when the two are inseparably interdependent, interlinked, and co-evolved in the real world. Thus there is a profound educational reason to substitute biological parks for zoological parks and botanical gardens. "Zoo" works well as an abbreviation, but "bio" does not; hence "biopark."

The biology, the evolution, the prehistory, and the history of our species are integral parts of the biopark theme. Since our origin, and particularly since we "invented" agriculture, our interactions with the rest of the living world have had profound, in fact terramutant, effects. And, of course, much of the artistic activities of humans, including graphic art, dance, music, and literature and poetry, have been inspired by nature and have often reflected it. Museums of natural history have collected specimens revealing the details of the changes in life over time, the geological history of our planet, our place in the universe, and the structures and physiology of living organisms. These subjects all belong together in the biopark's educational and expository features. The limits are not ultimately imposed by our imagination, but by considerations of space, resources, and particularly budgets. Thus the motive for the new entity is educational, and any enrichment of urban environments is a secondary—but perhaps sublime—consequence.

There is an element of déjà vu in my view of urban environments. In 1985, just over a year after I became director of the Smithsonian National Zoological Park (NZP), fresh from nearly 20 years spent in the tropics as a research biologist, we held a symposium at the zoo on the subject of wildlife survivors in the human niche. Thirteen distinguished contributors, including biologists, veterinarians, educators, and architects, addressed a variety of themes. The papers, unfortunately, were never published in the proposed symposium volume. But they are in our archives, and they refresh my failing memory. They illuminate many of the issues that I address here.

In fact, my own paper enunciated all the themes of the biopark idea, at least a year before I remember having such ideas, and also stated: "We are planting a wide variety of flowers . . . to bring back into the park all kinds of animals . . . that may have disappeared during urbanization." There was even a paper titled "Amphibious Assault on the Suburbs Successful."[4] Parks help enrich wildlife, but any kind of park is a built and in some sense artificial environment. Parks are, to repeat, artificial assemblages containing many "natural elements." When stating this as a negative we must remember that we ourselves live in totally artificial built environments, in villages, towns, cities, and conurbations amounting to megalopolises. We have replaced the kinds of habitations used by pre-civilized humans with functional substitutions instead of natural elements. In the terminology of classical ethology, such substitutions are "supernormal." (Think of Astroturf versus grass.) The steps that led to unnatural concentrations of human living units and populations are worthy of major educational exhibits in the biopark. The starting point, which was not really a point but a process spread over time, is easy to establish: it was the profound shift from our ancestors' being food gatherers to their becoming food producers. This is what V. Gordon Childe called the "Neolithic Revolution." Agriculture transformed us from being ecologically indistinguishable from any other large primate, depending on the natural carrying capacity of the environment, into an agent of change, producing new and higher carrying capacities. Cultivation, domestication, the consequent anthropogenic alteration of huge areas of the Earth's surface, and the origin of civilization are all interconnected. History began with agriculture's triumph; everything before that was prehistory.

CONSEQUENCES OF THE BIOPARK IN THE URBAN ENVIRONMENT

THE BIOPARK AS AN URBAN OASIS

The parks, gardens, and outdoor bioexhibits that present-day cities enclose are islands or peninsulas of a semi-natural world. They may only be scaled-up versions of domestic gardens and city allotments, but their scale can and usually does affect the richness of their natural history. Small open spaces may lack the ambiance and/or the resources that are needed by wild animals. Two groups of organisms may be found in such urban settings. The first group comprises animals and plants that exploit urban environments because they are in a sense pre-adapted as "opportunists" to be colonizers. These organisms are adaptable, and those that are animals are resource generalists. They have effective distance dispersal mechanisms and broad ethological and/or physiological tolerances. Rats, mice, opossums, squirrels, pigeons, sparrows, and starlings are common city commensals in North America. Cockroaches have been house guests around the world since we created semi-tropical conditions in our kitchens. In other parts of the world urban animals include similar types but different species. In India, for instance, crows and kites are prominent urban scavengers, and monkeys often exploit the favorable scrounging situation. In Alaska polar bears invade towns, but are not established there, while in other parts of the world the common mynah takes over from the starling, or sparrow, as a common passerine.[5] Just as the accidents of zoogeography determined the regional range of early domesticates, so it is with urban colonizers. On the plant side, hardy undemanding species with efficient seed dispersal (often windborne) are typical urban wasteland weeds. The second group of organisms comprises those that humans have chosen to plant or to otherwise introduce into the cityscape. Trees that line streets or are planted in squares and other "civic" areas belong to this category. These and other plants were, and still are in many cities around the world, chosen for their resistance to pollution. Our atavism for being surrounded by plants is reflected in the extraordinary range of garden cultivars that are grown by those householders who possess gardens (in addition, of course, to house plants). These constitute an often exotic element in the urban environment, including plants originally from distant continents or even from obscure oceanic islands. Introduced, unrestrained animals are comparatively rare. (Cattle in Indian cities are an exception.) Most urban introductions are house pets, and their feral by-products constitute a nuisance.

In relatively unrestricted large parks, bio- or otherwise, extensive areas of trees, shrubs, and herbaceous plants and ornamental or natural ponds, lakes, and streams attract fauna from nearby wild and semi-wild habitats. Birds can invade these areas more easily than flightless animals. A large "rookery" of black-crowned night herons nest at the National Zoo every year, flying to the Potomac to catch fish for their young. In Greater Washington, bird counts carried out by more than 100 volunteers recorded 91 breeding species, with an estimated 115 migrants passing through. Within Washington, the highest species diversity was found in parkland and the lowest in commercial and high-density residential areas.[6] The number of migrants is

FIGURE 1
One of the National Zoo's ponds. (National Zoological Park)

probably at least three times the number of resident birds. The grounds of
the National Zoo contain more than 150 tree species. Since plant species
richness sustains that of animals, it is no surprise to find rich faunas. Certainly flower-dependent hymenopterans (bees, wasps, and ants) flourish
where there is a rich diversity of ornamentals, as in many parks. It is also
true that butterflies, because of their relative conspicuousness, are easier to
count than most other insects, but their diversity probably reflects that of
less conspicuous insects. E. O. Wilson, the great evolutionary biologist, told
me that his interest in ants was whetted by his boyhood experience of collecting in Rock Creek Park. With its interconnected open areas, which
include the National Zoo, this is one of the largest urban parks in the country. The zoo is visited from the park by white-tailed deer, foxes, raccoons,
and other mammals. The park also sustains a population of feral domestic
cats.

Within 25 miles of Washington, George R. Zug recorded 58 species of
amphibians and reptiles. These included 11 newts and salamanders, 15 frogs
and toads, 5 lizards, 18 snakes, and 9 turtles. (A 25-mile radius extends

beyond the District of Columbia limits by 13 miles N-S, and more on an E-W axis.) The National Zoo's ponds sustain toads, frogs, newts, and salamanders. Bullfrogs have been observed, as have ringneck snakes, black rat snakes, DeKays snakes, and Eastern box turtles. These creatures, particularly the amphibians, are probably not as mobile as most invading mammals, and they are certainly not as mobile as birds.

THE BIOPARK AS DELIBERATE ENHANCEMENT, ACTUAL AND POTENTIAL

These examples of natural history richness exist as a result of the park syndrome of habitat creation. The human-contrived components of a park may be non-indigenous in plant species composition, in layout, and even in topography, but they exist in a climate that is normal for the area. They are usually planned, designed, nurtured, and weeded. The main aim of these operations is not, usually, to encourage the presence of wild animals and plants. If they do attract "wild" elements, as is argued above, this is usually the unplanned consequence of creating "green" areas for human recreation. A biopark, existing primarily to educate about life, should aim to enrich its biology. Such enrichment can be done in numerous ways, many of them largely dependent on the interest and dedication of staff at many levels and on processes that are not intrinsically expensive. For instance, at the National Zoo we have planted certain species as nectar sources and larval food to attract butterflies; such plants are also frequently beautiful and often fragrant. This policy seems to have worked, since large numbers of swallowtails are now seen, along with other less striking but interesting butterflies. Caterpillars of some moth species are attractively "furry," and these can be encouraged by the right plantings and by tolerant gardeners. Flowers with long corolla tubes attract bumblebees and many species of solitary bees and wasps.[7]

Many insects construct nesting burrows in sandy areas, while others construct nests of moist clay. Providing both these resources can encourage insect richness. We have established a Dragonfly Pond, which is working well, and many of our other ponds are conducive to colonization by a wide range of aquatic organisms, some large, some small, and some microscopic. Bird species richness can be enhanced by planting to provide autumn seeds and fruits, by providing nest boxes, particularly for hole nesters, and by provisioning feeders. We have planted thistles for goldfinches, a glorious sight in autumn. Our wetlands exhibits attract large numbers of migrating water-fowl. They are attracted to the water, and also to the on-demand feeders that offer duck food to our exhibit animals. One of our ponds, close to Rock Creek, attracted a muskrat, which we encouraged to become resident by providing it with a nice den and a supplementary diet. Attracting mammals is a more complex matter, but houses for bats and flying squirrels are a good possibility. Frogs and newts can be seeded into our ponds. Immigrants can be a mixed blessing, since they are not quarantined and they may introduce pests and diseases.[8]

THE EDUCATIONAL EXPLOITATION OF BIOPARK NATURAL HISTORY

Many a zoo fails to use the educational assets of the park in which its buildings are set. Such assets are largely independent of the zoo's horticultural style, which may range from "imperial" formal (as in the great Tiergarten Schonbrunn of Vienna) to the very different formalism of some Japanese zoos. In contrast to such stylistic approaches are the relative informality of the otherwise disciplined gardens of most US zoos, the tropical splendors of the Colombo Zoo in Sri Lanka, and the great impeccable gardens and real forests of the Singapore Zoo. Despite the spectrum of so many styles, species of plants, and climates, with the consequent differences in wild animals that they attract, large and small, all zoos have a great potential to be used to tell great biological stories. Even those zoos that are islands within cities can tell stories of urban natural history. The gardens are a good starting point from which to begin the evolution from zoo to biopark, since they are the least expensive places to change. (Built exhibits are difficult to change quickly.)

Species richness is useful in natural history education. At the National Zoo, we have developed two series of graphic panels under the themes "Science in the Park" and "Backyard Biology." These deal, respectively, with research carried out by Smitsonian researchers and with plants and animals in the park. "Science in the Park" is only indirectly related to natural history, although studies of the squirrel population conducted by one of the Department of Zoological Research scientists will eventually be highlighted. Other planned "Science in the Park" signs cover a wide range of biological topics, including lactation in humans and other mammals, a subject that we feel will be of great interest to visiting mothers.

Black-crowned Night Herons
Backyard Biology
These birds choose to nest at the National Zoo

More than 400 wild black-crowned night herons summer at the Zoo. They choose, ironically, to nest in trees around the Bird House. You may see juveniles practicing their hunting skills in our wetlands ponds.

Black-crowned night herons are messy, eating almost any aquatic animal. Under the trees where they nest, you'll find fish bones, half-digested frogs, feathers, and crab shells.

Night herons are also noisy, filling the air with a variety of interesting sounds.

YIP, YIP
(young begging for food)

QUOK
(while flying)

PLUP-BUZZ
(courtship)

KAK-KAK
(threatening)

The flock arrives in March. Chicks hatch, several days apart, and parents bring them regurgitated food. Later, they are fed whole small prey which is sometimes larger than the chick.

Night herons preen and sleep during the day. At dusk, they head for Rock Creek, and the Potomac and Anacostia rivers to hunt.

The species has a vast range—almost worldwide. Our night heron flock probably winters in Central America. During this time, their nests sit empty, waiting for the flock to return.

FIGURE 2

A "Backyard Biology" sign. (Jessie Cohen, National Zoological Park)

At present the "Backyard Biology" series is the more extensive of the two. One set of signs deals with pollination (particularly the interactions between plants and their pollinators) and with human allergies to wind-borne pollen. Other signs cover butterfly gardening, Victoria water lilies, dragonflies, web-building spiders, and squirrels and seed dispersal.[9] Near our former wetlands exhibit (now wild) is a wildflower meadow. Extensive graphics on migratory birds prepared by the Smithsonian's Migratory Bird Center have been installed.

The zoo's Grasslands exhibit includes 52 species and varieties of grasses. Two pavilions give details of the worldwide importance of grasslands, and prairies in particular, as well as information on the major plants in the adjacent gardens.

AN URBAN NATURE CENTER AND A WATERWORLD

Among the new things planned for the next few years are an Urban Nature Center and an exhibit currently referred to as Waterworld.

Intended to promote "public appreciation of the local flora and fauna in the metropolitan area," the Urban Nature Center will be an outdoor classroom. It will have a pond stocked with aquatic plants and a variety of gardens designed to attract butterflies, hummingbirds, and other small wild animals. Bird feeders, viewable nest boxes, and bird baths will help to attract local and migratory birds. A former animal barn will be refurbished and equipped with learning stations and audio-visual interactives. In a restricted-admission area, formal and informal activities will be greatly enhanced and facilitated, particularly for school classes and organized groups of all ages.

Waterworld (a tentative but descriptive title) is an exhibit planned around multiple themes of immediate global relevance. The overriding concept is water as the essential basis of all terrestrial life, its cradle and sustainer. Other themes will be the earth-sculpting effects of liquid and solid water; water as the basis of industry, agriculture, and civilization; and water's uses in human recreation. Water-conservation issues will be highlighted, partly by reference to the hydrology of the local watershed from Rock Creek to the sea. The exhibit will be built around the three ponds of the zoo's former wetlands-and-wildfowl exhibit.

CONCLUSION

Urban environments are greatly enriched by the presence of parks of all kinds, whether they are simply open areas of unmodified countryside, conventional, eccentric, zoological, or biological. Where, as in my concept of a biopark, wild species richness is enhanced by careful planning, this can give rise to more complex and exciting natural history experiences than parks that are managed simply as urban green spaces, or as islands of clean air. Such an enhancement, can, of course, be created by the same means in parks not intended to function as educational entities. This should be encouraged. Aesthetic pleasure, stimulation of curiosity, and the advancement of bioliteracy all are urgently needed, now and forever.

NOTES

1. F. Lutz, "The Use of Live Material in Museum Work," *Museum News* 6 (1930), 7–9.

2. See e.g. "Biodiversity, Bioparks, and Saving Ecosytems," *Endangered Species UPDATE* 10 (1993), 52–57; "The Biopark Concept and the Exhibition of Mammals," in *Wild Mammals in Captivity,* ed. D. Kleiman (University of Chicago Press,

1996), 161–166; "Multimedia in Living Exhibits: Now and Then," *Museum News,* July-August 1997, 37–43, 67–67; "Multimedia in Living Exhibits: Now and Then," in *The Virtual and the Real Media in the Museum,* ed. S. Thomas and A. Mintz (American Association of Museums, 1998), 37–55.

3. I think that this attraction is preponderantly an atavism dating back to our hominid past, which was overwhelmingly spent as barefoot biologists, as hunter/gatherers. It is also reflected in the present day phenomena of keeping pets and nurturing house plants.

4. NZP Office of Public Affairs, Archives 1985. On wildflowers, see J. Gilmour and M. Walters, *Wild Flowers* (Collins New Naturalist, 1973).

5. An important set of behavioral traits is involved in the tolerance of novelty in habitats/environments. It affects tendencies to invade urban areas. Fear of novelty has been called neophobia so its opposite should be called neophilia. Russell Greenberg, who has carried out ingenious experiments using cages with novel (artificial) habitats of differing complexity to compare levels of neophobia in birds, has found that closely related species differ markedly in exploratory behavior.

6. John Hadidian, John Sauer, Christopher Swarth, Paul Handly, Sam Droege, Carolyn Williams, Jane Huff, and George Didden, "A Citywide Bird Survey for Washington, D.C.," *Urban Ecosystems* 1 (1997), no. 2, 87–102.

7. For background, see B. Heinrich, *Bumblebee Economics* (Harvard University Press, 1979); Christoper O'Toole and Anthony Raw, *Bees of the World.* (Blandford, 1991). For a review of trap-nesting species of bees and wasps with instructions on making artificial inducements to their nest-building, see K. Krombein, *Trap Nesting Wasps and Bees, Life Histories, Nests, and Associates.* (Smithsonian Press, 1967). On wasps, see H. E. Evans, *Wasp Farm* (Doubleday Anchor, 1973). For descriptions of devices to attract hymenoptera, see B. A. Cooper, "The Hymenopterist's Handbook," *Journal of the Amateur Entomologist's Society* 7 (1943), 1–160.

8. On attracting a wide range of animals, see S. W. Kress, *The Bird Garden* (National Audubon Society, 1995); M. Schneck, *Your Backyard Wildlife Garden* (Rodale, 1992); M. Tekulsky, *The Butterfly Garden* (Harvard Common Press, 1985); Xerces Society and Smithsonian Institution, *Butterfly Gardening* (Sierra Club Books).

9. Examples of "Backyard Biology" signs:

Cicada Killer: A good provider. In late summer, female cicada killers hunt cicadas to feed their young. They're easy to spot; they grow to almost two inches long. Look for these mighty wasps lugging their heavy prey to burrows in this area. The cicada killer's needlelike stinger injects poison into the cicada, paralyzing it. Clutching the large victim, she flies to her underground nest. The wasp backs down the burrow, pulling in the cicada. She lays a tiny egg on the cicada's body. When the young hatches, it eats the cicada. Wasps also feed their young grasshoppers, beetle grubs, and caterpillars. By preying on insects that destroy crops, wasps do humans a great service.

Black-crowned Night Herons: These birds choose to nest at the National Zoo. More than 400 wild black-crowned night herons summer at the Zoo. They choose, ironically, to nest in trees around the Bird House. You may see juveniles practicing their hunting skills in our wetlands ponds. The flock arrives in March. Chicks hatch, several days apart, and parents bring them regurgitated food. Later, they are fed whole small prey which is sometimes larger than the chick. Night herons preen and sleep during the day. At dusk, they head for Rock Creek, and the Potomac and Anacostia rivers to hunt. The species has a vast range—almost worldwide. Our night heron flock probably winters in Central America. During this time, their nests sit empty, waiting for the flock to return. Black-crowned night herons are messy, eating almost any aquatic animal. Under the trees where they nest, you'll find fish bones, half-digested frogs, feathers, and crab shells. Night herons are also noisy, filling the air with a variety of interesting sounds.

Pollen in the Air: People and plant mating. When it's time to reproduce, some plants fill the air with their male sex cells, pollen. The pollen sails the wind to fertilize female cells in far-flung flowers. These plants produce millions of grains to increase their chances of success; yet only one grain among millions reaches its target. Are you a victim of plant sex? When pollen lands on some people, their bodies' immune systems go into overdrive—runny eyes, coughing, sneezing, suffering. Flowers that are pollinated by animals have bright colors, showy petals, and seductive scents. But flowers pollinated by wind have a different look. Their fluffy flower clusters are exposed to the wind, and tend to be small, pale, and odorless. Wind needs no seduction.

PORTRAIT OF INNOVATION: JON C. COE

MARTHA DAVIDSON

"The creative process is a foggy journey," begins a poem by Jon C. Coe, a landscape architect and one of the world's most innovative zoo designers. Fog is an image Coe often uses in talking about creativity. Creation is an oscillation, he explains:

It's an inward-outward oscillation. It's also an oscillation between what I call the fog and the clearing. . . . Sometimes you come on this hilltop, and you can see where you're going, but not the road that gets you there. To get where you're going, you have to go back in the fog. In the fog, you don't know where you are, and you're trying out all kinds of things, and most of them aren't working. To be a successful designer, you have to be able to be completely comfortable in the fog. . . . You've got to be comfortable enough to carry your client through it.

It's not surprising that Coe chooses fog as a metaphor, since his interest in zoo design did not come primarily from a love of animals, but from a more profound appreciation of nature in all its forms. "I'm passionate about animals," Coe admits, "but it's more: plants, too, and rocks, canyons, water-falls, and peoples; cultures, old and new, past and future." His approach to design, as it is to life, is a fusion of poetry and science.

Coe was born in Logan, Utah, into a family of peripatetic, creative indi-vidualists, many of them artists or independent contractors. He and a fra-ternal twin brother were the youngest of six siblings. His mother was a nurse, his father a professor of horticulture. Professor Coe also had an impulse to help other people and was open to and respectful of other cul-tures. Under his father's influence, Jon Coe acquired a knowledge of plants and strong humanitarian values.

When Jon was 6 years old, the family moved to Germany for 3 years while Professor Coe was involved with reconstruction efforts after World

War II. Exposure to German at an early age gave Jon a facility with spoken languages that proved useful later in his life. With written language, he had more difficulties. He has dyslexia, a neurological condition that interferes with interpretation of letters and numbers. Although the condition has been linked to creativity, it was not well understood at the time. During his school years in California, Jon was shuttled between mainstream classes and those for slow learners.

His oldest brother taught him to make landscape models and inspired him to read both *Scientific American* and science fiction. Jon's curiosity about the natural world also led him to spend hours looking at photographs in *National Geographic* while lying on the cool floor of the local library where he sought refuge on hot summer days. Two boyhood camping experiences left indelible impressions. On a Boy Scout trip in the desert, he spent hours alone tracking a bobcat. His first camping trip to Yosemite made so strong an impression that the images still come back to him, often resurfacing in doodles and sketches that he calls "landscapes of the mind."

In 1961, while attending a community college, Coe wrote a paper on naturalism in landscape design, advocating the use of native California plants and landscapes, a practice that did not become popular until several decades later. Pursuing this interest, he entered a program in landscape architecture at the University of California at Berkeley. In 1963, he explored the idea of a biopark in a plan for San Francisco Bay's Angel Island. Coe won several awards at Berkeley and in 1964 received a bachelor's degree with honors in landscape architecture. It was at Berkeley, too, that he met Susan Webster, a landscape architect in the class of 1965; they married in 1966.

Coe continued his training at Harvard University's Graduate School of Design. From the start, he disagreed with the prevailing philosophy of the School of Design, which in the 1960s extolled Modernism, the stark International Style of Le Corbusier and others:

In my opinion, it was international elitism. It was the concept that a few elite individuals know what's best for people around the world and that the International Style, Modernism, works equally anywhere in the world. . . . I see myself as a populist, a regionalist. I feel that the architecture and the landscape of each different part of the world should be unique to itself, and each group of people, each culture, has their own expression. This variety should be celebrated!

Coe came upon his thesis topic while sketching at Boston's Franklin Park Zoo. Entering the elephant house, he saw three elephants fighting while restrained by heavy chains. The scene is as vivid for him today as it was at that moment:

They were trumpeting and flinging themselves at each other. They couldn't really reach each other because of the chains, but they were really upset, and the feeling in the air was just terrible. I asked the keeper "Why are they fighting?" He said "Because they are chained." I said "Why are they chained?" "Well, because they fight."

He knew then that he wanted to explore behavior as a basis for design. His faculty advisor was not encouraging, saying that one could not make a career in zoo design and that, in any case, no one on the Harvard faculty had expertise in that area. Not dissuaded, Coe found a professor of anthropology, Irven DeVore, who provided him with an extensive reading list. Coe spent the next nine months working on his own and reading extensively. He submitted a seventy-page thesis consisting almost entirely of quotations from animal behavior experts. In the thesis, he also proposed organizing zoos by biomes (bioclimatic zones). Although unconventional, his thesis and other work at Harvard won him respect, and in 1966, along with his master's degree in landscape architecture, the School of Design presented him with one of its three top awards: the Jacob Weideman Fellowship for study and travel abroad.

Later in 1966, unable to find a job designing zoos and unwilling to fight in Vietnam, Coe joined the Peace Corps with his wife. "I believed in national service," he says, "but I wanted to be involved in something constructive, not something destructive. And at that point in the war, they allowed deferments for the Peace Corps. So that was the direction I went."

The couple had hoped to go to Africa to work on housing, but the Peace Corps had just closed its African training program. Instead, they learned Portuguese and were sent to Brazil to organize 4-H Clubs. "As soon as we got there we jumped ship," Coe recalls, "and we started working on public housing, which is an area I had some knowledge about. So we moved into a favela, a slum area... and worked with local people, helping rebuild a dam, design a school, and plan a small city." They also did the first site survey of Glass Falls, the world's seventh highest waterfall, in the interior highlands of Bahia. The Coes gave their data to Diogenes Reboyas, a university professor active in the national parks movement in Brazil, who used it to win national park status for Glass Falls. They completed their Peace Corps service late in 1968 and spent the rest of that year traveling throughout South America on Jon's Weideman Fellowship, looking at national parks and historical cultural landmarks in eight countries. Among their most memorable experiences were brief visits to an African settlement in a remote, arid area of Ecuador and to an Indian fishing village in Chile's Tierra del Fuego. Jon wanted to understand how the confluence of geology, topography, climate, and history made each place unique. He kept notebooks of sketches and observations as he learned firsthand about the biological and cultural zones of South America. Those notebooks, along with other volumes of notes, sketches, and poems chronicling his travels since then to twenty countries on six continents, have served as a resource for design ideas.

After seven months of hard travel, the Coes were ready to settle down but not yet ready to return to the United States. They responded to an offer from two old friends who had started a firm in Calgary. Wanting a change from the tropics and eager to learn about the far northern latitudes, they moved to Alberta and worked for the Lombard North Group for 4 years.

Living at the base of the Canadian Rockies, Coe expanded his experience with biomes though work on projects on the prairies, in the mountains, and in the Northwest Territories. One of his first projects was a site plan for a community college. He persuaded the college to add to the plan several gardens that were re-creations of local native habitats. They included a dune with native prairie grasses and a black spruce bog. It was the first time Coe had taken a flat site and actually created regional habitats instead of conventional ornamental gardens, anticipating a trend that has since become popular. Environmental science classes at the college helped plant the bog and used the areas as learning laboratories.

Coe also played a leading role in two innovative housing projects. For Rundle Lodge, a home for the elderly, he collaborated with the architect Jack Long in using behavior as a basis for design. For example, hearth areas were created to serve as gathering spaces or as places for solitude. Garden plots were raised so gardening could be done from a bedroom window seat. Decorative elements recalled the rural communities where residents had spent most their lives, and the large building itself was designed to appear small from any viewing angle, on a scale with the small towns of their childhoods. In the second project, a low-income housing complex, Coe and Long substituted irregular, crescent-shaped arrangements of townhouses for the traditional grid, to allow a range of opportunities for socialization or privacy. The plan was influenced by Coe's experience in the favelas of Brazil. Both projects won many awards and still had two-year waiting lists for occupancy 10 years after opening. These projects were early examples of what he has come to call "affiliative design"—design that encourages positive interactions among people, among animals, or between people and animals.

affiliative design

While at Lombard North, Coe also helped develop an environmental analysis for a thousand-mile highway along Canada's Mackenzie River to the Arctic Circle, and he is proud of the fact that the report helped to dissuade the government from building it. Coe had become an ardent and aggressive conservationist. He helped lead a successful campaign to prevent development of a massive commercial ski resort in Banff National Park, and his comments at public hearings on the future of the Canadian Rocky Mountains National Parks were published in the Canadian edition of *Time*. Coe's voice was being heard on a regional and on a national level. These activities only increased his deep appreciation of the wilderness, yet he still had not found an opportunity to test his theories or use his professional skills in the design of wildlife parks or zoos.

By 1973, after 4 years in Alberta, Coe was ready to return to the United States. He had kept in touch with a Harvard friend and classmate, Grant Jones, who had his own firm in Seattle and wanted Coe to join him. The firm, Jones & Jones, was developing an outstanding reputation for quality and innovation in architecture and landscape design. Coe accepted the offer. Two weeks after he joined Jones & Jones, they landed their first zoo project: a master plan for Northwest Trek, a 600-acre wildlife reserve in Eatonville, Washington.

In designing Northwest Trek, Coe and Jones drew on the expertise of David Hancocks, a British zoo designer who was serving as a consultant to Seattle's Woodland Park Zoo. The zoo had raised $60 million for renovations, but had been forced to reject a problematic master plan developed by another firm. Now Woodland Park Zoo approached Jones & Jones, asking them to submit a new master plan, with Hancocks serving as the zoo's representative. The noted ecologist Dennis Paulson would be scientific advisor. An early draft of the master plan developed by this team so impressed the Seattle City Council that it was immediately approved, and construction of exhibits began before the master plan document was finished.

The plan was revolutionary. Instead of following the Modernist model of most mid-twentieth-century zoos, with tile-lined animal cages or concrete grottos arranged taxonomically or by continent, Coe and Jones organized the animal exhibits by biomes and emphasized naturalism and visual authenticity in re-creating characteristic landscapes for each species. They coined the term "landscape immersion" to describe this approach, which immersed both visitors and animals in natural settings, separated only by moats or other hidden barriers. It was not homocentric, but biocentric: humans, rather than controlling and dominating the space, were just one element of the landscape. By putting animals back in their natural settings, the designers also enhanced the possibility of meeting the animals' real needs.

Not until construction was underway was the Woodland Park Zoo's long-range plan completed. Coe was the principal writer and designer, but it was truly a team effort. The long-range plan, published in 1976, soon became the bible for zoo design, used all over the world. It describes all of the Earth's biomes (such as steppe, chaparral, and rain forest), explains relationships of climate zones, and provides a model for conceptualizing zoo exhibits, taking into account everything from vegetation and soil to animal service areas and viewing. The book—which had been out of print, but never out of demand—is being updated by Coe's present firm, CLRdesign, for re-publication.

The Woodland Park Zoo, which has won four major awards in its field, established Coe and Jones and their associates as zoo designers. Landscape immersion quickly took hold, and they worked on projects for zoos all over the country. Another talented architect and landscape architect, Gary Lee, joined Jones & Jones and worked closely with Coe on a number of those projects.

Coe was enjoying life in the Seattle area. He and Susan had a daughter, Alyssa, born in 1973. They lived on a farm on Bainbridge Island. After milking his goats in the mornings, Coe commuted to work by kayak and ferry. But as he traveled with increasing frequency, around the United States and internationally, he began to feel that he could live anywhere. "Home" was wherever he was immersed in a culture or a project, wherever he was at that moment. He was ready for a change.

Taking a leave of absence from Jones & Jones in 1983, Coe accepted a position as adjunct associate professor of landscape architecture at the University of Pennsylvania and moved with his family to the Philadelphia area.

While teaching at the university, Coe began to get calls from East Coast zoos. He resigned from Jones & Jones—an amicable parting, with mutual respect—and was able to build up his own practice, initially setting up a studio in his garage. In their country home, the Coes raised basenjis, barkless dogs of African origin that they had first acquired in Alberta; Susan has become an international expert on the breed.

It was not long before Coe needed a design partner. Gary Lee, with whom he had worked at Jones & Jones, came to Philadelphia to join him. Coe and Lee's role in the amazing rebirth of Zoo Atlanta in the mid 1980s lead to national publicity, including a feature piece in *The Atlantic*. Coe has since been quoted in *USA Today,* in the *New York Times,* and in the *Wall Street Journal.* Coe was named a 1988 Honoree by *Esquire,* and in 1993 he was made a Fellow of the American Society of Landscape Architects.

In 1995, John Rogers, an architect and a former classmate of Gary Lee, was brought in as a partner, and the firm's name became CLRdesign inc. The offices were moved to Old City, a trendy section of Philadelphia where art galleries, studios, and cafes now occupied former industrial buildings.

While creative collaboration is central to their work, each of the partners plays a distinct role in the firm. Coe is the philosopher, the naturalist, the public voice of CLRdesign. He does most of the writing and publishing, and he works closely with zoo curators and animal caregivers. Lee, a brilliant conceptual thinker, designer, and artist, also has a keen eye for business. He often represents the firm in dealing with zoo board members, directors, and senior administrators. Rogers, a pragmatist with a background in architecture and urban planning, handles much of the business management as well as design and client contact. The three principles are supported by an interdisciplinary team of architects, landscape architects, exhibit and interpretive designers, model makers, interns, graphic designers, and administrative assistants.

The office reflects a combination of professionalism and whimsy. The white walls are covered with photographs and huge graphics from past and current projects. By making the plans very large and colorful, CLR helps clients visualize and understand the ideas presented. Scale models are also used both in developing new concepts and in communicating them to clients. Yet along with the plans and models are playful animal sculptures, including a life-size, fake-fur-covered stuffed gorilla named Coco who sits near the receptionist. Coco is a favorite of CLR staff members' children, who frequently visit the office.

Coe's design process often begins with a story line or scenario, a description of the exhibits as if they already existed. A concept is suggested, and then all the component parts—animal facilities, viewing positions and paths, ground cover, vegetation, water features, interpretive signs, acoustical features, and mechanical systems—are developed to fit the scenario. Ideas and images emerge partly from Coe or other CLR designers, partly from the client's program. "There are a lot of people making this happen," Coe explains. "It's very much a team effort. By the time a project is actually built, it might have had two or three hundred people involved. It's not just having a certain vision, but it's having the communication to get it built the way you envisioned it."

Although his name is identified with landscape immersion, Coe does not feel confined by that concept. He continues to expand and refine the immersion experience. For example, he has advocated placing animals on higher ground than human viewers, arguing that the physical act of looking up will unconsciously engender in visitors an attitude of respect for wildlife. But he has also created innovative zoo exhibits that take an entirely different approach. In a "Primate Reserve" for the Philadelphia Zoo, CLR wanted to bring visitors closer to the monkeys and apes and allow people to watch behind-the-scenes interactions with caregivers that are hidden in immersion exhibits. The CLR team came up with a scenario of an abandoned lumber mill in the tropics that had been converted to a primate research and conservation station. They built a new $14 million structure that looks convincingly like a rehabilitated industrial building. Within the building, they used affiliative design principles to minimize visible barriers between visitor and animal spaces. There are even "howdy crates" where animals and visitors can come face to face through a glass partition. Coe believes such zoo concepts can become models for real wildlife rescue and conservation facilities around the world.

Coe now regards landscape immersion as "just one tool in the tool kit." Recently, he has been exploring the use of other important tools, such as operant conditioning training and behavioral enrichment, which in turn affect design. Operant conditioning, developed by Karen Pryor and other marine mammal trainers and based on ideas of the psychologist B. F. Skinner, is a method of controlling behavior through positive reinforcement. By this means, even wild animals can be trained to follow caregivers or to cooperate in their veterinary treatment. Behavioral enrichment involves expanding the range of choices and experiences available to the animal within its environment. Coe suggests that animals should even be allowed—by use of motion detectors or other means—to control the lights, fans, or temperature of their zoo environments. "Why can't the monkeys run the monkey house?" he asks, pointing out that primates in the wild have learned to meet their needs in a much more complex and dangerous habitats. He further explains:

To use these enrichment features requires some training and a real integration of those features into the immersion design. Activity-based design and management is the integration of all these things into one whole. Instead of us designing the exhibit, and then once it's open, the staff designing the husbandry system, we're saying they should all be designed together. Training, enrichment, presentation, education and entertainment, habitat and horticulture, all gets done together.

Coe sees many challenges for the next generation of zoo designers and believes that creative responses can be cultivated and channeled. Most important are exposure to varied, powerful experiences, to a range of cultures and ideas, and to people working in different disciplines. Creativity and honest dissent should be rewarded, and respect for others should be instilled by example. Aspiring designers must learn to really listen to clients and to work with their feedback. Not least important is a willingness to remain anonymous. "Remember," Coe cautions, "it's their project, not yours."

If there is a guiding principle at CLR, it is that design should be innovative:

As a firm, we try to emphasize innovation on every project. . . . But the subject of that innovation is driven by the client, not by us. In other words, when we start, we'll suggest all kinds of different ideas and new directions. If any of them find fertile ground, that's the one we'll pursue.

The firm is not proprietary about new ideas or technological improvements it develops. It provides a service, not a product. "We try to encourage our people to understand that the creative muscle gets better with use, that no idea is so precious you can't give it away and come up with a better one tomorrow. If you start defending it, maybe that's when you set up blocks," Coe says. "The universe is full of ideas. All you have to do is access them." Coe elaborated on his work process in a poem titled "The Journey":

The creative process is easily learned.
It can be like a family trip.
Know who you are,
where are you going?
And why?
What are your resources,
time, wealth, friends, family?
What obstacles are expected?
What routes seem best,
highway or byway?
What milestones, what havens?
How will you know
when you arrive?

But once the trip begins,
chance and change interfere.
Travelers change the journey.
The journey changes the travelers.
Roads are clear
or fog bound.
Sometimes you have
to get out and walk
and walking can be
the best part.

For most
the journey is the means,
necessary, unavoidable.
But for some,
navigators of the creative process,
joy is in the journey,
arrival is only a break
between trips.

In 1926, Fritz Lang's film *Metropolis* offered a bleak vision of cities of the
future. People lived in class-defined strata where opportunity and a fulfill-
ing life were restricted to the rich and powerful. Workers and the machines
they operated became indistinguishable from one another. Technology was
elevated to the role of co-star.

Lang, trained as an architect, held a strikingly different view of the
"techno-city" than others in the 1930s, as the historians Arthur Molella and
Robert Kargon point out. Defining techno-cities as planned communities
developed in conjunction with an industrial enterprise, the two historians
compare Norris, Tennessee (a product of the Tennessee Valley Authority)
and Salzgitter, Germany (established in connection with the Hermann-
Göring-Werke armaments factory). The two towns are examples of "high
modernism," evoking a faith in technology, science, rational design, and
control over nature. These "techno-cities," Molella and Kargon assert, "were
created by visionaries in each country as exemplars for the environmental,
economic, and moral regeneration of the nation." In the end, though, both
Norris and Salzgitter failed as viable communities—for reasons beyond
their founding philosophies.

The architectural visionary Paolo Soleri also addresses the idea of
planned communities. Labeling the single-family house "the classical tip of
the iceberg of wrongness," Soleri advocates combining living, working, and
service establishments into unified megastructures. As the art historian
Harry Rand explains, "Soleri wished to conserve the city's space rather than
allowing it to spread horizontally." Rand places Soleri's call for the death of
cities into context by contrasting it with the work of the artist and architect Friedensreich Hundertwasser. "If Soleri has tried to unify the city to
save it from sprawl," he notes, "Hundertwasser has tried to bring the village,

and village life, into the city so that every apartment [in his Vienna apartment building] is visually articulated and can be distinguished from street level."

Yet another perspective on the nature of cities and urban planning is found in the work of Erick Valle, a proponent of the "New Urbanism." This town planning movement combines traditional forms with technological innovation to create distinctive towns that reflect their inhabitants' heritage and strong sense of community. Central to Valle's approach is community involvement in the planning process.

ENVIRONMENTAL PLANNING FOR NATIONAL REGENERATION: TECHNO-CITIES IN NEW DEAL AMERICA AND NAZI GERMANY

ARTHUR MOLELLA AND ROBERT KARGON

From its beginnings, the Industrial Revolution aroused concerns about environmental degradation and its harm to society. The First World War and the ensuing worldwide depression refocused attention on the need for comprehensive approaches to economic development without environmental blight. The 1930s, for example, was an era of dramatic visions of the future. Mixed in with often virulent ideologies were bold attempts to devise inventive solutions to long-standing problems. Powerful activist regimes were established in autocratic Italy, Germany, and the Soviet Union as well as in democratic America. The European nations faced not only the great world-wide economic depression but also the challenges of securing their authoritarian rule and building their military capacity for the eventual showdown with their enemies. As confidence in centralized planning was immense, to address these problems they established new planned cities in connection with technical enterprises. These "techno-cities" were created by visionaries in each country as exemplars for the environmental, economic, and moral regeneration. First, all these visionaries advocated the decentralization of industry. "Back to the soil" provided a motto that was at once geographic, economic, and morally uplifting. Second, despite the nostalgia for a pre-industrial bond between land and people, these attempts to transform the nation paradoxically extolled the role of science and technology in building a new future. Theorists such as James C. Scott have termed this complex of ideas "authoritarian high modernism."[1] In an important sense, these techno-cities were *inventions* aimed at implementing the nation's planned tomorrow.

This essay compares two such endeavors: Norris, Tennessee (established in connection with the Tennessee Valley Authority) and Salzgitter, Germany (established in connection with the Hermann-Göring-Werke, an armaments factory).

NORRIS, TENNESSEE: A NEW TOWN FOR A NEW ERA

The Tennessee Valley Authority town of Norris, Tennessee, was conceptualized, planned, and built in the 1930s as a home for employees of the TVA. But its reach was intended to extend much further. Its planners envisioned Norris as a model for the nation, pointing the way to a better, richer, more satisfying life for its residents. Norris was to exemplify what rational and humane planning could do for Americans as a people. Genetically linked to utopias of the nineteenth century, it was a product as well of the technological optimism so common in the United States. Using technology appropriately, the optimistic view held, Americans could protect themselves from the worst excesses of the rapid growth of the nation: unchecked industrial capitalism, untoward urbanization, and rampant expansion of the population.

Perhaps surprisingly, our discussion of Norris begins with the automobile manufacturer Henry Ford. In the immediate aftermath of the First World War, Ford—like his friend Thomas Edison—was considered a genius who contributed to the technological marvel that was the modern age. In July 1921, Ford offered the US government $5 million for its hydroelectric dams and nitrate plants along the Tennessee River at Muscle Shoals in Alabama. This offer brought to a head decades of debate over public versus private ownership of these resources. Senator George Norris, a Republican from Nebraska, opposed the plan. Norris pointed out that the plants had cost $90 million to build, and he submitted a bill to Congress for government operation.[2] In January 1922, Henry Ford astonished the nation by coming out with an expanded version of his vision of the future of Muscle Shoals, a vision that greatly increased his popularity and changed the debate over Muscle Shoals forever. On January 12, the *New York Times* ran a headline: "Ford Plans a City 75 Miles in Length." The article, which went out on the Associated Press wire, outlined Ford's plan for a new regional development model for the Tennessee Valley, including a city 75 miles in length for the Muscle Shoals area. "It would be made up," the article stated, "of several large towns or small cities. This is in line with the manufacturer's view that men and their families should live in small communities where benefits of rural or near-rural life would not entirely be lost."[3] Ford's vision set off a frenzy of enthusiasm in the region and a wild boom in the real estate market. Edison publicly announced his support for the plan and advised Congress to accept it.[4]

The "Seventy-Five Mile City" was the subject of a long, laudatory *Scientific American* piece in September. The new Tennessee Valley industrial center would depend on the establishment of linked "hydro-driven plants." Between these factories would be the "farm-homes of the factory workers." An employee, it was asserted, could "be a food-producer and salary-earner at the same time."[5] The idea was "factory and farm close together, yet co-operation between them."[6] Shortly thereafter, in an interview with the magazine *Automotive Industries,* Ford envisioned "a great industrial city on the banks of the Tennessee, which will rival Detroit."[7]

Ford's plan was ultimately rejected by Congress. The opposition was led by Senator Norris, an advocate of public power. Though Ford bitterly blamed "the international Jews" for the failure of his plan, he did make one claim that rings true: "If we haven't done anything else, we have shown what Muscle Shoals are worth."[8] Ford had indeed changed the public's thinking about Muscle Shoals. No longer would the question be about fertilizer and hydroelectric power alone; it would increasingly focus on the larger vision—a regional plan to uplift the area from backwardness to leadership. The Tennessee Valley could be a great utopian experiment in reversing some of the excesses of the Industrial Revolution, which Ford himself had done so much to advance. Ford was not a proponent of urbanization. He decried the movement of farmers from the land into the cities. "Factory and farm," he said as early as 1918, "should have been organized as adjuncts of one another, and not as competitors." The city, in his view, had been a mistake.[9]

THE COMING OF THE NEW DEAL

These ideas resonated with those of another public figure from another side of the political spectrum: Franklin Delano Roosevelt. One historian writes that within a year of the signing of the Tennessee Valley Authority legislation President Roosevelt invited Ford to the White House to discuss "getting people out of dead cities and into the country."[10] Ford and Roosevelt shared a commitment to "the land." Roosevelt, the Hudson Valley patroon, had a visceral distrust of high-density urbanization. And, like Ford, Roosevelt was seeking a new way to mitigate the worst excesses of industrialization.

For Roosevelt, unlike Ford, one way out was through planning, and especially regional planning. Roosevelt became interested in regional planning, by his own testimony, through his uncle Frederic Delano. Delano, who had been a sparkplug in the creation of Daniel Burnham's Chicago Plan of 1908, went on to chair the planning committee for the New York Regional Plan of the 1920s.[11] Before World War I, Delano introduced Roosevelt to the City of Chicago Plan. "I think from that very moment," Roosevelt wrote in 1932, "I have been interested in not the mere planning of a single city but in the larger aspects of planning. It is the way of the future."[12] Upon his nephew's election to the presidency, Delano moved to Washington as an advisor, and became head of the newly created National Planning Board.[13]

While governor of New York, Roosevelt solidified his concern with planning and, as the historian Paul Conkin has written, his enthusiasm for "preserving scarce resources, for moving as many people as possible back onto the land, for making cities as orderly and country-like as possible."[14] In an address on state planning delivered in June of 1931, Roosevelt defined his philosophy: "Government, both State and national must accept the responsibility of doing what it can do soundly, with considered forethought and along definitely constructive, not passive lines." One area of concern was land utilization. "Hitherto," he said, "we have spoken of two types of living and only two—urban and rural. I believe we can look forward to three rather than two types in the future, for there is a definite place for an intermediate type between the urban and the rural, namely a rural-industrial group. . . . It is my thought that many of the problems of transportation, of overcrowded cities, of high cost of living, of better health for the race, of a better population as a whole can be solved by the States themselves during the coming generation."[15]

Roosevelt's interest in regionalism doubtless was behind the invitation to be keynote speaker at the July 1931 Roundtable on Regional Planning at the University of Virginia. Important members of the New York-based Regional Planning Association of America (RPAA) were there: Clarence Stein, Lewis Mumford, Henry Wright, and Benton MacKaye. Stuart Chase, Howard Odum, Charles W. Eliot II, Frederick Newell, and others made presentations. Roosevelt forcefully asked a receptive audience: "Isn't there a third possibility [between urban and rural], a possibility to create by cooperative effort some form of living which will combine industry and agriculture?"[16] This important theme, a third way between city and rural area, was repeated often during the years leading up to Roosevelt's successful campaign for the presidency and during the first years of the New Deal.

Over and over again, he would denounce "the profligate waste of natural resources" and the "gigantic waste" that industrial advance has entailed.[17] After his election and before his inauguration as president, Roosevelt made a point of traveling to Muscle Shoals, where he made the following extemporaneous remarks:

I am determined on two things. . . . The first is to put Muscle Shoals to work. The second is make Muscle Shoals a part of an even greater development that will take in all of that magnificent Tennessee River from the mountains of Virginia down to the Ohio and the Gulf. . . . Muscle Shoals is more today than a mere opportunity for the Federal Government to do a kind turn for the people in one small section of a couple of States. Muscle Shoals gives us the opportunity to accomplish a great purpose for the people of many States and, indeed, for the whole Union. Because there we have an opportunity of setting an example of planning, not just for ourselves but for the generations to come, tying in industry and agriculture and forestry and flood prevention, tying them all into a unified whole over a distance of a thousand miles.[18]

Within weeks of his inauguration as president, Roosevelt moved to establish the Tennessee Valley Authority, combining Senator Norris's interests in power, the agriculturists' concerns with fertilizer production, and his own vision of regional planning in an all-encompassing bill, which he signed on May 18, 1933.[19]

Roosevelt turned his attention next to selecting a director for the TVA. He had consulted Arthur E. Morgan, president of Antioch University, a flood-control engineer, and a scholar of the utopian writer Edward Bellamy. Morgan had shown a serious interest in community redevelopment while at Yellow Springs, Ohio.

A TECHNICAL UTOPIA

Morgan's interview with FDR reinforces the notion that Roosevelt was interested in the TVA primarily as an exemplar of regional planning and not merely as a source of power and fertilizer. Morgan later reported: "He talked chiefly about a designed and planned social and economic order. That was what was first in his mind."[20]

Within a month of the signing of the TVA Bill, Morgan began planning a new community to be associated with a new dam at Cove Creek, Tennessee. It would be a permanent town, rather than temporary housing for

TVA workers, and it would, at Morgan's insistence, be called Norris, after the senator who had fought so hard for the TVA Bill.[21]

One of the people brought on board early was Benton MacKaye, a stalwart of the RPAA, a founder of the Wilderness Society, and the conceiver of the Appalachian Trail. In an important article in the May 1933 *Survey Graphic* titled "Tennessee—Seed of a National Plan," MacKaye extolled the breadth, scope, and depth of the Roosevelt plan for the Tennessee Valley: "President Roosevelt has spread it out from a dam to a river to a region. ... He has done more—he has related a local project to a national emergency; he has sown the seed of that "national planning" announced in his inauguration speech."[22] MacKaye was ultimately concerned with the protection of what he termed "the basic settings"—wilderness, community, and wayside. To stem the tide of the "metropolitan slum," he urged the "townless highway" and the "highwayless town." Highways bypass the town, but are connected to it by spur roads. Surrounded by green spaces, the town is to the region as the cul-de-sac is to the main road.[23] MacKaye concluded with a peroration: "The Tennessee Valley project sows the seed of a national plan for the country's redevelopment.... Further steps ... must in due course carry on the national evolution conceived in the Roosevelt statesmanship."[24]

The actual planning and implementation of Norris were in the hands of Earle Draper, director of the TVA's Division of Land Planning and Housing, and his assistant director, Tracy B. Augur. The town was put on the fast track. The site was picked in July 1933, and housing construction started in January 1934. The idea remained "high concept." In an article published in December of 1933, Earle Draper noted:

To serve the entire community a complete town center has been laid out adjacent to a 14-acre public recreation ground.... Here will be grouped the public hall and administration building, a small hotel, stores, public market, bus station and service garage and other community features as the need arises. Centered on the main axis of this group will be the public school, away from traffic.... The utilities, including electric distribution station and steam laundry are relegated to nearby but unobtrusive locations. ... [Norris] will demonstrate that the unduly congested, insanitary, matter-of-fact ugliness and the usual haphazard growth ... can be avoided inexpensively.[25]

Tracy Augur, director of planning for Norris, drew a direct line from Ebenezer Howard's garden city ideal to Norris. According to Augur, the

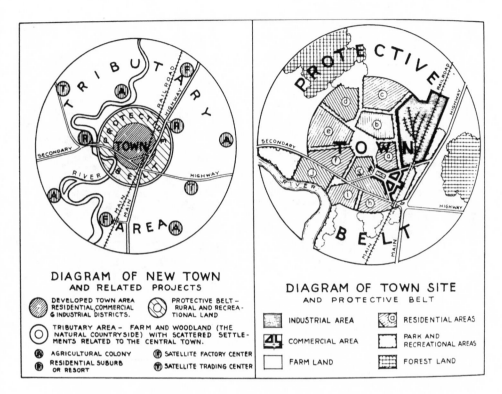

DIAGRAM OF NEW TOWN
AND RELATED PROJECTS

⊘ DEVELOPED TOWN AREA
RESIDENTIAL,COMMERCIAL
& INDUSTRIAL DISTRICTS.

◉ PROTECTIVE BELT –
RURAL AND RECREA-
TIONAL LAND

◯ TRIBUTARY AREA – FARM AND WOODLAND (THE
NATURAL COUNTRYSIDE) WITH SCATTERED SETTLE-
MENTS RELATED TO THE CENTRAL TOWN.

⊛ AGRICULTURAL COLONY   ⊛ SATELLITE FACTORY CENTER

⊛ RESIDENTIAL SUBURB   ⊛ SATELLITE TRADING CENTER
OR RESORT

DIAGRAM OF TOWN SITE
AND PROTECTIVE BELT

▭ INDUSTRIAL AREA   ▣ RESIDENTIAL AREAS

▣ COMMERCIAL AREA   ▢ PARK AND
RECREATIONAL AREAS

▢ FARM LAND   ▦ FOREST LAND

FIGURE I

Tracy Augur's plan for the resettlement of families displaced by Norris Reservoir, 1934. (*The Tennessee Planner*)

town had three "focal points": a community center, a construction camp site, and a business center devoted to attracting future industry. A population of 1,000–1,500 families was desired. The houses would be easy to maintain and would make maximum use of electricity for heating and cooking.[26] The town had a protective "greenbelt" around it and, consonant with Roosevelt's rural-industrial ideas, would have a home instruction center (offering classes in cooking, child care, budgeting, and furnishing), a trades and engineering center (with instruction in auto mechanics, aviation, plumbing, wrought iron work, and electrical and mechanical skills), and, in the greenbelt, dairy and poultry farms where Norris workers could engage in part-time agriculture. Workers would thus be educated to become foremen and managers, to provide an agricultural-industrial basis for everyday life,[27] to attract industry to balance the town's rural character, and to enable residents to partake of a rural-industrial community, or what

FIGURE 2
Model houses in Norris Village. (Walter Creese and Earle S. Draper Jr.)

TVA literature described as "a community based upon the orderly combination of industrial work and subsistence and farming."[28] According to Augur, "the fundamentals of the plan were never sacrificed—a recognition of the underlying purposes of the community—a sympathetic treatment of the site, abundant open space for children's play and adult recreation, attractiveness in all things big and little, from the iron bracket of the street signpost to the roadway's gentle curve and the school's straightforward architecture, simplicity, economy, a place designed for pleasant living and convenient work."[29]

The reality fell far short of the dream. The costs of housing were higher than those projected. As a result, the professional staff of the TVA were attracted to the original houses in Norris, while workers found themselves able to afford only the cinder-block houses in the southeast corner. The small population and rural location (approximately 25 miles from Knoxville) rendered Norris unattractive to businesses and industry. By 1936, when the major phases of construction were over, workers began to leave Norris, replaced in the housing units by outsiders on a waiting list.

However, the original élan of Norris was undermined, and it began to look more and more like a bedroom suburb of Knoxville. Moreover, the TVA itself was undergoing great changes. The two additional directors appointed by Roosevelt (David Lilienthal and Harcourt Morgan) consistently out-voted Arthur Morgan and turned the TVA increasingly into an agency for economic development.[30] By 1937 the TVA decided to sell Norris, having long before abandoned its regional development ideas.[31]

SALZGITTER, GERMANY: AS IN A DISTORTED MIRROR

While the Roosevelt administration developed its city of the future in Tennessee, Hitler's Reich was making similar plans for a *neue Stadt* called Salzgitter about 20 kilometers to the south of the Lower Saxon city of Braunschweig. Taking its name from one of the original villages in the region, it was built to house workers for a vast new mining and iron- and steel-making complex known as the Hermann-Göring-Werke. Although having quite different proximate causes, the National Socialist New Town and the TVA project shared some of the same environmental and ideological goals. Both experiments sought to reinvent the environment in response to impending industrialization. Both identified with technology, but expressed a profound ambivalence toward it—a desire to mitigate the effects of modern technological civilization with rural values and an appreciation of nature. And, as with Norris, Tennessee, the fate and fortunes of Salzgitter serve as a trail along which we can follow a fascinating process of invention and reinvention. But, while the Reich's project mirrored the TVA in significant respects, it was as if its reflection were seen in a distorted mirror, twisted by the extreme social, political, and economic conditions of the Nazi era.

ENVIRONMENTAL TRANSFORMATION: THE CITY OF THE HERMANN-GÖRING-WERKE

In the words of its architect, Herbert Rimpl: "With the rise of the [Hermann Göring] Werke the villages and the farm towns of the Salzgitter hills were awakened from their peaceful existence."[32] A farming area of 55,000 acres, approximately 20,000 people, and some thirty small towns became, in just a few years, "one of the largest concentrations of industrial might in the world."[33]

The agent of this astonishing transformation was the Reich's secret rearmament drive in the 1930s. Having ceded its main iron- and steel-producing regions in Alsace-Lorraine after the First World War, Germany looked for alternative sources of ore. Extensive deposits in the Salzgitter hills, known to exist since ancient times, had remained dormant because of the inferior quality of the ore. For centuries Salzgitter had been known as a bucolic agricultural district of wheat and sugar beet fields, its only claim to fame being mineral baths frequented by the princes of Braunschweig.[34]

But an invention changed all that. A newly patented chemical process provided an economical method for enriching Salzgitter's low-grade ore, thereby making its ore deposits usable. Hermann Göring, the powerful field marshall responsible for making Germany resource independent, incorporated Salzgitter in his Soviet-style four-year plan for putting Germany's economy and natural resources on a war footing.[35]

Construction began in 1937. A huge industrial complex of mines, foundries, molding plants, forges, chemical and electrical facilities, and other supporting installations was underway. With the region lacking a pool of skilled industrial labor, the project imported workers from all over Germany and Europe. When labor still fell short of demand, a concentration camp was erected nearby in 1941 to supply slave labor for the mines and foundries.[36] Salzgitter exploded in size, ranking as the fastest-growing and the most densely populated region in Germany.[37] With the huge influx of workers and their families, Göring was immediately confronted with a housing crisis. An overall plan for new housing construction was needed.

THE SALZGITTER SIEDLUNGSBAU

The Reich's approach to the housing problem was more than a pragmatic solution, however; it reflected concerns deeply rooted in German history and culture. Salzgitter's planners resolved to avoid the mistakes of Germany's industrial revolution, whose precipitous onset—the most rapid in Europe—had caused extreme social and economic dislocations. (The anxieties from rapid modernization are well documented in such German writings as Spengler's *Decline of the West*.[38]) It was feared that the development of the Hermann-Göring-Werke would inflict on the region the same environmental and human blight as had Germany's initial industrialization—flight from the land, urban overcrowding and suburban sprawl, polluted air and skies, the "ghettoizing" of workers in undesirable districts, and unhealthy living conditions.[39] Of even greater concern to Germans of this generation was the destruction of community, of *Gemeinschaft* (whose distinction from *Gesellschaft* was so critical to Ferdinand Tönnies and to the sociologists of the Frankfurt school).[40] If environment was a shaping force of society, as they believed, then who could predict what moral and spiritual damage would result from depriving the people of the open air and healing light of the countryside, natural connections to the land, and the joys of rural community? Such concerns, coupled with Germany's long history of anti-city sentiments, framed the planning for Salzgitter.

Yet the Reich was not willing to forgo the power afforded by technology, the foundation of economic, military, and political strength for the modern nation state. In inventing Salzgitter, therefore, it pursued a seemingly paradoxical course: aiming to preserve the environmental benefits of traditional rural life while building up the nation's industrial and military muscle.

THE GERMAN ENVIRONMENTAL IDEOLOGY

Incorporating ideas from Britain, Italy, and the United States, the German *neue Stadt* was a retranslation of the New Town concept in terms of National Socialist values. The solutions to housing, industrial, and environmental problems were wrapped in a "blood and soil" ideology that found peculiar expression in a paradoxical Nazi world view, combining progressivism and atavism—an awe of modern industrial might with a nostalgic pre-modern vision of the *Volk,* a romantic myth of small-town agrarian Germany (a cultural contradiction that Jeffrey Herf has labeled "reactionary modernism").[41]

This contradictory ideology was the basis for Germany's famous *Siedlungs* (settlement) program, which besides Salzgitter built thousands of new towns during the 1920s and the 1930s.[42] Salzgitter was intended as a model project of the type. The program, an expression of German anti-urban sentiments that go at least as far back as the seventeenth century, was a direct offshoot of Ebenezer Howard's turn-of-the-century Garden City movement, which had found an immediate and enthusiastic following in Germany. Howard's invention of the green-belted city, strictly limited to a population of 30,000 and combining industry and country in a new organic relationship, had German parallels in the contemporary anti-urban,

New Town ideas of the anti-Semitic propagandist Theodor Fritsch.[43] Relocating city dwellers into smaller settlements at the edges of cities or out in the countryside, the Reich's program of decentralization—also referred to as "internal colonization" or "repatriation"—produced new housing projects around the country, usually in conjunction with industrial sites.

According to Nazi propagandists, resettlement would remove downtrodden city dwellers from corrupting urban environments and bind them spiritually and morally to "Mother Earth" in a new form of rural industrial town. In actual practice, the decentralization program became a tool for social control and forced migration of undesirables; it emphasized the resettlement of minority groups and workers—a policy with deadly implications.

NAZI IDEOLOGIES OF RESETTLEMENT

The *Siedlungs* program's framing ideologies can be seen in the views of two high Nazi officials: Richard Walter Darré and Gottfried Feder. Darré, Hitler's agriculture minister and one of the Nazi Party's most powerful figures, brought a fierce anti-urban and anti-technology bias to National Socialist ideology. In his "blood and soil" philosophy, he categorized Europe's original peoples as either "settlers" or "nomads." Darré idealized the non-urban, land-loving peasantry—the settlers—as precursors of the "Nordic race" and condemned the nomads, citified purveyors of godless technology, as the progenitors of all other races, especially the "Semitic" and the "oriental." He called for "repatriating" urban populations to the soil as the only way to restore Nordic racial values. His racist mythology took a deadly turn when Heinrich Himmler made him head of the Race and Settlement Office within the S.S., which implemented Himmler's resettlement policies, and he became deeply implicated in the execution of "the final solution."[44]

Gottfried Feder, the Reich's *Siedlungskommisar,* launched a broad program of invented cities that sought a compromise between urban/technological and rural/agrarian values. He had a powerful role in shaping the Reich's policy of decentralizing urban populations and in carrying out the ideology of "blood and soil." Not sharing Darre's anti-technology views, he envisioned "green" towns of around 20,000 people that combined agriculture with industry. Feder compared the new city to an organism, hierarchically organized with lesser parts linked to greater, much as cells in the body constitute ever larger systems and finally the whole.[45] His ultimate goal was "the dissolution of the metropolis, in order to make our people be settled again, to give them again their roots in the soil." "The reincorporation of the metropolitan populations into the rhythm of the German landscape," he asserted, "is one of the principal tasks of the National Socialist government."[46]

REINVENTING SALZGITTER: HERBERT RIMPL

Nazi resettlement and environmental ideologies converged and crystallized in the planning for Salzgitter, one of the Reich's most ambitious attempts to design an environment integrating industry and agriculture, town and country. Awarded the contract for the overall planning of the Salzgitter *Werke* and of the housing and administrative facilities was Herbert Rimpl, Hitler's chief industrial architect and, after Albert Speer, the leading German architect of the Nazi era. Rimpl's company employed some 700 architects and had branches all over the Reich.[47] He had worked often for Göring, building installations for the Luftwaffe chief; Salzgitter was to be one of his biggest commissions.

Rimpl as yet has no biographer, but we know something of his thinking from his published reflections on architectural theory and practice, written in the philosophical mode often affected by celebrated architects.[48] His writings reveal a love of nature reminiscent of the romantic fantasies of Goethe and Rousseau. "Yearning for nature" [*Natursehnsucht*], he wrote, "logically follows technology"—an assertion that echoed Nazi antiindustrial sentiments.[49] "Already in 1905, after the epoch of the crassest materialism, the Garden City of Letchworth was built as a logical reaction." Rimpl's admiration for Letchworth, Ebenezer Howard's original new city, revealed his debt to the British Garden City movement. He praised the new domestic styles from England, where the house "is oriented toward the sun and bound with nature" and "the garden is a part of the house, whose inner spaces it extends outward, over terraces or meadows, over flower or vegetable gardens." He belived that "the basis for the rise of this kind of dwelling was the hated industrial city." Rimpl's love for nature verged on worship: "The inclusion of the breadth of landscape in cities, the opening up to nature afforded by glass houses, the yearning for green, for gardens, the sun, water, the mountains, undisturbed forests, all of these are visible signs of the embodiment of a pantheistic point of view."

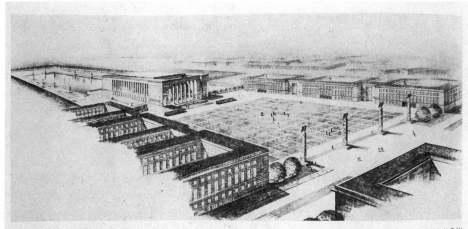

DIE STADT DER HERMANN-GÖRING-WERKE · DER HAUPTPLATZ   ARCHITEKT HERBERT RIMPL, H.G.W.

FIGURE 3
Rimpl's rendering of the main plaza for Hermann Göring Works City ("Die Kunst in Dritten Reich," *Die Baukunst,* April 1939)

In a Nazi-sponsored architectural journal, at the behest of Hitler, Rimpl laid out his plans and justifications for the Salzgitter project (figure 3).[50] Commissioned explicitly to preserve the rural character of the area as well as a sense of the natural environment within the heavily industrialized landscape, Rimpl faced an ultimate challenge to the tension in National Socialist ideology between nature and technology.

The architectural styles Rimpl used in Salzgitter tended to be eclectic, selected for pragmatic reasons rather than consistency. Thus, the principal public buildings—the *Volkshalle* and the Nazi headquarters (figure 4)—were in the symbolic Speer style of neo-classical giantism; the factories adapted the efficient glass-and-steel construction of the International Style (Rimpl's firm employed many former *Bauhäusler*). The workers' homes, 15,000–20,000 of which were planned for the overall area, exemplified the traditional vernacular style of German housing.[51]

What unified the whole was an overarching organic concept, reflecting the holism and organicism at the core of Nazi ideology.[52] Like Speer, Rimpl regarded architecture as both a practical and a symbolic art. Hence, his plans for the Nazi technology town can be read not only as a literal blueprint but also as a set of signs and symbols. Specifically, Rimpl's plans embodied the notion of the "body politic," a metaphorical comparison of the town plan

DIE STADT DER HERMANN-GÖRING-WERKE, ARCHITEKT HERBERT RIMPL, H.G.W.   DIE VOLKSHALLE AM HAUPTPLATZ DER STADT

FIGURE 4
Volkshalle and Nazi headquarters. (Rimpl, "Die Kunst in Dritten Reich," April 1939)

to the body. A tradition that goes back as far as Plato's *Republic* and that found ideological reinforcement in contemporary German cell theory, the "body politic" concept was intended to justify social arrangements by declaring them "natural."[53] As an organic entity, Salzgitter was to be a "green city"—a city at one with its natural surroundings, in the tradition of Howard's Garden City.

Both to separate it from and to connect it to major urban and industrial centers, Rimpl nestled the technology town in the heart of an existing network of rail and autobahn routes, to which was added a canal for industrial access. In the choice of physical location, environmental considerations—the amount of clean air and sun, the visual landscape, soil, water, and health factors—figured prominently. The town was located in a valley that was north of the foothills of the Harz Mountains (where the ore deposits lay) but west of the foundry areas (so prevailing winds would carry pollutants away from the town).[54]

Salzgitter was to serve as a hub for existing towns of the district as well as for the new settlements that were to be built to accommodate workers. The town proper was projected to have a population of about 130,000,

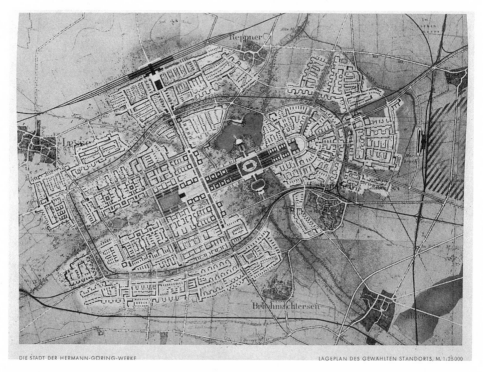

DIE STADT DER HERMANN-GÖRING-WERKE   LAGEPLAN DES GEWÄHLTEN STANDORTS, M. 1:25000

FIGURE 5
Rimpl's plan for Salzgitter. (Rimpl, "Die Kunst in Dritten Reich")

with the whole region, including the mines and industrial sites, reckoned
eventually at 250,000.[55]

Salzgitter was defined by the confluence of two rivers, the Flothe and the
Fuhse, and their green banks. Rimpl invoked the "body politic" metaphor
quite literally. The Flothe formed the 2-km backbone of the town, while
the convergence of the two rivers defined the skeleton of the body. The
joining of the green valleys provided the setting for a sport and health complex—the town's heart. The green areas, conducting cleansing mountain
winds through the town, were the lungs through which the town breathed.
The whole effect was to "give the new industrial city the character of a city
in the country."[56]

The transportation node at Salzgitter's northeast end formed the legs and
arteries of the town. A second symbolic point, the site of the Volkshalle and
Nazi headquarters, was the head that directed the organic functions of the
body. Order was of utmost importance, and hierarchy was to be preserved

at all costs, another reminder of the authoritarian subtext of Nazi organicism and holism.

The town's residential areas operated on the same hierarchical principal whereby the whole ruled the parts. The apartment buildings and single-family dwellings constituted the cells of the organism, and the roads winding around among them (as opposed to rectilinear major arteries) formed its capillaries.[57] Green gardens and commons around and among the houses and apartments kept the system well sunned and oxygenated. The cells were arranged in subsystems of small communities, each with its own schools, groceries, and other facilities.[58] At every level of organization, multi-story buildings were gathered at the center to preserve the sense of order and control from the center and above.[59] The main transportation routes were strictly separated from the residential areas, embedded in a greenbelt separating the town from other population centers and defining the city's skin.

The body politic of Salzgitter, however, never matured fully. Because priority had been given to the industrial sites and to workers' housing, Rimpl's grand scheme lay uncompleted at the end of the Second World War, realized only in the form of visionary architectural renderings and models. With Germany's defeat, the *Werke* shut down, ceding their role back to the revived Ruhr and other iron districts. Salzgitter lived on, however. During the Cold War, West Germany maintained Salzgitter and, to a small extent, its industries; in competition with nearby East Germany, it did not want to show an increase in unemployment. However, the experiment—the environmental dream that was Salzgitter—was over.

CONCLUSION

Both Norris and Salzgitter offer insight into perceptions of the urban condition as we enter the twenty-first century. Both were interesting early attempts to come to grips with problems left by the worst excesses of the Industrial Revolution and unplanned urban sprawl. Both displayed enormous confidence in planning for the future. Moreover, planners in both the United States and Germany tended to see those problems in regional and ultimately in national terms, foreshadowing later attempts, in the 1980s and the 1990s, to take up these issues.

In the end, of course, both Norris and Salzgitter were failures, whether judged by their own initial goals or by today's standards. Each failed owing to contingencies. Germany lost the war and was disarmed. The Hermann-

Göring-Werke ceased to function, and Salzgitter lost its purpose. After the war, the nationalist and *Volkisch* ideology of Salzgitter fell into disrepute; the German people wished to forget. Norris was practically stillborn. Just as it was beginning to develop, its political backing was cut loose from it. The New Deal, confronted by a stubbornly resistant depression, turned away from visionary planning and toward practical, ad hoc solutions. The pragmatists at the TVA, David Lilienthal and Harcourt Morgan, ejected Arthur Morgan, and the TVA turned mainly to power production and distribution. Beyond contingency, however, both experiments were ultimately victims of an intrinsically untenable concept. The inner contradictions of the founding idea—a massive industrial complex within a "green" environment—may have doomed them to failure from the start. In *Seeing Like a State,* James Scott termed the impulse behind this kind of planned city "high modernism" and characterized it as an ideology combining faith in scientific and technical progress, rational design of the social order, and control over nature.[60] Salzgitter and Norris reflect all these, plus a characteristically pre-World War I nostalgia for the rural and a distrust of the industrial city.

The legacy of the 1930s techno-cities is precisely this: they remind us once again of questions of the limits and strengths of planning, of looking for the optimum ways to deal with sprawl and congestion through the Garden City idea, and of the role of visions of the future, the utopian thrusts, and their dangers.

NOTES

1. James C. Scott, *Seeing Like a State* (Yale University Press, 1998), 89, 90.

2. Judson King, *The Conservation Fight* (Public Affairs Press, 1959), 98.

3. "Ford Plans a City 75 Miles in Length," *New York Times,* January 12, 1922.

4. Preston Hubbard, *Origins of the TVA* (Vanderbilt University, 1961), 40.

5. Littell McClung, "The Seventy-Five Mile City," *Scientific American,* September 1922, 156.

6. Ibid., 214.

7. "Ford Tells What He Hopes to Do with Muscle Shoals," *Automotive Industries* 47 (1922), October 19, 753.

8. Ibid.

9. Allan Nevins and F. E. Hill, *Ford: Expansion and Challenge 1915–1933* (Scribner, 1957), 227.

10. Reynold Wik, *Henry Ford and Grass-Roots America* (University of Michigan Press, 1972), 193.

11. Walter Creese, *TVA's Public Planning* (University of Tennessee Press, 1990), 38, 39.

12. Franklin D. Roosevelt, "Growing Up by Plan," *The Survey* 67 (February 1, 1932), 483.

13. Creese, *TVA's Public Planning,* 38, 39; Donald Krueckeberg, "Norris and Environmental Tradition," paper presented at conference "A Planned Community: Norris Tennessee after 50 Years," 1983, 7, 8.

14. Paul Conkin, "Intellectual and Political Roots," in *TVA: Fifty Years of Grass-Roots Bureaucracy,* ed. E. Hargrove and P. Conkin (University of Illinois Press, 1983), 24.

15. F. D. Roosevelt, *Public Papers and Addresses* (Random House, 1938), 486, 487, 494.

16. Roosevelt, quoted in Creese, *TVA's Public Planning,* 51.

17. F. D. Roosevelt, "Address at Oglethorpe University," May 1932, 642.

18. F. D. Roosevelt, "Informal Extemporaneous Remarks," January 21, 1933, 888, 889.

19. Frank Friedel, *Franklin D. Roosevelt: Launching the New Deal* (Little, Brown, 1973), 351.

20. Ibid.

21. Roy Talbert Jr., *FDR's Utopian: Arthur Morgan of the TVA* (University of Mississippi Press, 1987), 115.

22. B. MacKaye, "Tennessee—Seed of a National Plan," *Survey Graphic* 22 (1933), 251.

23. Ibid., 293.

24. Ibid., 294.

25. Earle Draper, "The New TVA Town of Norris, Tennessee," *American City and County* 48 (1933), 68.

26. Tracy Augur, "The Planning of the Town of Norris," *American Architect and Architecture* 148 (1936), 19–26.

27. Creese, *TVA's Public Planning,* 258, 259.

28. Charles Stevenson, "A Contrast in Perfect Towns," *The Nation's Business* 25 (1937), 19.

29. Tracy Augur, quoted in National Resources Committee, *Supplementary Report of the Urbanism Committee, vol. II* (National Resources Committee, 1939), 72.

30. Thomas McCraw, *Morgan vs. Lilienthal: The Feud within the TVA* (Loyola University, 1970), 36; Richard Lowitt, "TVA 1933–45," in *TVA,* ed. Hargrove and Conkin, 44, 45.

31. Daniel Schaffer, "The Tennessee Transplant," *Town and Country Planning* 53 (1984), 316–318.

32. "Die Stadt der Hermann-Göring-Werke," in *Die Baukunst, Die Kunst im Dritten Reich,* April 1939, 140. Translations are ours unless otherwise indicated.

33. Winfried Nerdinger, *Bauhaus-Moderne im Nationalsozialismus* (Prestel, 1993), 172. I am grateful to Dr. Nerdinger, Director of the Architecture Museum at the Technische Universität München, for his insights and for leading me to numerous German sources on Nazi architecture and town planning.

34. By far the best source on the development of the Salzgitter region is Christian Schneider, *Stadtgründung im Dritten Reich, Wolfsburg und Salzgitter* (Heinz Moos, 1978).

35. See Schneider, 55–57.

36. Jean Chardonnet, *Métropoles Économiques* (Armand Colin, 1968), 101; Hans Günter Schönwälder, Werden und Wandel des Industriegebietes Salzgitter, Ph.D. dissertation, University of Hamburg, 1967, 81.

37. Schönwälder, 81–83. The population was initially projected to grow to 300,000 but never actually reached that size. See Chardonnet, 113.

38. On Spengler's and the National Socialists' reaction against the modernist city and industrial state, see Carl E. Schorske, *Thinking with History: Explorations in the Passage to Modernism* (Princeton University Press, 1998), 52, 53.

39. On Nazi ideological antipathy to cities and industrialization, see Barbara Miller Lane, *Architecture and Politics in Germany, 1918–1945* (Harvard University Press, 1968), 155; Schneider, 9, 11.

40. On Tönnies's distinction between *Gemeinschaft* and *Gesellschaft* in response to industrialization, see Arthur Mitzman, *Sociology and Estrangement: Three Sociologists of Imperial Germany* (Knopf, 1973).

41. Jeffrey Herf, *Reactionary Modernism: Technology, Culture, and Politics in Weimar and the Third Reich* (Cambridge University Press, 1986).

42. Among them were edge cities near Düsseldorf, several in industrial areas of Berlin, and elsewhere. By 1941, some 355 million Reichsmarks had been spent to create a reported 184,000 *Kleinsiedlungen* (new mining, steel, and agricultural towns) and more than 275,000 dwellings, apartments, and *Volkswohnungen* for middle-class, working-class, poor, and unemployed Germans. One of the most interesting examples of a new German industrial town (recalling places like Pullman, Illinois, and, to some extent, Norris, Tennessee) was the city of "KdF (Strength through Joy)-Wagens," the Volkswagen factory town, also near Braunschweig, built under Reich auspices by Ferdinand Porsche, a great admirer of Henry Ford. See Schneider, 63.

43. On Howard's Garden City Movement, see Lewis Mumford, *The City in History: Its Origins, Its Transformations, and Its Prospects* (Harcourt, Brace & World, 1961), 514–522. On the rapid adoption of the movement in Germany, see Schneider, 10; Ute Peltz-Dreckmann, *Nationalsozialistischer Siedlungsbau* (Minerva, 1978), 43–45, 203f.

44. Lane, 154–156.

45. Heavily influenced by Ebenezer Howard's Garden City ideal, Feder sketched a plan that established the template for towns like Salzgitter in "Versuch der Begründung einer neuen Stadtbaukunst aus der sozialen Struktur der Bevölkerung." Feder's proposals are outlined in Peltz-Dreckmann, 43–45, 193–204.

46. Quoted and translated by Lane, 205, 206.

47. In December of 1937, "Wohnungs A.G." was incorporated in Braunschweig to build and manage the housing for the employees of the Reichswerke (Schneider, 62).

48. Although published nearly a decade after the end of the Salzgitter project, Rimpl's tract, *Die Geistigen Grundlagen der Baukunst unserer Zeit* (Callwey, 1953) reveals the ideological essence of his thought.

49. All quotations from ibid., 6, 134–137.

50. "Die Stadt der Hermann-Göring-Werke," in *Die Baukunst, Die Kunst im Dritten Reich*.

51. As Winfried Nerdinger has pointed out, Rimpl's firm was known for hiring modernists, including a considerable number from the Bauhaus and Gropius's office. *Bauhäusler* assumed leading positions in Rimpl's giant Salzgitter operation, making it "the largest reservoir in Germany" for the modern architects who did not flee the country after the rise of the National Socialists. For photographs of the workers' housing, see Schneider, 79, 80.

52. See Anne Harrington, *Reenchanted Science, Holism in German Culture from Wilhelm II to Hitler* (Princeton University Press, 1996).

53. See especially Book IV of *The Republic,* where the components of the state are compared to those of the individual. On the German biological theories underlying the notion of the state as organism, see Harrington, 59.

54. Rimpl, "Die Stadt der Hermann-Göring-Werke," 148, 179. Rimpl justifies the choice of site and layout in terms of its harmony with the natural environment and topography.

55. Ibid., 140, 141. The projected size of Salzgitter fell well beyond the population guidelines of Nazi planners like Feder, who envisioned an ideal *Mittel-Stadt*—a mid-size city of approximately 20,000—large enough to be self-sufficient and to avoid the backward conditions of small German villages but small enough to avoid dependence on special modes of transport and other disadvantages of big cities. See Peltz-Dreckmann, 194, 197.

56. Ibid., 148.

57. Among the dwellings were some 300 experimental homes, using such new materials as "gas-concrete" (i.e., aerated concrete blocks) and novel building techniques. See Nerdinger, 172.

58. See Chardonnet, 114. In this respect, his designs showed the influence of the "Neighborhood" concept developed by American city planners in the 1920s. See Schneider, 67.

59. Rimpl explains (ibid., 149): "The architectonic Gestalt leads from the green outer areas in the form of settlers' houses, and the single-storey row houses and detached homes through the two-storey apartment areas to the closely adjacent city center whose dominating structures give the general impression that the city was designed by an overpowering Will." It has been remarked that the overall organization of Salzgitter represented the hierarchical organization of the Nazi party, reminding the people graphically of the precedence of the social and political order (Nerdinger, 67).

60. Scott, *Seeing Like a State,* 89.

PROSPECTS AND RETROSPECT: THE CITY OF HUNDERTWASSER AND SOLERI

HARRY RAND, WITH A STATEMENT BY PAOLO SOLERI

Certain notions so deeply rooted in our thinking resemble instincts and aspects of personality more than ideas. Among these, general cravings toward form are a rare class of diagnostic self-perception which historically present themselves in few instances. These central notions may not be universal yearnings, although their manifestation seems to have been the driving force of much of what is called culture and almost everything called civilization. Some of these longings seem more like natural tropisms that underlie technological evolution rather than culture-specific urges toward the creation of distinct inventions that derive from convention. Among the great locomotives drawing humanity forward through history, a very few notions were realized in historical times and some even within the memory of living people. That is, after millennia of yearning—and probing nonlinear experiments of more-or-less value—within the last few generations some of these central urgings have been requited. Eye-witness testimony to a great shift in the way our species adapts is rare, but recognizing the satisfaction we experience at having achieved these goals helps visualize the pride early civilizations felt in overcoming the first great beckoning ambitions—none more primary than the others.

The airplane gave us flight, for which we had envied the birds since the time when bipedal humans could look up to the clouds. Letting us travel faster than the wind or seven-league boots, our wonder at flight, and the godlike convenience it bestows, is either keenly appreciated from our window seat or forgotten in the aisle seat. In the nineteenth century the telephone brought far-off voices and conversation instantly to us, withering space and making Marco Polo's distant report as easily obtained as a routinely punched-in country-code; or, a telephone call can be unspeakably poignant, more than the words uttered over distance. But even this invention was not as momentous, in terms of assuaging human longing, as the breakthrough innovation of the movies, by which life could be captured beyond its moment and place and preserved past the lifetimes of those pictured—a kind of immortality was added to verisimilitude, both of which had driven art from the first time a hunk of charcoal began to be applied to a cave wall. The movies succeeded where the Pharaoh's mummies failed: to have the dead walk and talk among the living. That desire may date from Neanderthals' (supposed) burials and every other subsequent intimation of immortality. Such innovations tell us how long-ago people triumphed over the taunting of time and space. While these technical victories can still be savored, and are visible against the background of other, no longer dramatic technologies (of fiber, agriculture, pyrotechnics, etc.), one of the greatest early accomplishments is now all but invisible, absorbed into our assumptions.

More distantly in the past than most things we recognize as culture, the city was accomplished with the same drama as these other inventions except that the wonder of that achievement has passed from memory, and sunk into the oblivion which today only poetry can rescue from complacency. Familiarity with cities adds background awareness of a certain urbane drama, an option to be exercised at will—by visiting a city, or, if you live in one, by leaving it. Urban excitement is very old, dating back to shortly after the contrivance of the first city—which was, assuredly, an invention.[1]

Quickly and universally the idea of the city spread. Perhaps like the alphabet, it might only have been invented once since contact with it convinced of its superiority to other forms.[2] And, like the alphabet which has only one form—adapted by various peoples to different writing instruments, surfaces, and speech patterns—there is one central word for city, just so quickly did the idea spread.

"City" (meaning a town of notable size, usually incorporated with a municipal government and defined boundaries) derives from the Indo-European root for "couch" or "lie," and even generates such words as "cemetery." Through a quirk of English's development we must use the word "city" to refer to a far deeper realm of meaning. In ancient Egypt the heavenly city was called "per" Anu.[3] At some distance, in South Asia, in Sanskrit, "pur" is only slightly different but means the same thing, city; and the great cities of Mahamalapuram or Yasodharapuram carry that word within them.[4] (Ancient Etruscan confounds the leading scholars, who "do not entirely agree with one another on the meaning of such [a] basic Etruscan

[word] as *spura,* 'city.'"[5] And the Etruscan language does not seem related to very much else on the European mainland.)

The unvoiced "p" of *per* or *pur* equates to a voiced "b" in the German *burg* or English suffix *burgh* or "borough," and some English speakers pronounce the latter simply as "burr."[6] During the post-glacial period the sound has hardly changed since the dispersal of the 'Semitic' ancient Egyptian and Sanskrit, all the way through into modern English. Far from England, on the Pacific Ocean is Singapore, which name is formed from "Singh" = lion + "por" = city. What a perdurable word-idea-cluster this is, how central to human self-definition. It must have come into people's heads at the dawn of time and represented our mastery of the environment in some deeply satisfying way.

Retaining the initial "p" sound, if the final "r" is sounded as an "l"—as so many languages do conflate these two sounds—a whole other batch of examples comes into view. The Greeks have their "pol" in "polis" which spawned the modern words derived from that root (police, policy, polite), all distinguishing the social refinements and necessities of city life.[7] Even further afield, the Greek "pol" might have been enunciated as "bol,"[8] which could yield an alternate pronunciation as "vol," the likely predecessor of "ville" (which seems as reasonable an ancestor for this word as "villa," a kind of palace). If the initial consonant was softer—or, being unvoiced, was almost silent—then other, seemingly unrelated words may emerge from the root. The Bible records that Abraham came from the Chaldees' city of Ur, which can be presented orthographically as "(p)ur" or "(p)er." Nor is this citation the only such transliteration of the sound. The Hebrew for city is simply "er"—again, basically the same word as all these others. Moving from the Semitic to the Indo-European languages, we recall that Latin's "ur(b)" is the root for words like urban, meaning city, and urbane—how a city dweller carries him or herself to adapt to the man-made situation.[9] Although Roman metropolises, or at least Latinate ones, arrived on the horizon five to eight millennia too late to have been invented in Italy the root word for this planned or accumulated construction remains the same. The linguistic traces are everywhere.

"In cuneiform writing, prefixes called determinatives usually define the characteristics of the following word but are not themselves pronounced or translated. However, the use of determinative in the [dispatches from the senior administrator Abdi-Heba to Pharaoh concerning] Jerusalem in the [fourteenth century B.C.E.] Armana letters is different from that in other letters from Palestine. Town names in the other letters are usually preceded by the determinative URU," as in Urusalim, as Jerusalem was then written.[10] (This indicator remains nested in the name Jerusalem.) In the South of India, where the Hindu personify every aspect of the world (mountains can be divinities and rivers alive), Tamil villagers distinguish between the map-delimited town in which they live and its spirit, "ur."[11] The ur is the town's personified space, the soil from which its particular consciousness arises and its spirit, its locally distinct traits which interpenetrates the villagers' bodies. (In the abstract this sounds foreign, and as coldly clinical as do many anthropological descriptions, but the same can, and is, said by and of, New Yorkers and Parisians, and many others, as in "You can take a New Yorker out of New York but you can't take New York out of a New Yorker.")

In whatever its form—bur, pur, pol, per, er, uru, or ur—the city's imposing vigor was expressed by a single word root. This urban idea must have been so impressive that—like radio or powered flight today—each nation vied to claim primacy of having invented and built the first city; hence the hometown of Gilgamesh is Erech, which might mean "er" (city) "ech" (first).[12] There must be a reason for this near universality of word and concept, which might arise from the diffusion of an original invention—just as there are relatively few words for modern inventions, like the telephone or the airplane. Inventions that originate at a single source tend to be known by one word.

Just how deeply seated the idea and attraction of the city was (and remains) must be confronted in any discussion of resistance to, or fundamental modification of, the city. From the oldest civilizations came the urban ideas that inform the present. The geographic reach of this primal idea of urbanism was vast, resistance to demoting its intrinsic appeal intractable, and the urban idea remains universal in, and fundamental to, industrialized cultures and the societies that aspire to become industrialized. Yet clearly the modern city is sick, perhaps because it is so old, perhaps rickety as a notion to confront the behemoth of industrialism. There are those who would profoundly alter its constitution. One of these is Paolo Soleri, who believes, essentially and with good reason, that the city is a neolithic invention that has run its course and been choked by outracing its logistical possibilities.

Born in Turin on June 21, 1919, Soleri received a doctorate in architecture from the Turin Polytechnic in 1946. He then made his way to Taliesin West, in Arizona, to work as an apprentice of Frank Lloyd Wright from

1947 to 1949. Soleri became disillusioned as he discovered, to his dismay, Wright's suburban predilections for stand-alone houses (or, eventually, attention-grabbing and self-aggrandizing monuments like the Guggenheim Museum). Soleri fled back to Italy, but returned in 1955. He settled in Scottsdale, Arizona, assuming the post of professor of art and architecture at Arizona State University. While he has built many distinguished buildings, Soleri is one of the best-known utopian city planners of the twentieth century. His reaction to the current urban situation is hostility, or aggravation with the city, and his solution—the annihilation of the city as we know it—contends with one of the deepest-seated of cultural inventions. Naturally, resistance to his ideas outstrips his actual proposal, which, if implemented, would only enhance the lives of the vast majority of city dwellers.

Soleri wishes to conserve the city's space rather than allow it to spread horizontally along the ground. The megastructures he envisions are designed both to perpetuate the natural surroundings by allowing them to remain self-sustaining, even under agriculture, or be planted as park-like recreational gardens, while within his giant buildings support networks would intensify the humanly social activities of living aesthetically while fostering creative work. These noble and apparently non-controversial ideals fly in the face of urban reality as its has developed from its ancient roots. For a very long time the basic idea of "the city" has appeared less like an artifact, subject to deliberated adaptation, and more like a value to be accepted or rejected but not fundamentally discussed. Indeed, any city's shaping and evolution (with the increasingly rare plastic influence of a Haussmann upon Paris) has become as estranged from its dwellers as their control of the other aspects of their lives. Yet this was not always the case, as the centrality of the urban idea recalls a constructed milieu intended to rival nature with a sculptural density and an outline against the sky, a hive of activity that must have been envisioned thousands of years before it was achieved. It offered hope of control, a technology to alleviate present threats (beasts of prey and marauders) while enhancing the human social potential (crafts, commerce, art). The great urban justification, the price of these benefits, may be becoming exorbitant, as Paolo Soleri recognized earlier than many.

Beginning in 1959 with his designs for Mesa City (a desert city housing 2 million people), Soleri investigated the possibilities, mainly with thought experiments, of condensing cities spatially—his beloved notion of "miniaturization." The resulting integrated, total environments, Soleri dubbed "arcology," a term he invented by joining *arch*itecture and ec*ology*. The term refers to Soleri's utopian blend of light industry supporting a complex that would include schools, parks and greenhouses—all within constructions that he delineated in drawings of great beauty and imagination. The vision is an old and pleasant one, of the city amid public gardens, untroubling of its surroundings though reliant upon them. Dating from antiquity, this conception was illustrated by Pol Limbourg in his "May" page for the book the *Trés Riches Heures de Duc de Berry*. Although this picture of Paris around the year 1416 presents the old ideal—the organic and harmonious vision of mankind amid nature—the densely clustered beautiful towers rising from park-like land could have been tendered by many a twentieth-century architect.[13] The benefits of urbanism's intensely rarefied culture shines amid a benign, and somehow reciprocal, relationship to the land which must supply the city. This ideal and idealistic, though fundamentally practical relationship, was considered by Soleri when he anticipated inserting a hugely dense population into the undisturbed, or otherwise depopulated, countryside. Soleri visualizes very densely inhabited megastructures dotting the landscape linked by mag-lev trains.

His drawings and models of such images (minus the then undreamt mag-levs as regional connectors) were presented as a traveling exhibition in major American cities in 1970 and brought his ideas widespread notice. Concurrently, Soleri's 1969 book *Arcology: The City in the Image of Man* provided an overview of his aspirations, which might have remained a dream.[14] In 1970 Soleri began to build a version of Mesa City, a prototype center which he called Arcosanti, a community that would eventually have a population of "only" 5,000. In Arizona, 65 miles north of Phoenix and halfway to Flagstaff, he began Arcosanti as an experiment for a city in the middle of nowhere, purposefully aloof from other centers, so as to free him and his collaborators from the pressures of urban land-developers. Within or adjacent to an extant city, his slow-moving experiment could never have survived its infancy. This project, not conceived in the ambitious scale of many of his drawings, was to serve as a laboratory, mediating between grandiose self-contained cities, which being theoretical inferred many unanswered questions, and everyday lived reality for the inhabitants of such gigantic structures.

His enormously inventive hand-building techniques could be applied to much larger structures using only moderately more advanced technology. Set upon a mesa that rises from a 4,000-foot-high desert, this impressive

FIGURE I

Pol Limbourg, "May" page from Trés Riches Heures de Duc de Berry, c. 1416, (Réunion des Musées Nationaux/Art Resource, New York)

experiment in community dynamics, construction technique, and urban socialization, Arcosanti is itself a commanding sculptural presence striving toward an organic inner life to match its shell. The urge toward a natural systemic, sustainable and vital, was noted by Naomi Bloom in the March 1991 issue of *Science Digest:* "Paolo Soleri is an architect who designs cities the way nature designs the Universe." Yet the form did not recall bygone pastoral ideals. More than one skeptical visitor recognized that "amid the tumbleweed and cactus, a futuristic space colony materializes in the desert."[15] The unquestionable aridity of the spectacle in its wilderness setting imparts a science-fiction flavor advertently or inadvertently associated with megastructures—they seem part of somebody's future. Like other captivating accoutrements of science fiction, we are not sure how to acquire these artifacts, which, nevertheless, seem appealing, but only for somebody else at the moment. (Before they were achieved, cities must have been talked about and wondrously described by storytellers sitting around the campfire, just as movies today show future megastructures. The narrative association is both a curse and a blessing; a blessing because even non-theoreticians understand what a megastructure is while some theorists sniff at a pop-culture idea, and the population is terrified, not by Soleri but, by the vulgar presentations of these unified cities which serve as the backdrop for horror stories.)

When completed, Arcosanti may house its 5,000 people in a dwelling 25 stories high, surrounded by light and air, set in 14 cultivated acres amid an 860-acre tract of greenbelt. The work, by unpaid students, proceeds slowly; today the project is not nearly done and is partially financed by the sale of the ceramic and copper wind bells Soleri produces and by the tuition of students and advanced professional architects who come to Arcosanti to apprentice with Soleri for short periods. They, along with many theorists, architects, and enthusiasts around the world share his salvific notion of how the city must change or perish. (See Soleri's expression of his vision at the end of this essay.)

An alternative way to frame the question of the cause of the death of cities assumes their fundamental viability and people's deep attachment to them as an idea and way of life, and, though counterposed to Soleri's ideas, this alternative arrives at a solution oddly hospitable to Soleri's. This is the proposal by the Austrian artist Friedensreich Hundertwasser.[16]

Born in Vienna on December 28, 1928, of a Jewish mother and an "Aryan" father, Hundertwasser barely survived the Hitler period in which most of his family was murdered. Even as a child who drew and painted, he was piloted in his development by the direction of Egon Schiele's painted architectural citations; Hundertwasser was attracted to portray buildings and was a generation ahead of his time as an ecologist. At last, after 30 years of painting fantasy ideal environments which attracted a huge audience and which goaded authorities with depicted taunts that a better coexistence with nature was possible, Hundertwasser was given the chance to build. While Hundertwasser had been preparing throughout his life to erect the marvelous structures he imagined in his paintings, finally on August 16, 1983, ground was broken for his apartment block on the corner of Löwengasse and Kegelgasse; the City of Vienna had challenged him to actually erect one of his buildings instead of merely criticizing the academic and municipal post-modernists.

Today Hundertwasser's apartment complex, the third-most-visited attraction in Vienna, is 600 percent oversubscribed. The people who dwell there log less sick leave than other workers, as Hundertwasser had predicted. They report that they are happier than other workers. Their children are more creative. The building cost essentially as much to raise as a traditional structure. Hundertwasser's building contains the same number of apartments as prescribed in the original public works program—he was given no dispensation for "art"—and was constructed with basically the same budget as the municipality allotted for this project before Hundertwasser assumed it; his success drove the professional architects crazy in anger. He has gone on to erect many other buildings. Yet, for all his ameliorating of the urban situation, Hundertwasser does not believe the city must be abandoned as a concept. Here we face the two extremes, the alpha and omega of good-willed questioning of urbanism's basic assumptions. Paolo Soleri, trained as an architect, produces and plans giant sculptural forms while Hundertwasser, an artist compelled to realize his forms in flat graphic space—as, initially, no one would entrust him to realize such creations—produces buildings. In fact, as an example of Hundertwasser's unartistic practicality, during 1988–89 he reworked the exterior of one of ecology's real villains, a building all modern planners love to loathe: an urban incinerator. (In Vienna, Hundertwasser retrofitted the modern Great Satan of urbanism on strangely Romantic, aesthetic, and oddly pragmatic terms: the city must still have incinerators; this was the most advanced and least polluting model; it had to be near the city to reduce the trash's travel, consequently it is visible from the city center; therefore Hundertwasser sought to make it "beautiful" and whimsical, as good a neighbor as possible.) These two figures, the Italian-American Paolo Soleri and the Austrian Friedensreich Hundertwasser, seem opposed, yet they respectfully and mutually coexist if we do not insist that true progress move in lockstep toward some dogmatic future.

Cities only came into being gradually, realized with planning and probing that was no more a straight-line progression than any other technology exhibits. Undoubtedly blind evolutionary alleys and failed attempts litter urbanism's pre-history, along with discarded notions and occasionally an idea so good it was adapted and repeated. Sometimes urban successes were forgotten (the great Indus valley cities), or were merely predecessors, not genuine cities (Çatal Hyuk in Asia minor) or enduring pre-urban forms (the Amerindian Southwestern pueblo)—all anticipated the truly achieved city. The earlier forms too were driven by a striving toward efficiency, a need to gather together trades and the specialized information represented by so many occupations, and to defend a way of life.[17] Urban culture resulted. This mass of human-borne knowledge quickened the evolutionary step of each successive cultural advance; indeed, as a colonial animal the city expressed the cultural externalization of biological functions that exist only in the aggregate. If technology is the amelioration of the environment by means of human intervention, then, that mediation between nature and the culture-magnified faculties (tools) finds its greatest and most complete expression in the city. The city does not intercede with nature on behalf of any one faculty or sense—as tool-making substitutes or extends one sense, faculty, limb, or organ at a time—but on behalf of perception itself. The caloric price for this achievement has always been steep: dependence upon the surrounding countryside, and lost privacy. Yet there are gains too. New Yorkers, who travel by the millions through tubes in the ground eschewing a personal car and who live in boxes surrounded on six sides by neighbors (not often of their choosing or preference) use something like 40 percent less energy than their rural fellow citizens[18]; the mythically liberal, suspiciously effete, gourmandizing, and fashion-conscious New Yorkers are subsidizing their conservative, rural, and independent-minded political adversaries, who enjoy nothing better than to pillory city dwellers as immoral. (The equalizer is a cash, not a caloric, economy.) Cities draw heavily on the countryside for resources, but use those resources more efficiently than the rural population that supplies them and therefore the countryside prospers at the city's expense.[19] Not the other way round.

a city a technology

Now, some like Paolo Soleri argue that we are decades past the need for this urban achievement and that humans and their aggregated artifact should part ways. It hardly matters to his opponents whether the city was intended as a designed environment, however sloppy or cumbersome, or if the city was achieved as the sum of vernacular conglomeration, an architectural medley feeling its way blindly to the sun with none of the natural elegance of a coral reef. Viewed as a manifestation of biologic function expressed by technology (not a metaphor), the city comes in two flavors: planned and unplanned.

The planned city, can be designed, like Baghdad which was mapped out from its inception (though overgrown today beyond the recognition of its commissioning organizers and patrons); or schematized (as was Boston along functional zones).[20] The unplanned city may grow upon a spot that offers some unique natural circumstance: a ford in a river (Oxford) or a deep-water harbor (hence the "pool" in Liverpool); access to some mineral or other exploitable natural resource (the salt at Salzburg, the Indian-dug wells that preceded Phoenix); a rare defensible position where otherwise is found only open space (ancient Paris on its two islands or Venice on its many).

Like the unplanned city, in its subsequent life the planned city too assumes organic patterns as it ripens. It grows neighborhood by neighborhood filling the most easily colonized zones available along and within ready lines of commercial intercourse. To describe this process I would like borrow a term from music: *durchcomponieri,* German for "through-composed," meaning that a song may be set so that the music for each stanza is different (unlike strophic folk songs). The resulting form of a "through-composed" song is not necessarily determined by the form of the poem upon which it is based, but the music may respond expressively to the changing tone, concepts, images, and situations in the verse. Likewise, as each urban neighborhood grows it may or may not conform to the pre-existing overall plan for the city, in building size or type, street width and character. The through-composed city is alive and responsive.[21] Yet the felt need to consolidate the city, to make it a single imposing thing rather than the sum of neighborhoods, is a predilection evident from the very earliest cities.[22] Perhaps more than a sky-challenging mentality, this consolidating impulse underlies stories like that of the Tower of Babel (whose illustration by Breughel so resembles one of Soleri's arcologies, with unintended irony contributed by the viewer). The hubris of mankind and the vision of an architect can be disentangled, even when the architect is planning large. Soleri's campaign to integrate functions and concentrate facilities springs from wholly salutary intentions. The perceived need to unify a city arises from a sense of civic organization (if not fairness), military efficiency, esthetics, and the economies of scale that technology can contribute.[23] Yet the arcologies are not intended to overwhelm the inhabitants, however great the impression may be from the outside that the dwellers would be reduced to drones in an enormous hive.

Interior scale and living space within an arcology would be quite pleasant and consistent with spaces available in urban and suburban dwellings today. In fact, an arcology is intended to combine the efficiency of high-rise living with the square footage available to suburban dwellers.[24] Likewise, Hundertwasser has introduced numerous amenities into his buildings while, within the components of a single living unit, intensifying the through-composed character of the city below the level of the neighborhood's personality or even of one building. If Soleri has tried to unify the city to save it from sprawl, Hundertwasser has tried to bring the village, and village life, into the city so that every apartment is visually articulated and can be distinguished from street level. In his "Hundertwasser Haus" the identity of each apartment is expressed externally on the building's face. The inhabitant of a Hundertwasser building, fighting the city's anonymity, can point up and say "I live in the blue apartment on the fourth floor." In contrast to this simple, but immediately palliative approach, Soleri is asking much deeper questions that will resonate far into the future. For example, how will denizens of an arcology live as neighbors?

Soleri plans population densities—215 people per acre projected in Arcosanti—much higher than typically encountered in cities (New York 33 and New Delhi 72 per acre) that rely on horizontal space-gobbling commutes that foul the air, waste time, degrade quality of life, and, in the United States, every 3 years, directly kill the population equivalent of San Francisco in cars. Yet, because of how Soleri's constructions stack, the effect should be pleasantly communal. The vision is breath-taking for some and terrifying for others, age and imagination being only two factors of predisposition. The chorus of support for Soleri's arcologies includes not merely the cheerfully idealistic jeans-wearing undergraduate, but hard-headed industrialists who recognize the benefits of economies of scale. Eschewing any idealism, industrial interests recognize that resources like natural gas, electricity, phone, water, newspaper delivery, groceries, etc.—anything needing transmission to a point of consumption—can be conveyed to one arcology with much less investment than to the 2,500 or so homes it replaces. The economies of scale only increase with the size of the arcology. Hundert-wasser, too, has considered density, but in a typically idiosyncratic way that distills into his proverb, "the vertical belongs to man, the horizontal to nature," by which he means that people can build as high as they want as long as the resulting upper surface is planted so that an airplane ride above it would not reveal a city, only the primal forest transposed many stories into the air, to regenerate and cleanse water, add shade, disperse the wind, contribute oxygen, provide a home for wildlife—all so the city is not cut off from nature. However improbable this may sound, wherever Hundert-wasser has implemented the theory it has worked. In contrast to injecting tangibly vibrant nature into the city, "Soleri believes that for maximum efficiency the city should adopt a three-dimensional form, as biological organisms have."[25] One is a metaphor, the other a praxis.[26]

No one knows how people will dwell in an arcology. People may tend to live segregated by income and social status (nearly the same thing), or by job description/type, or by some other propensity. Will an arcology have neighborhoods? (For that matter, once constructed and colonized will the arcologies exhibit homeostatic maintenance of a planned polity or will they exhibit unplanned growth? No one knows.) The present social experiments at Arcosanti are probing some guidelines. Dwellers in an arcology may be allocated different volumes of living space, but will this assignment be made on the basis of need or wealth? Clearly two systems of value will struggle to apportion space. Today no one knows if it will be more desirable to live on the ground floors of an arcology—with seemingly endless garden views out the window and no street noise or soot or smoke—or will it be deemed more desirable to live in the upper floors (as from the bridge of a ship or a skyscraper's penthouse), with an imperially symbolic view of the landscape? No one knows what money will buy in the way of preferential living within such a community. In contrast to these theoretical questions testing the essentials of what people want from their cities, Hundertwasser, building within the through-composed urban structure, tried to recreate a small town of stacked cottages. In addition to the usual conveniences of modern life (laundry rooms and storage space, etc.) his buildings are full of amenities: indoor winter gardens, cafés, children's playrooms, roof gardens with access from several floors, colorful and variegated interiors, and imme-diate entry to the city's grid plan by which to commute or play.

Within an arcology it cannot now be known if people will be predisposed to commute by foot or by elevator. His theorizing has not all been conjectural, but Soleri has begun to apply these ideas to an actual building where commuting could be solved or abolished, and energy used more efficiently; in mega-structures, solar energy will power moving walkways which, combined with elevators will make every part of the arcology accessible with a vertical commute and clusters of living, working and leisure activities in one structure, along with schooling, medical and some agriculture. Even Soleri does not know if hypothetical inhabitants of an arcology will wish to live near a school, or will some other vector influence social cohesion? While numerous such questions must be addressed by Soleri, Hundertwasser relies on existing civic textures (and sometimes rural settings) into which his buildings are injected.

Formerly, cities were unified by a limited variety of available materials (usually locally obtained) worked with a limited number of technologies.[27] To exceed these resources displayed conspicuous wealth.[28] In proto-history and the chacolithic, the available technologies pretty well saturated entire societies without regard to social rank. The pyrotechnology of ceramics was used throughout the reaches of society from highest to lowest; textile work would have varied mainly by delicacy of the fiber, how labor-intensively it was worked and how much woven work was displayed (to soften hard surfaces and as insulation), but the sources of the fiber were pretty uniformly distributed. (Certain expensive dyestuffs were symbolically segregated, like royal purple in the ancient Mediterranean or Imperial yellow in China).[29] Likewise, woodcarvings and basketwork would have been found in palace and hut; the degree of workmanship, the choice of metal, and number of metal objects were the surest signs of social class. But aside from the workmanship of the stone in a palace, the permeating technologies and resultant styles afforded a visual uniformity that must have been harmoniously agreeable—when it was not monotonous. The classical world's style permeated architecture, fabrics, metal and wood work, painting and sculpture. Uniformity of appearance, as a threat in the modern world, is accounted and parried by Paolo Soleri and Friedensreich Hundertwasser as a peril to be seriously regarded for the community's health.

The entire built environment of the classical world harmonized in a way that was enviable to the Medieval and later occidental mind. This visual consonance bespoke a polity in unison—regardless of whatever the factional and partisan realities at any past moment. In the classical world the

[margin note: vectors of social cohesion]

[margin note: classical uniformity/ harmony of design]

same motifs could be found on buildings, garments, eating utensils—food, clothing and shelter all resonated together. During the subsequent industrial age diverse efforts were made to re-integrate the visual environment. In several labor and craft movements the alienated worker's creativity was re-united to his product. Likewise, the citizen was re-united with his home that had become nothing but a compartment to house a worker; streetscape and city were coaxed to become, again, more than a giant, apparently uncontrollable machine. Much of this campaign was understood a century ago when various programs were proposed to remedy the new urban and industrial malaise. Attempts to unify the built environment in modern times have arisen, basically, in three waves and each of these has its resonance in the present situation.

At the end of the nineteenth century, Art Nouveau, basically a decorative program with different national expressions, employed the new industrial technologies in a holding action against drably oppressive Victorian clutter. Employing craftsmen to ornament every imaginable surface from buildings to bookbindings, Art Nouveau was practiced with great success in Scotland, Bohemia, France, Belgium, Austria, Italy, Spain, and the United States. This commercial movement had no manifesto or intellectual underpinning, but spontaneously informed a deeply felt reaction against the jettisoning of generations of accumulated knowledge of craft and the richness of the visual environment which that craft was prone to enliven. What Art Nouveau produced was not for the masses, was not inexpensive, and, though visible as typography everywhere, was not generally accepted— with the notable exception of Thonet's bentwood designs which are manufactured to this day as well as his other furnishings in the Austrian version of Art Nouveau, called *Jugendstil*. This easy-going disregard of the masses while creatively advancing industrial production (as exampled by Thonet or by Tiffany) was not true of Art Nouveau's successor.

Begun in 1919, the Bauhaus employed up-to-date technological and mass-production procedures and systems, in conjunction with the modern materials of the industrial age, and unified these materials and techniques under the rubric of "design." The detachment of the formal concept of a work from its execution may have been among the first examples of conceptual art, but the social program of the Bauhaus and its grimly earnest doctrine were as organized (and Germanic) as carefree Art Nouveau was a mercantile product for the middle and upper classes. Nevertheless, despite their brilliant compositional, motific, and stylistic triumphs, neither Art

Nouveau nor the Bauhaus won the public's affections though the Bauhaus was striving to rescue the public from the bleak industrial world into which the mass of people were being submerged—and had been, increasingly, for nearly two centuries, in cities without soul, greenery, light, or clean air or water.

Only one urban movement achieved popular success—the one without a plan or backing, with neither manifesto nor philosophy, without a spokesman or even a name until well into its maturation. Art Deco promised to deliver an urban situation veined with the atmosphere of the garden.[30] It sentimentally mingled the fruits of industrial progress with modernity as an aesthetic, but also doted upon nostalgia. Its politics were a mess and therefore never closely examined because its product, exhilarating cities and buildings, were completely intoxicating, and still represent the standard for the modern urban experience. It is the style of the twentieth century and in it, after 8,000 years, the city had discovered its fullest expression. Yet Art Deco's exhilarating mixture never specified where its nostalgia pointed, and, therefore, this same heady and imprecise blend could be found in the capitalist West, in Hitlerian Middle Europe, and in the pre-Stalinist East of early Communism.[31] No one could misuse the progressivist Bauhaus; that was instantly apparent to Stalin, who suppressed all vestiges of Russian modernism, as did Hitler in Germany. But Art Deco was another matter.

It would be easy to mistake Art Deco for another manufactured popular culture, but it arrived unbidden, unlike the merchandise of pop culture, which Clement Greenberg noted, "is a product of the industrial revolution which urbanized the masses of Western Europe and America."[32] Soleri strives toward a high-tech Italian hill town and Hundertwasser wants a colorful medieval streetscape—these are not kitsch solutions though they each rely on the amiableness of living in circumstances that neither fight the greater environment nor prove uncongenial to the intense cultural activities cities foster and (self-servingly) celebrate. These two approaches each invent and acclaim the pastoral as an escape from the city. (Farmers, despite their love of the land, failed to devise, then ignored the pastoral mode, which is deliverance from pressing cares of the city's pace. Now that e-mail can deliver those cares to our desks at light speed, it is to be seen whether the "e-pastoral" will be low tech or no tech.) Elemental assumptions of all three—Art Nouveau, Bauhaus, Art Deco—are found in Soleri's and Hundertwasser's work. They cannot be understood in purely technological

workman-ship

terms as their social-architectural solutions are framed within historic time's limited choice of possible responses.

From Art Nouveau both Hundertwasser and Soleri have inherited issues of workmanship. Soleri uses low-tech construction that can be divided into many sub-sections while Hundertwasser employs skilled craftsman to "over endow" a building as a post-industrial labor sink. In an "excessive" labor investment, Hundertwasser's craftsmen are given almost total freedom to work liberated from the architect's constraints. If wastage were reduced in the modern consumer economy the investment in labor-intensive, post-industrial, crafts and skills could award buildings a much longer life span. Against the onslaught of uniformity the alienated worker could be re-invested in his product by assimilating excess labor into construction, so that the city becomes a "creativity sink" and the citizen in his newly hab-itable city becomes, again, more than a scurrying denizen amid huge machines for living, or a renter flitting without consequence through un-noticing spaces.[33] From Art Deco's counterpoise of nature and building come two approaches: Soleri puts the city into a garden; Hundertwasser puts the garden into the city. In either case, somewhat surprisingly given its wholesome ambitions, the Bauhaus emerges the enemy. The problem is one of focus, of assembling an environment from well-designed artifacts or inducing design from the entirety of the city.

waste

Though a multiplex object, with as many parts as a living being, the city was always viewed as a single entity envisioned as both sculpturally fixed (a visual impression, for example, a skyline) and dynamic (as a living being), a thing and a process, a noun and a verb. This may seem a modern notion, something that grows from concepts of city planning and contemporary musings about how to assign function, but the idea is very old. The book of Jonah describes the metropolis Nineveh.[34] In this characterization, really a very brief inventory of things, a great composite object is indicated— inferring many houses, lanes, shops, bazaars, walls and gates, temples, palaces, gardens, districts perhaps set off by larger roads, military strong points, fountains, assembly places, etc. Yet the same text says that "Nineveh was an exceeding great city of three days' journey" (Jonah 3:3) and that, accordingly, "Jonah began to enter into the city a day's journey" (Jonah 3:4). Here time was exchanged for space. The city was being measured as so many days' pedestrian journey across. We still measure that way: some one asks how far a certain point is, and we reply "half an hour away." Space is equated with distance traveled at an assumed rate—the most efficacious and usual way to travel, whatever that may be. Energy is equated with time and compared to space. The relationship sounds so sophisticated and modern, but the Bible was doing it first, describing cities that way. An object became a process became a distance expressed as time.[35] At a deep level, Soleri and others try to resolve the tension that exists as a function of civilization between thing and action, weighing each with relative value. It is easy enough to recognize that the dynamic of the modern city arises from the commute. Time, energy, comfort, cost, and social considerations (fear, detestation, disregard) for surroundings all come into play as the commuter pierces and eventually infiltrates the city with the values of the commute. The tempo of city life, though intangible but nevertheless real, is affected, rarely for the good.

The existential effect of this struggle is felt by every citizen who wanders a city stopping to look first at one then another building; to the stroller, each structure evidences the thought of a single architect or patron. As long as that artist's work is visible as a distinct contribution the spectator's experience of architecture remains intact as a string of disparate constructions each with its own qualities. The city, however, has become so dense, and the automotive pace of viewing so rapid that the experience of architecture is becoming increasingly rare—hence the ever-more frequent attention-grabbing architectural follies of post-modernism. Only by inflecting the entire built environment, or by creating some outlandish monument can the architect practice a distinct art. So, three choices loom for the planner. Anonymity through teamwork, such as Hundertwasser advises, beckons to certain idealists of a practical and humanitarian nature.[36] To others the Romantic gesture of the free-standing building validates not only their own identities, but, by association, the viewer is caught up (as in the heroically irrational construction of Wright's Guggenheim Museum, or Gehry's Bilbao museum). Finally, Soleri offers the opportunity for traditional architecture to dissolve into the fabric of the city by becoming the city so that architecture and urbanism unite. This is not obvious.

Architecture is usually experienced one building at a time while urbanism or city planning is the experience of maps or long walks. Soleri drew up the plans of a series of gigantic urban centers that extend the city plan vertically. Thus, Soleri returns the city to its Neolithic origins as the coherent product of deliberation within a small community, even while, correctly, claiming to have abolished the city. At least as the city is currently assembled—as a maze of pretty spigots for utilities, buildings fashioned after playful arrangements of form—the architect's traditional role must be diminished. (Zoning has begun to limit this effect, except for the prima donnas, builders who are excused due to some perceived overriding good.) Inventing for the urban environment, Soleri would argue, must be taken as seriously as technology employed elsewhere. The history that informs urban development, ancient and modern, can be absorbed, and, perhaps, much of it discounted as new building materials and societal issues render old assumptions a field of obsolete thinking. Wherever real issues of mortality and efficiency are already perceived to be at stake personality and self-expression take a distant second. We see no follies or idiosyncrasies in good engineering; public discontent would not allow senseless and dangerous construction generally, and we have laws and oversight to enforce those laws regulating civil engineering. Naval architecture tolerates no frivolity; lives, weight, speed, and freight are at stake, none are risked lightly for the cause of the designer's self-aggrandizement. The space vehicles of whatever nation, and space stations planned, are not symbols—they must work. These fields offer a standard of performance that is translatable. If we wish to pre-serve the ancient benefits of urbanism to humankind, city design will have to be taken equally as seriously as naval architecture or aerospace environ-ments. The stakes are even higher, but the inertial greed more widely dis-persed. To overcome perceived, but incorrect, notions of self-interest, and to oust the ancient and deeply rooted pride of the city, something new will have to be tried. As was the case whenever necessity over-ruled fashion, the architect may have to give way to the artist like Hundertwasser or to a newer profession, such as Soleri represents.

STATEMENT BY PAOLO SOLERI

I

While positioning myself in various domains like society, politics, technol-ogy, sex, environ, and so on, I have adopted a device I find useful. I call it the Bubble Diagram. As is true of all expedient devices, its existence is justified by its usefulness, not by its intrinsic meaning. In fact, it does not per se have a meaning; it is a virtual landscape subdivided into four domains. The bor-der line of each domain is porous to let a trespassing from one to another.

In what follows, I will try the Bubble Diagram through the four perspectives suggested by it—Political Correctness, Historical Fitness,

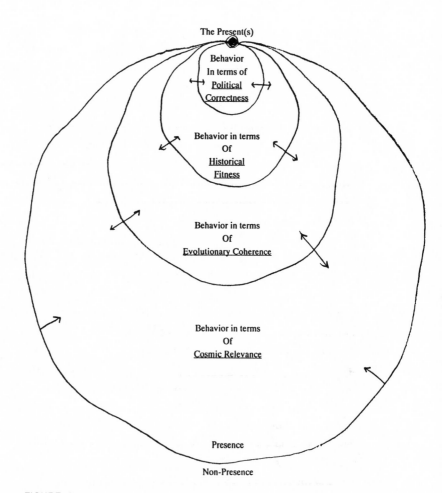

FIGURE 2

"BUBBLES: The Political Correctness bubble is as big as the USA today. The Historical Fitness bubble is as big as the summation of human history. The Evolutionary Coherence bubble is as big as the sum of all biospheres peppering the cosmos up to date. The Cosmic Relevance bubble is as big as the cosmos to date." (Paolo Soleri)

Evolutionary Coherence, and Cosmic Relevance—in relation to the habitat problems.

II

Before that, I would like to remind myself that, of all human interventions upon the planet, two are unavoidable. Farming and habitat: to eat and to be sheltered. As such, they have, or they should have, precedence over everything else. To fall on the wrong path in pursuing feeding and sheltering ourselves may mean impending catastrophe.

It has been my conviction that in the shelter domain our pursuit of improved wrongness is synonymous with impending catastrophe. I refer to the habitat I call sub-exurbia.

A second premise: For 3 million years or so, we have been captives of the animism we invented with great imagination. It is now time to abandon ship and go it alone on our own, away from the tutelage of our gods and their apostles.

The epoch of animism should be over. Its presence has been a protracted exorcism against the unknown. Now that such unknown has and is being "scientifically" investigated, we are discovering the astonishing richness of nature, the magic of nature. The magic of nature has no need of a surnatural composer-creator or of a conductor, let alone an absentee one. Reality is a harsh democracy of sorts and can no longer suffer under hypothetical despots, cruel and absent minded.

Via bootstrap virtue, space—the stuff of reality—has been making itself from a very elementary beginning (Big Bang) into a monster of a beast punctuated by excruciating episodes of life and now (3 or so million years on this planet) endowed with consciousness and reflection. The hypothesis sketched and inferred here may come in handy when the reflections on habitat play within the Evolutionary Coherence and Cosmic Relevance bubbles.

We have to keep an eye on the most fundamental fact: we are the "stuff of stars," and as such we will make it or we will not. That is why eventually it is in the Cosmic Relevance bubble we will find resolution, that is, genesis concluded or dissolution, that is, final nemesis.

From here to genesis concluded, it will take as many eons as is necessary. From here to possible nemesis, it could take just a minor astronomical event, a sizable asteroid, a protracted ice age, or something strictly of our own making: suicide, voluntarily or involuntarily by the whole human genome.

III HABITAT AND BUBBLES

THE POLITICAL CORRECTNESS BUBBLE

Fully invested as the Americans are by the storm of production-consumption, we are carpet bombing the land with the single-family house. The single-family house is the classical tip of the iceberg of wrongness. Wherever the single-family-house bomb is dropped, an avalanche of commitments and goods follow. A shrapnel-like chain of "events" investing the underground, the landscape, and the atmosphere. The commitments are the rules, regulations, mortgages, the covenants among neighbors, policing, law and order, etc. The goods, delivered directly or indirectly by the rubber wheel, pour in in ever larger clones of itself.

Political Correctness demands that the sub-exurban pattern be the tangible proof of the "American Dream" and that the glamour of life be in an escalating momentum: more is better. Consumption is good, hyperconsumption better. I call this the unarrestable engine of a better kind of wrongness.

This dream made into a planetary reality would do in China, India, Africa, South America, and all. Just as a pointer, the tip of the iceberg again, the number of cars, trucks, buses, and similar rubber-rolling stock. Ten planets would probably not deliver sufficient disposable biosphere to satisfy our hyperconsumptive felicity.

THE HISTORICAL FITNESS BUBBLE

Waste, pollution, environmental disruption, and segregation have been the shadow projected by our presence in search of food, shelter, and dominion. But for most of our presence on the planet, let's say 3 million years, the impact of such presence has been absorbed by the vastness of its biosphere.

No longer so and mainly because of two agencies: population and howness. Population growth has been the direct consequence of howness: how to make something into something else in gender and in size. The scientific and technological revolutions have been nothing short of magic, putting to shame the shaman and the witch doctors of animism.

What has emerged at the very beginning of the creational process of reality is howness. With the quasi-miraculous tools of depth vision, larynx

sound making and the versatility of the hand (from tool making to caress-
ing to piano playing), in their double task of executing the brain directions
and as maker of the brain itself, our species has opened the safe of the universe to find in it the dim but indispensably rich ancestry of our presence.

In terms of Historical Fitness, we should carefully calibrate our endeavors in ways that seem to work for and not against the progression of more
and more mind being extruded from matter, a theotechnology of sorts
(technology is present at the very beginning of things, of the universe itself,
that is).

The price of civilization with the extraordinary fruits it has generated,
science and technology as we regard them now among others, has been
high, too high, especially if we now think of plunging into the rematerial-
ization of life promised by hyperconsumption. The historical fit we may be
seeking seems to be related more to the not-so-noble drives history has
showered upon civilization and culture: personal opportunism, greed,
hypocrisy, envy, ill will, hatred, the enemy at every crossroads, racism, and
xenophobia.

Civilization was the invention of cities in an almost runaway mouth-to-
tail cartwheeling of the mind and the toil of generations succeeding generations. The urban effect is not really the invention of man from zero
point. All organisms are urban effects, the unbelievable coming together of
the miniaturized and the complex. Man as hyperorganism is the coming
together of trillions of events, all spatial, by the way. Without an implosive
miracle of such extravagant nature, no organism would exist in the biosphere (the microchip is now attempting a similar feat on silicon matrix).

Getting away from the city is not only counter culture and civilization,
it is counter life. One has only to imagine what the global hermitage generated by unleashed cyberspace would deliver. Three billion enclaves
"served" by an impossible logistical grid, in perpetual gridlock. Sub-urban
and ex-urban sprawl are promising the Garden of Eden and are in the
process of delivering despair: physical absurdity and mental anguish. We
must halt the folly.

THE EVOLUTIONARY COHERENCE BUBBLE

It was with the first sparks of consciousness millions of years ago when tiny
flakes of the noosphere began to manifest an utterly new presence on the
planet. Pierre Teilhard de Chardin was the first man to use the term "noosphere" and the electronic envelope now being knitted around the globe is

a powerful presence. Cyberspace produces not only a great wealth of information, but also great financial wealth. How this new source of wealth is put to use will greatly influence the destiny of the species and of the planet.

What is going on now is paradoxical.

The computer, an ephemeralizing instrument by definition; i.e., it generates intellection via miniaturization, is, on the side so to speak, generating a new powerful wave of materialization. A large chunk of it is poured into the form of habitat that shamelessly blankets our "beloved" nature with what I call the global hermitage skin. This ex-urban metastasis taxes life, human and not, greatly. It is a devolution into the sink of a materialism that might turn out fatal for body and mind.

Eventually, the sooner the better, we must turn to the city, one of the most crucial inventions of *Homo sapiens,* and seek in it the guidance toward a transcendence of the hedonistic dream constructed on the delusion of the Garden of Eden. A Garden of Eden stuffed with all the magic gadgetries, the super-shopping mall, virtual or not, thrown relentlessly at us from infancy to death.

A congested civilization is, by physical laws, a civilization of gigantism as our so-called cities testify. The one million or so Phoenicians' metastasis is occupying 400 square miles of desert, and "growing." Such gigantic pseudo-structures are logistical nightmares, wasteful, pollutant, and segregational. There is no evolutionary coherence to be found there. There is delusional "well being" draining away the core of the best gift we are in the process of creating for ourselves and for the reality from where we come: the animation of matter.

The Evolutionary Coherence bubble shows wrinkles, its pressurization is defeated by the hole we might punch in it, in our hyperproduction-hyperconsumption drive. "Slow down and ponder" could be a good slogan for the first steps into the new millennium.

THE COSMIC RELEVANCE BUBBLE

With the gusto we exhibit in improving the wrong thing, that is, encouraging and improving wrongness, we are digging for ourselves personal and luxurious graves and the cosmos could not care less. In fact, closer to us, the biosphere itself could not care less. Mindlessness has no cares as it has no aims unless we define as aim the population opportunism of the species. Absurd as it might sound, only reality filtered by the brain-mind might eventually give aim and meaning to a reality foreign to even the notion of them.

If ever brain-mind reaches that power point, the cosmos will be infected by the bug of transcendence. From there on is the Mystero Tremendo. Only the urban effect can take us from here to there, the singularity of intellection becoming a cosmic event. The cosmos could be transfigured by such intrusion.

We know that the planet as a source of life is doomed. No gods nor spirits will do anything about such nemesis. They are after all only simulations constructed by our anguished minds. What can carry on the tasks of consciousness is its potential power of pervading the cosmos and even prevail on it, as senseless as it is, to reveal itself to itself in an astonishing act of self-revelation. We, all of reality, would be there in an esthetic and equitable fullness. "Religion" (*religo*) fully concluded.

L'homme c'est une machine à faire des dieux, as Henri Bergson once wrote. Star dust, us, making stars into grace.

NOTES TO "PROSPECTS AND RETROSPECT"

1. Something of the wonder of early cosmopolitanism can be savored in the following Biblical passages: "And all countries came into Egypt to Joseph for to buy corn; because that the famine was so sore in all lands." (Genesis 41: 57) "And Joseph's ten brethren went down to buy corn in Egypt." (Genesis 42: 3) "And the sons of Israel came to buy corn among those that came: for the famine was in the land of Canaan. And Joseph was the governor over the land, and he it was that sold to all the people of the land." (Genesis 42: 5, 6) A crowd as varied and colorful as a United Nations cocktail party is conjured by the phrase "all countries came into Egypt," and all that varied humanity ("among those that came") was gathered into one central administrative zone ("he it was that sold to all the people of the land"). Far more recently Charles Baudelaire (in his essay "The Painter of Modern Life") became, perhaps, the first modern to notice the excitement of moving among the crowds that were only then appearing as a daily feature of cities, providing a kind of theater with sets on two sides and one-dimensional movement down the sidewalk with self-presentation as the principal form of expression.

2. See Hans Jensen, *Sign Symbol and Script* (Allen and Unwin, 1970). When the development of an alphabetic script, with consonants and vowels, was completed, "Further evolution is conceivable still in only two directions: 1. in the direction of greater accuracy . . . and 2. in the direction of greater simplification," 53. Plainly, all notations are not alphabetic, and scientific, musical, mathematical, etc. and other notational systems abound. Likewise, all true cities are the same and differ.

3. The hieroglyphic symbol for city (which Wallis Budge translates as "house," although recognizing it as a component of and reference to Heliopolis, "the

metropolis of the XIIIth Nome of Lower Egypt"—see E. A. Wallis Budge, *The Book of the Dead* 1920/1960 n. 1 24), is a wonderfully compressed pictogram, a circle circumscribed around a cross. This hieroglyph illustrates the two principal functions of a city. The cross represents the crossroads—the junction where travelers meet. At such intersections can begin true commerce, not merely exchange but planned and centralized commerce from which, by extension, can grow all organized industry (not local crafts traded to third parties) and business. The circle stands for the surrounding wall, not only protection in a physical sense (a pen for cattle or a wall against marauders), but a border between the city and non-city. This circular barrier, a palisade, demarcates the city-slicker from the hick or hillbilly, one who is "beyond the pale." Throughout his writings Charles Olson made the point of the city as embodying these characteristics and having been first established at spots to which early nomadic peoples felt constrained to return periodically—where they buried their dead. Hence, he stipulates, a third quality: necropolis preceded metropolis.

4. For example, we learn that when the putative "Aryan" invaders of ancient India encountered the indigenous population's citadels, "for these cities the term used in the Rigveda is *pur* meaning 'rampart', 'fort', or 'stronghold'" (Sir Mortimer Wheeler, "Harappan Chronology and the Rig Veda," *Ancient India* no. 3 (1947), 78; *Ancient Cities of the Indus,* ed. G. Possehl, Vikas Publishing, 1979, 291). See note 3 above on how close the idea of 'palisade' is related to 'fort', indicating the central function of the bastion in earliest cities.

5. Ingrid Rowland, "Etruscan Secrets," *New York Review,* July 5, 2001, 16.

6. My mother-in-law, for example, who has lived her whole life in western Pennsylvania, pronounces the word "borough" as "burr," which is understood by her similarly accented neighbors without correction to be a complete word. Linguistics legitimately offers the opportunity to use and rely on very small samples of native speakers.

7. See, among many such sources treating the Indo-European word root "pele" (sometimes thought to mean "fortified high place"), Joseph T. Shipley, *The Origins of English Words* (Johns Hopkins University Press, 1984). A broader view of this word root is taken by Mario Pei in *The Families of Words* (Harper, 1962). Yet even in Pei's more embracing construction all of the urban roots remain.

8. Ezra Pound delights in the transcription *"Brododaktylos."* (For a discussion of this apparent aberration see Hugh Kenner, *The Pound Era,* University of California Press, 1971, 56.) Kenner notes, citing the original manuscript source, that *Brododaktylos* as Sappho used it in her dialect did not refer to the Homeric "rosy-fingered dawn.") That dialect B, incidentally, shows up as an important link as it unites—by means of the b/v substitution as found in "brodo"—the otherwise unattested

English word "Rose" (a common enough word, but, nevertheless, a word with murky sources) with its distant cousin in Hebrew "Vered," meaning rose. If the V/B is dropped, as it usually was in *Brododaktylos,* the remaining root is a cognate, which, again, indicates that proto-Indo-European was affiliated with other language groups, many of which today share a common word for the urban core.

9. Even "suburb" has a long history that wends it way through proto Indo-European and the Semitic languages so that the French *purlieu* shows the core word for city, but, at least according to one scholar, so does the Hebrew words for suburb and wilderness. See Isaac E. Mozeson, *The Word* (Shapolsky, 1989), N.B. "Barrio," 28.

10. Nadav Na'aman, "It Is There: Ancient Texts Prove It," *Biblical Archaeology Review,* July-August 1998, 43. Na'aman, continues that, "all of the names of places with these determinatives mentioned by the Jerusalem scribe should be translated as the names of towns."

11. D. Dennis Hudson, "Madurai: The City as Goddess," in *Urban Form and Meaning in South Asia: The Shaping of Cities from Prehistoric to Precolonial Times,* Howard Spodek and Doris Meth Srinivasan, eds. (University Press of New England for National Gallery of Art, 1993). Hudson (p. 125) makes a distinction between how Tamil townspeople "perceive their village as two different ways: as a map (*granman*) and as a person (*ur*). As a *granman* the village is a legally defined space, but as an *ur* it is a personal space."

12. Interestingly, the *Epic of Gilgamesh* begins with the invocation of a city, indicating that all of the ensuing journey-narrative contrasts to the city that is left behind. The "Prologue" begins "O Gilgamesh . . . was the king who knew the countries of the world," meaning that he was widely traveled. A few lines later the "Prologue" ends: "Climb upon the wall of Uruk; walk along it, I say; regard the foundation terrace and examine the masonry; is it not burnt brick and good? The seven sages laid the foundation." (translation by N. K. Sandars, Penguin 1960, 59) These claims are, and remained ever after, the two most important assertions in city planning. The lofty beauty of the city was constructed of the best indigenous materials, fired brick (not sun-dried), and the seven circumpolar stars were used to perfectly align the city along a propitious north-south axis. Good materials were wedded to good design in this advertisement.

13. This ideal showcase for architecture is hardly Le Corbusier's monstrous *Voisin* plan for Paris of the mid 1920s, a perfect piece of Stalinist oppression. Although I cannot digress for such a discussion, it must be noted that modernism neither implies nor is best represented by the worst excesses committed in the name of the "international style"; therefore, the defense of modernism is hardly the point of the present article, although thinkers and builders as various as Soleri and Hundert-

wasser each rail against modernism (of whose progressivism they liberally partake), while insignificant, though well-paid, cunning, and famous, minions of "post-modernism" will gather allies anywhere outside the modernist camp.

14. Soleri's writings include *Arcosanti: An Urban Laboratory?* (third edition: Cosanti, 1993); *Technology and Cosmogenesis,* co-authored by Scott M. Davis (Paragon, 1986); *Paolo Soleri's Earth Casting: For Sculpture, Models and Construction* (Peregrine-Smith, 1984); *Space for Peace* (Cosanti Foundation, 1984); *Fragments—A Selection from the Sketchbooks of Paolo Soleri* (Harper and Row, 1981); *The Omega Seed: An Eschatological Hypothesis* (Doubleday Anchor, 1981); *The Bridge between Matter and Spirit Is Matter Becoming Spirit* (Doubleday Anchor, 1973); *The Sketchbooks of Paolo Soleri* (MIT Press, 1971); *Arcology: The City in the Image of Man* (MIT Press, 1969); and *Archetipi Cosanti—Paolo Soleri Architetto* (booklet, 1956).

15. Ralph Blumenthal, "Futuristic Visions in the Desert," *New York Times,* February 1, 1987.

16. For a complete exposition of the following argument see: Harry Rand, *Friedensreich Hundertwasser* (Taschen Verlag, 1991; abridged edition 1993). A short bibliography of materials dealing with this architectural position: *Das Hundertwasser Haus Österreichischer* (Bundesverlag/Compress Verlag, 1985); Michael T. Turkiewicz, *Ein Haus* (M.T.T. Eigenverlag, 1987); Hametner/Melzer/Spiola, *Das Hundertwasser-Haus* (Orac, 1988); *Friedensreich Hundertwasser: Gemeinsam einen Schritt zu Gott Sankt Barbara: Die Kirche in Bärnbach/Steiermark* (Róm.-katholisches Pfarramt, 1989); Pierre Restany, *The Power of Art: Hundertwasser—The Painter-King with the 5 Skins* (Benedikt Taschen Verlag, 1998).

17. On the role of trade, see Jane Jacobs, *Cities and the Wealth of Nations* (Random House, 1984); N.B. "Faulty Feedback to Cities."

18. For a good survey of the history of underground urban railways, see Benson Bobrick, *Labyrinth of Iron* (Holt, 1994). There has yet to be written a sufficient essay on the effect of building these mazy complexes that network domestic neighborhoods with business districts, connect shopping zones with consumers, and, more recently, lace together nodes of regional and air transportation with local travel.

19. For example, street lighting is far more efficiently utilized in a dense urban situation than along a country lane or a town center. For a fine examination of the efforts to achieve this lighting see Wolfgang Schivelbusch, *Disenchanted Night* (University of California Press, 1983 and 1988).

20. For a discussion of Boston as a dynamically planned city, see John Brinckerhoff Jackson, *American Space* (Norton, 1972), especially chapter 4.4.

21. When New York's subway workers went on strike and the entire system shut down, it was fascinating to see the enormous metropolis reduced to forty or fifty neighborhoods, each with its own small shops and a vitality circumscribed by easy walking distance. That is, New York City as currently experienced is not a political entity at all; I would maintain it is now a function of its mass-transit system. No metaphor to bodily circulation is required to defend this point once it was demonstrated by the transit workers' job action.

22. The aesthetic implicit here (or politico-managerial outlook) was best spelled out by Gilbert Sorrentino (*The Perfect Fiction*, 1968, 29), whose poem of a single triadic stanza I quote in its entirety: "Something plus something is not one thing. / An insufferable vicious truth. / One lives with it or one dies. Happy."

23. There is, to be honest, a paternalistic manner in approaching this issue which has been characterized by Alexander Tzonis as the "paradox that most theories of architecture whose task was rationalization have in common their preoccupation with visual order, the look of the product. There is a greater concern that the building should *look* rational rather than that rational methods should be employed in its design." (*Toward a Non-Oppressive Environment,* i press, 1972, 87) While Soleri clearly wishes to anticipate and organize every aspect of society within his giant colonies, the visual effect of his planned buildings is not so much rational as heroic sculptural while Hundertwasser is merely anti-rational in appearance while he employs ultra-rational methods.

24. It would prove guileless to ignore that "many people like suburbia" (Robert Venturi, Denise Scott Brown, and Steven Izenour, *Learning from Las Vegas*, MIT Press, 1977, 154). On the other hand, many people like to eat fattening foods, smoke, remain sedentary, wallow in thoughtlessness, drive SUVs in cities, abuse their spouse, vote foolishly and scorn whatever is alien or novel. What people 'like' hardly matters when considering the possibility of societal self-extinction. The mythical sybarites were as happy as the modern ones. On the other hand, a censorious tone is not mandatory; many Americans live in suburbia because, the central cities having been eviscerated, the vital urban experience must be gathered on vacations when suburbanites gladly fly thousands of miles to walk streets where people work *and* live *and* play *and* shop.

25. Naomi M. Bloom, *Science Digest,* March 1991.

26. Hundertwasser has also developed his theory (first exhibited as part of "Triennale di Milano," 1973) when he planted twelve "Tree Tenants" through windows at Via Manzoni. These trees each occupied an apartment (rent-subsidized by a grant). Subsequent tree tenants have been placed in Vienna, at Alserbachstrasse, 1980, and, to marvelous effect on the roof of Kunsthaus Wien where they form a forest, colo-

nized by local wildlife. (Seven stories above the street I witnessed a mother duck strutting ahead of a string of ducklings who had been born in the wooded thicket atop the KunsthausWien instead of the nearby Stadtpark.)

27. These limiting variables, praiseworthy for what they can contribute to social unity, are the foundation of the approach advocated by Prince Charles in his successful revitalization efforts through the Prince's Trust.

28. Exotic goods were used sparingly for palace or temple, but never for common homes: "And Hiram king of Tyre sent his servants to Solomon; for he had heard that they had anointed him king in the room of his father; for Hiram was ever a lover of David." (1 Kings 5: 1) "And Hiram sent to Solomon, saying, I have considered the things which you sent to me for: and I will do all you desire concerning timber of cedar, and concerning timber of fir. My servants shall bring them down from Lebanon unto the sea: and I will convey them by sea in floats unto the place that thou shalt appoint me." (1 Kings 5: 8, 9) Having already written about this once before, I see no reason to try to reformulate my position in less efficient expressions: The materials of which things are made can and do, by themselves, carry associations. Extraordinarily fine components, rich and exotic materials, indicate the princely mode (to acknowledge Oleg Grabar), a form of "conspicuous consumption" (to acknowledge Thorstein Veblen, *The Theory of the Leisure Class,* 1899). Things have cost, and the price of things is relatively well-known to people in a given society, and the object carries that implicit cost in labor and materials, whether barter is appropriate or not. Things evidently or subtly exhibit the technique by which they were made, and high artisanship implies a tremendous caloric investment by the entire culture as the artisan is being trained, and thereby his or her labor and creativity are withheld from other, capital-forming, work. Craft represents a mixture of practicality and culture-specific aptness, a symbolic precipitate of which every object is the residue. This symbolic component has a history, as much as the evolution of functions, so that style can be divisible from function, but only to an outside observer. Together, all these considerations convey a powerful social message that completely transcends the ostensible use to which a thing is to be put by the user. To the beholder things mean something else entirely than to their putative or actual owners. The deliciousness of the situation arises from all of us being both users and observers. (Harry Rand, "The Uses of Things," *Things* 5, winter 1996)

29. See Isaac Herzog, *The Royal Purple and the Biblical Blue* (Keter, 1987).

30. For a discussion of the origin of this style's name, see Bevis Hillier, *Art Deco* (Studio Vista, 1968), 10. For a discussion of how one of the most successful artists of this period interpreted the unstated assumptions by which Art Deco suffused a society that had already solicited its birth, see Harry Rand, *Paul Manship* (Smithsonian Institution Press, 1989).

31. The cocktail of hi-tech and Romantic mythology perdures in the high/low-tech visions of *Star Wars,* where feudalism and chivalry abide next to tomorrow's hardware.

32. Clement Greenberg, "Avant-Garde and Kitsch," 1939, in *Art and Culture* (Beacon, 1961), 9. Later in the same essay Greenberg notes, importantly for the current discussion, that "kitsch had not been confined to the cities in which it was born, but has flowed out over the countryside, wiping out folk culture." Kitsch is the self-consciousness of marketing (akin to the hyper-consciousness of modernism); it is the "high art" of commerce, capable of obliterating the organic connection between high and folk art, and thereby undermining the vital moral correlation between apparently amoral science and liberality.

33. This intemperate abuse of the environment was noticed from the beginning of the Industrial Revolution. William Blake, in his 1794 poem "London," wrote "the chartered Thames does flow," by which he dourly observed lordly nature being turned into a resource to be exploited.

34. "Nineveh, that great city, wherein are more than sixscore thousand persons that cannot discern between their right hand and their left hand; and also much cattle." (Jonah 4: 11)

35. The implicit quality of such statements engaged the not-too-excessive time required. We might say that a commuter should, ideally, spend no more than 45 minutes traveling to work, although in such a normative we do not assume the mode of transportation or even the intermodal changes that may involve pedestrian travel, bicycle, bus, elevator, escalator, street car, or heavy rail—in any order.

36. Admittedly, this construction of the argument skirts the question of vernacular or anonymous architecture wherein such questions may be moot for, as Bernard Rudofsky has pointed out, "vernacular architecture does not go through fashion cycles. It is nearly immutable, indeed, unimprovable, since it serves its purpose to perfection" (*Architecture without Architects,* Museum of Modern Art, 1965, frontispiece caption).

PORTRAIT OF INNOVATION: ERICK VALLE

MARTHA DAVIDSON

As a proponent and a practitioner of the New Urbanism—a town planning movement that seeks to enhance community life through a revival of traditional principles and forms—the architect and urban designer Erick Valle looks to the past, drawing on memories of the plazas and public spaces of his native Costa Rica. As an innovator on the cutting edge of computer applications for planning and design, he is a visionary with an eye on the future.

Born in Costa Rica in 1961, Valle moved to Chicago with his parents when he was 5 years old. Valle explains that his parents came to the United States "to give the American Dream a shot" and that "both worked hard to provide me with opportunities they did not have back at home," having come from "modest family backgrounds." They chose Chicago because that was where job opportunities and wages were greatest at that time; Valle's aunts later joined them there. Valle's parents did not have other children, and they devoted all their resources to giving their son a good start in life. "My mother simply worked alongside my father to provide me with an education and moral values," Valle says.

Although he had no siblings, Valle grew up surrounded by cousins, both in Chicago and in Costa Rica (where he spent many summers). "I remember the different little towns they lived in, going to the squares, looking at the churches, visiting their schools. . . . And in Chicago . . . I walked to school every day. I had a corner store where I got my candy in the morning. That's the kind of experience I remember. . . . I have a very clear vision in my mind of what these places are about. And my work has gone back to that. It's a big circle."

Even as a child, Valle was drawn to design. He distinctly remembers his first set of blocks, which he used to build all kinds of objects. Later, as an adult, Valle learned that one of his great uncles was an important architect in Costa Rica who had designed the main railroad station in San José, as well as other civic buildings.

When Valle was 16, his family moved to Florida. His father, a second-generation printer, started a business in Miami. Valle began assisting in the print shop as a summer job. By 1979, when he was 18, he was the company's graphic designer. But the move to Florida was not an easy one for Valle. "Miami was a shock after Chicago because it was less urban. There is a museum of science, but it's nothing like the Museum of Science and Industry in Chicago. There are little art galleries around, but you don't have art centers like you do in Chicago. . . . Certain experiences weren't there any more." Valle did learn to appreciate Miami's open spaces, but he remains at heart an advocate of more traditional urban environments, with the cultural and social opportunities they offer.

Another experience when he was 18 affected Valle profoundly: "I had a close tragedy that changed my outlook on life. I was diagnosed with cancer in my arm and it was going to be amputated. After lots of praying and two long operations, my arm was saved. More importantly, since I lived in the hospital for several months and saw all my bed fellows die, I do not take life for granted. I am clearly self-motivated because I know how valuable life is and how short a time we have in this world. I believe an individual can make a difference."

After his recovery, Valle attended Miami-Dade Community College before enrolling in the University of Florida, where he earned a Bachelor of Design degree in 1984. While an undergraduate, Valle worked as an intern in the offices of several Miami architects, drafting and making presentation drawings for building projects. The hours spent bending over drafting tables were difficult and led to another discovery: "I started realizing I had a real problem with my back. I couldn't be leaning over all the time on those drafting tables. . . . It was around 1982, that time when computers were just beginning to be used, and I saw [computer technology] as an alternative for me to be able to continue in what I was doing. Like everything else, I took it on like a passion. . . . As technology evolved, it became my palette for expression and exploration."

He continued his training at the University of Miami's School of Architecture (UMSA), receiving a Bachelor of Architecture degree in 1986. At UMSA he pursued a course of independent study, investigating aspects of design that particularly interested him. He quickly developed expertise in computer applications. "Nowadays, digitizing images and altering them is commonplace," Valle says, but it was not so at that time. Although computers had been used by architects and designers since the early 1970s, Valle's approach to the technology was highly innovative. He developed a project "that looked for the first time at using video and CAD [Computer Aided Drawing] to simulate before-and-after conditions. I captured video of an existing condition and then using CAD overlaid a wire-frame image of an architectural proposal." The simulated environment was a tool that could help developers and government officials visualize a planner's concept, facilitating the decision-making process.

CAD for before & after

Valle's studies at UMSA included a course taught by Andrés Duany, a faculty member who, with his partner Elizabeth Plater-Zyberk, was among the foremost pioneers of the New Urbanism. The basic principles of the movement resonated with Valle's own sensibility about urban communities. According to Valle, the essential ideas of New Urbanism are the following:

new Urbanism

• the establishment of public space—plazas, squares, and green spaces—where people can come together
• a return to a pedestrian scale, where goods and services are within walking distance for most residents and where broad sidewalks encourage walking and mingling, in contrast to the dominance of the automobile that is part of contemporary suburban sprawl
• an emphasis on diversity through mixed usage and mixed housing types, so that residential, commercial, and recreational needs can all be met within a neighborhood, with people of different economic levels living in proximity as members of the same community
• creation of a strong sense of identity for the neighborhood, to distinguish it clearly from surrounding neighborhoods and communities.

Urban Codes

In 1987 Valle worked with Duany and Plater-Zyberk to computerize the urban codes of Seaside, a town they had designed in the 1970s as one of the first manifestations of the New Urbanism. "Urban code" is a term used by the New Urbanists in preference to the standard term "zoning law." Valle explains that, while zoning laws are associated with restrictive measures, isolating neighborhoods to one income level, one building type, or one function, urban codes "are about creating places that have a mix of incomes, building types, and uses." The distinction in connotation is important, and the New Urbanists hope that a general evolution in thinking about communities will accompany the evolution in language as the term "urban code" becomes more widely used.

Valle helped digitize the urban codes of Seaside by creating a database and a design program that included all the components of the town's codes (such as building heights, building types, and street types). With the information digitized and integrated, the designers could manipulate it more easily in refining their plan; the computerized information provided them with checks and balances as they worked.

Valle describes Duany as the "Pied Piper" of the New Urbanism movement, one who has done much to organize it and publicize it. He regards Duany as a brilliant individual and a role model. "My role models have simply been achievers, people who started literally with nothing and made a difference in their respective worlds or areas of expertise—people who keep trying and eventually win the odds in life." He has found similar inspiration in the writings of Ayn Rand and in the work of the Austrian town planner Camilo Sitte (1843–1903) and contemporary colleagues such as Leon Krier, a planner from Luxembourg, and William Mitchell, dean of the School of Architecture at MIT and an expert in the field of computer technology.

While Duany, Plater-Zyberk, and a small number of other architects and planners such as Peter Calthorpe and Peter Katz (whose book *The New Urbanism* offers a clear overview of the history, philosophy, and principles of the movement) are regarded as the first generation of New Urbanists, Valle is identified as a leading name in the second generation. The first generation did ground-breaking work in establishing the theory and principles of New Urbanism and in dealing with government policy and legal issues. Valle and others of the second generation have developed the methodology and the technology that make this type of planning distinctly different from other approaches to urban design. "Our urban codes, method of creating urban plans, and our architecture is intuitively more in line with the principles that govern this movement. We are dealing with more of the technical issues than the first generation. They have dealt with the regional issues, we are dealing with the issues that reconstitute the neighborhood."

In 1987, while still a graduate student, Valle founded and was named director of UMSA's Image Transformation Laboratory. In this capacity, he was able to continue his research on the merging of CAD with video to represent design solutions in a built environment. His work caught the attention of other professionals and of private industry. "One thing led to another," Valle says. "How can I put it? Computers and I . . . I am able to talk to them. I understand them. . . . I took this technology, merged it so that it got a lot of excitement . . . and there was a lot of pressure by private industry to finance me and to come out and do something with it. Basically, I got together with two colleagues, and the next thing we knew, we had $100,000 in our pockets and were off and running. A year later, we had seventy people and five offices . . . doing work all over the world. . . . All this computer simulation just really took off."

Valle's original partners were two other UMSA graduates, Victor Dover and Joe Kohl. Their company, Image Network, was based in Coral Cables, Florida. Its mission was to bring the benefits and applications of image processing to the constructed environment for the service of design professionals. By creating electronic simulations of development, Image Network helped architects, builders, real estate developers, and city agencies see how proposed changes in the built environment would look. The products they created were important tools for presentation and decision making, and Valle's chief responsibilities included publicizing their work and dealing with clients, as well as research and implementation of new technologies.

Image Network was so successful that Valle soon found himself "sitting in board meetings all the time, talking to investors. . . . It was too big an animal for us. Not something they necessarily prepare you for in school." Within 3 years, Valle and Dover were questioning their commitment to the business. "We said we weren't sure we were enjoying ourselves," Victor recalls. Just at that time, Jaime Correa, an architect/planner on the UMSA faculty, joined as a partner in the firm. Valle was enlivened by the connection with Correa, who also had worked with Andrés Duany. Correa became something of a mentor, encouraging Valle and Dover to look at historical models for communities and inspiring them to work along the lines of the New Urbanism.

During that period, Valle also met a talented designer who was an undergraduate at UMSA. Her name was Estela Lupe Garcia. Originally from Cuba, she had come to Miami with her family in the Mariel boatlift of 1980. After earning her Bachelor of Architecture degree, she worked for Duany and Plater-Zyberk at their firm DPZ. Valle found in her not only an architect who shared his interests and vision but also a partner in life. Married in 1989, they celebrated their wedding in one of Miami's beautiful historic buildings, El Jardin. They now have three children: Elizabeth, born in 1994, Eduardo, born in 1995, and Ellen, born in 1997.

Valle was awarded a Master of Architecture degree from the University of Miami in 1991, receiving a certificate for academic excellence and a special citation for being the outstanding student in his field. Completion of graduate studies prompted some career changes. Valle left Image Network and started his own industrial design firm, Valle Studio & Associates. Between 1991 and 1994, he collaborated with specialists in various disciplines in designing a variety of products and received patents on designs for a wine cooler and a toothbrush. Other product ideas, which he continues to develop from time to time, remain archived until there is time to patent them and put them into production.

Valle also joined the faculty of UMSA, first as an instructor in 1991 and then, in 1992, as an assistant professor of computing and design. On the faculty, his duties included research and the teaching of undergraduate and graduate students. In his advanced computer visualization course, students explored the design potential of existing software such as AutoCAD and 3D Studio. They learned applications of geographic information systems technology to spatial analysis and data management for urban planning and design. In Valle's media course, they were encouraged to create interactive information resources integrating text, photography, video, sound, and computer graphics. Students were challenged to apply their skills to real-world projects.

In response to the devastation caused in 1992 by Hurricane Andrew, UMSA created the Architectural Recovery Center (later renamed the Center for Community Design), which provided technical assistance to homeowners whose property had been damaged or destroyed and coordinated information and proposals from government agencies, civic organizations, and professionals to create plans for redevelopment of the area. Valle played an important role in the process through the visualization techniques and other applications of digital information that he had developed. His students were actively involved in gathering information and creating electronic databases. The simulated environments Valle and his students produced from the data helped citizens and government officials visualize alternatives for rebuilding their communities. In subsequent years, students developed databases for other purposes, such as for the creation or modification of zoning codes for Key Biscayne, Key West, and other communities.

design crit

Among the many innovations Valle made at UMSA was the introduction of an on-line design "crit" (a year-end review of student work by other faculty or outside professionals) in 1995. Valle posted a request for jurors on the Internet and received more than 100 responses from architects and planners on four continents. Valle selected a jury that included faculty members from Harvard University, Pratt Institute, the University of California at Berkeley, and universities in Hong Kong, Italy, and Colombia. Student presentations were made on a web site, and jury members responded to the work by e-mail over a four-day period. According to *Miami Magazine* (spring 1996), Valle's Internet design crit was the first such instance for any school of architecture in this country.

Also in 1995, he published the first volume of his series *American Urban Typologies*. In this series, Valle intends to present detailed analyses of American cities that have retained a distinct sense of place instead of adapting the kind of standardized urban environment that is so prevalent today. He believes that certain elements of architecture and design that make these places function well could be adapted for other communities. In each volume, he will examine the principles (building types, architectural and urban codes, and site conditions) that form the underlying structure for each city. So committed is he to this project that he funds the research himself.

For the first book in the series, Valle chose Key West, Florida. Among the features he admires are zoning laws that allow stores and offices in the corners of residential areas; heterogeneous neighborhoods with houses of different sizes and prices; a downtown area that is lively after business hours because residents live above the shops; and a block and sidewalk design that is pedestrian friendly. Much of the book is devoted to an analysis of varieties of the conch house, a distinct building type that is prevalent in Key West.

Watching his children growing up in the Miami suburbs, where everything from libraries to churches to motels can be found at shopping malls, all with a similar architectural style, gave impetus to the project and intensified Valle's commitment to traditional urban design. His leisure time and family outings are often devoted to research for his books: "We would take trips to Charleston, Savannah, New Orleans, and other places and realize what makes them so memorable. And then we took the time to actually go and measure, walk it, photograph it, document it, and try to understand it. To me it's very relaxing, and at the same time it's work. Fortunately, my wife loves it as well, so it's a team effort."

Working on the series, Valle explains, "is my way of understanding the principles that govern this movement known as the New Urbanism." "The urban ingredients I rediscovered which constitute those Southern cities has since become my measuring bar for our planning projects. The list of cities goes on to include other cities throughout North, Central, and South America."

His growing involvement with urban design led Valle to make another career move. In 1996 he entered into a partnership with his wife, Estela Valle, and with Jaime Correa. CVV (Correa, Valle & Valle) is an architecture and town planning firm that is committed to the principles of New Urbanism. To devote more time to the business, Valle gave up his faculty position at UMSA. Within the firm, the three partners collaborate on all projects, but have distinct roles and responsibilities. Correa is responsible for planning, Estela Valle is the lead architect and designer, and Erick Valle is the negotiator, handling "all presentations, negotiations, contracts, invoicing, lectures, research, and other issues." "I enjoy convincing my clients/audience of what is wrong with the built environment and presenting them with the alternative," he says. The three partners share a common vision, working harmoniously and usually in complete agreement, but Valle acknowledges they must often compromise when working with existing codes or with developers "who are not visionaries."

CVV has won commissions for architectural and urban design projects in North, Central, and South America. Because it can take as long as 10 years for a building project to be completed, much of their work is still in planning stages. By chance, the firm's first constructed design was for a civic building in Costa Rica—the library of the Instituto Centroamericano de Administración de Empresas, an important business school that was founded with assistance from Harvard University.

The three partners have made "cultural urbanism" a focus of their firm. "We are very interested in celebrating the cultural diversity found in our inner cities," says Valle. "We are currently working on projects that celebrate the cultural roots of the people that bring historical value. My firm's mind, heart, and soul are gearing up for the next wave of growth, which will demand memorable neighborhoods of mixed income, mixed uses, and mixed cultures."

One example of their approach to cultural urbanism is an independent study they did for Wynwood, a low-income Miami neighborhood with a population that is nearly 90 percent Puerto Rican. CVV developed a master plan that included public spaces and architecture that were based on traditional architectural styles and urban forms of Puerto Rico. The public spaces were not greens or parks, but paved plazas. The architecture was not bungalow-style, as in much of Florida, but buildings with courtyards.

Although Valle and his partners undertook the Wynwood master plan as a pro bono project to create a model for the kind of cultural urbanism they advocate, their plan captured the interest of two companies with roots in Puerto Rico: the Bacardi Corporation and Goya Foods. Bacardi and Goya offered to finance construction of the public spaces.

The Wynwood master plan does not call for the existing neighborhood to be razed and reconstructed; rather, like much of New Urbanism, it relies on "infill" (construction on existing empty parcels) to transform the area. The addition of new buildings influenced by Puerto Rican architectural forms would gradually give Wynwood a cultural identity distinct from the neighborhoods around it.

Other examples of CVV's cultural urbanism are Bahama Village in Key West and the neighborhood of Overtown in Miami. Bahama Village is a community whose population is largely from the Bahamas. CVV brought in consultants from the islands to create a plan for development and infill that would celebrate their culture. To define the edges of the community, CVV introduced a color palette typical of the Bahamas and encouraged residents to use it in painting their homes and shops. They also created a small workshop where people could learn traditional island crafts, such as carving, that they could sell from their doorsteps. Similarly, in Overtown, an African-American community, CVV drew on African architectural styles and design forms to give the neighborhood a distinct cultural identity.

Valle and his partners draw on extensive experience and research in finding solutions to the challenge of urban design. "You may see us with a hundred books opened up," Valle explains, "because we're exploring and looking at things that we saw at one point or experienced. Or we're looking at photographs for details that really touch us, that we think would really make a difference. Most of our design comes from either travel or books or even reading. . . . We draw from history consistently in everything we do."

Other important aspects of the work are understanding the technical and legal language of zoning laws, knowing how to deal with numbers, and being able to negotiate with clients and government agencies. Clients are drawn to Valle because of his ability to work outside prevailing trends. Valle credits his innovative approach to design partly to his own intuition, but largely to the inspiration and creative collaboration of his partners: "All of us understand what we're looking for—to us it's what makes common sense. Some of the things that have been built around us just don't make sense."

Valle's work has received national and international recognition, including honors for his contributions to urban design and awards for his innovations in multi-media. Recently, several projects by CVV were selected by an international jury of architects for "The Other Modern," an exhibition in Bologna representing alternative architecture and urban planning for the twenty-first century. Valle himself is constantly exploring new possibilities in design and the technology that supports it, like computer animation to show how a proposed community would actually function. For Valle, the next project is always the most exciting one. "My most important work," he declares with enthusiasm, "has not yet been done."

Rediscovering an old technology can be as innovative as creating something new, according to the sociologist Kathryn Henderson. She has been documenting, from historical and environmental perspectives, the revival of building with straw bales. In her essay, Henderson writes about the inventions for baling straw that made this type of construction viable in the late nineteenth century, especially in areas with limited timber supplies. She continues with a description of the cost and energy savings, as well as the community-building effects, of today's straw-bale building movement. Henderson's insights into this type of construction come from research and from hands-on experience.

The architect Marley Porter has also been active in the renaissance of straw-bale building. He asserts that a building must unite sound architectural practice and personal beliefs, which include respect for the environment. Condemning "tract houses with thyroid problems," Porter describes one of his buildings—the Wimberley House of Healing—as an example of the sort of "living architecture" he supports.

David Hertz's work as an artist, an architect, and an inventor combines the reinvention of old technologies and a visionary use of new materials. Hertz is the creator of Syndecrete, a concrete-like substance that incorporates materials from society's waste stream and is used in buildings, in furniture, and in other applications. To the three R's of Reduce, Reuse, and Recycle, Hertz has added Restore, Reinterpret, and Reincarnate.

The range of architectural innovation is remarkable, especially in the design of materials and construction methods. Behind the technology, however, is also an underlying philosophy of leaving the environment better than the architect found it.

STRAW-BALE BUILDING: USING AN OLD TECHNOLOGY TO PRESERVE THE ENVIRONMENT

KATHRYN HENDERSON

In the late 1800s, when Nebraska pioneers experienced a shortage of locally available building materials they were able to draw on the recent invention of modern hay-baling equipment, which enabled them to use hay bales like building blocks to build everything from churches to houses.[1] Hence, hay-baling machines were then and are now part of what has been termed in science and technology studies "heterogeneous engineering"[2] —the simultaneous co-construction of the network of people and things that is co-produced while it facilitates the development and production of a technology. Just as industrial design engineers must negotiate not only technical aspects but the whole heterogeneous network of social and technical relationships, so must those who build homes with straw bales. They must build a network that includes the farmers who grow crops that produce straw, the baling technology used to convert straw in the field to building-grade bales, transportation systems and moisture-proof storage for the bales, people to help put them together, designers to plan buildings, building codes that will allow them to use the technology, local knowledge of how to put all these elements together, and a motivation for going to so much trouble. The motivation may be aesthetic and/or may be a philosophy that articulates that this is worth doing for individual economic savings along and for the ecological good of the community and the planet. This essay examines the manner in which activists have built their tacit knowledge and their networks, the barriers they have faced, and the social and historical factors that have contributed to the renaissance of a way of building that is more than 100 years old.

This approach comes from the perspective in science and technology studies that views technology and its design as mutually socially constructed and as shaping society. Network analysis[3] more closely examines this shaping during innovation and development of new technologies, including artifacts as actors in heterogeneous engineering. My previous work and that of others has paid attention to the importance of visual representations and tacit knowledge in the heterogeneous engineering of technology.[4] Since the richest form of knowledge is that situated in practice,[5] sketching, drawing, and building are often the loci of technical knowledge.

"Tacit knowledge"[6] is an important concern in science and technology studies. This kind of knowing is action that requires skill such as that held by a carpenter who knows just how a certain kind of wood must be handled, or what type of joint will serve best, but could not put that knowledge into words. I am using the term here to signify knowledge that is not verbalized, in some cases because it cannot be but in other cases because it may be taken for granted or regarded as too trivial to warrant verbalization. The generation and/or elicitation of all types of tacit knowledge is intrinsically linked to practice. Collins states that "all types of knowledge, however pure consist in part of tacit rules which may be impossible to formulate in principle."[7] His study of newcomers' attempts to build lasers using documentary information reveal that even with access to accurate diagrams and blueprints these scientists could not build lasers without having participated in real laser building. This is the kind of knowledge that is being transferred in workshops on straw-bale building. While visual and textual documentation of straw-bale-building design and technique are widely available, participation is still crucial for acquiring thorough understanding, just as in the case of Collins's laser builders. In order to get at this kind of knowledge, I joined the straw-bale network.

METHODS

I conducted extensive ethnographic field work with the straw-bale-building community in central Texas, an area enjoying significant growth in ecology-oriented building innovation. I attended meetings of the Straw Bale Association in Texas and Green Building and Natural Building conferences and symposia in Texas and Maryland. I participated in wall raisings and finishings. I interviewed homeowners, contractors, designers, architects, and Department of Energy evaluators. I surfed the Internet. I conversed with members of the movement from all over the United States (in person and by electronic means) in order to find out how one becomes an "expert" straw-bale builder. Participating in hands-on building as a member of a local straw-bale community allowed me to observe firsthand how local knowledge is passed on, not only spreading tacit skills but also extending the community. There are now straw-bale buildings in every American state, in Canada, in Mexico, in Europe, and elsewhere.

Qualitative methods were particularly suitable for this study because participant observation provided the necessary rich base of data to track the passing of tacit or unarticulated knowledge between individuals against a background of relevant context. In-depth interviews helped add the individual dimension, revealing members' meanings and narrative accounts of experiences not observed by the researcher. A snowball technique in which each informant provided further informants was used not only for contacts but also to trace human networks. Written materials, including newsletters, workshop announcements, and printouts of electronic communications, were collected to trace communication patterns and networks. Photography was used to document work practices during construction.

Pseudonyms are used here for individual home builders and for some activists and architects. Those who are highly visible and are quoted for their expertise are identified by their true names, as are individuals who did not request pseudonyms.

A PRACTICE-BASED PHILOSOPHY

In today's world of rising building costs and concerns over deforestation and pollution, the technique of building with straw bales is enjoying a renaissance. Advocates maintain that the technique produces excellent thermal mass in walls, puts waste-product straw to use, and produces aesthetically pleasing and sustainable buildings. (Sustainable systems are those that could continue interminably without harm to the environment). At the same time, the technique of stacking large bales can utilize a large number of unskilled laborers, facilitating the community camaraderie that many feel is missing in our individualist late-capitalist society. Unlike timber, straw grows to usable size in less than a year and can be rotated annually or grown with other crops to maintain healthy soil. Straw can be grown in low-quality soil, certain strains being highly productive in soils with high salinity and alkalinity. The production costs of straw and straw bales also contribute to minimal energy consumption, avoiding much of the energy and waste usual to the production of common industrial building materials.[8] This energy efficiency is measured in terms of "R values," which designate the insulation value of the walls; the higher the number, the more efficient the insulation. The walls of a conventional wood home will have an R value between 18 and 24. A straw-bale wall between 18 and 24 inches thick will have an R value between 2.4 and 3.0 per inch, substantially higher than other building materials.[9] While building with straw bales is not new, the renaissance that began in the 1980s with a handful of owner-builders who were willing to experiment is now being carried on both at the grassroots level by progressive home builders and at the professional level by architects and contractors. Activists have worked to get building codes for straw-bale construction passed all over the country and are now working toward a national building code.

Straw-bale techniques have become sufficiently well known and accepted in some locales that anyone who so desires can use them, even if not committed to the philosophy of living lightly on the land and building community. However, in order to gain the detailed tacit knowledge based on experience that truly makes the technique viable, one must enter into and interact with the community. Once exposed to the community, the philosophy becomes integrated with the technique.

One couple had chosen straw-bale construction for their dream home purely for aesthetic reasons. They bought lakefront property and planned to simply hire a straw-bale-construction consortium and paid labor to construct their home, because they felt they were too old for strenuous manual work and could afford to hire professional builders. However, the straw-bale contractor refused to take the job unless the couple agreed to have a communal wall raising. Similarly, folklore in the Austin community tells of unnamed individuals whose building project failed because they intentionally remained isolated, not assimilating the experiential knowledge of the group.

Communication in this growing network often takes place over the Internet as well as in conferences, workshops, and newsletters, combining low-tech, regional, hands-on knowledge with the electronic age. Much of this building is taking place in the southwest, having originated in Arizona and New Mexico, but was also adopted early in the movement in central Texas. The network of humans and artifacts that has built the straw-bale community along with straw-bale buildings is one of the prominent sociological factors of significant interest here. Part of that network is the technology of hay baling, the development of which began more than 100 years ago. In fact, straw-bale building has three distinct histories: the history of the invention of the hay baler in the late 1800s; the history, linked to that invention, of the first straw-bale homes in Nebraska; and the history of the contemporary straw-bale building movement which started in Arizona and New Mexico in the 1980s, with its own distinct phases.

A BRIEF HISTORY OF THE HAY PRESS OR HAY BALER

The earliest hay balers, then called hay presses, were developed from cotton press technology as early as the late 1850s. These large, upright, horse-powered, stationary presses produced bales weighing up to 200 pounds. For this reason they were not especially successful; the large bales were difficult to transport. Between 1854 and 1870, P. K. Dederick developed various models of horse- and hand-powered hay presses from his Columbia Prize-winning design for a cotton press. In 1872 he patented a smaller horse-drawn and horse-powered portable press with a horizontal ram rather than a vertical one (figures 1, 2). It produced bales of dimensions close to those of contemporary balers: 18–24 inches in width and 36–48 inches in length. This technology was successful for a number of reasons. Not only could the finished bales be more easily manipulated (even though they were heavier than today's bales, because side tension as well as vertical pressure was

bales

FIGURE I
Dederick's Standard Bale Columbian Cotton Press. (Smithsonian Institution)

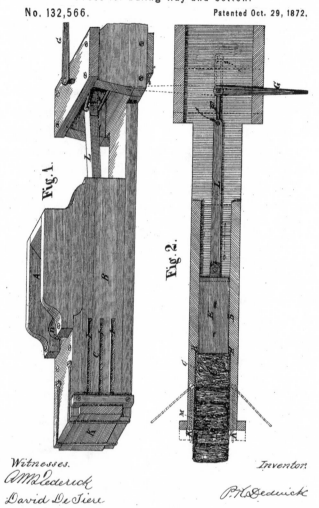

P. K. DEDERICK.
Press for Baling Hay and Cotton.
No. 132,566.   Patented Oct. 29, 1872.

Fig. 1.

Fig. 2.

Witnesses.

Inventor.

FIGURE 2
Dederick's 1872 patent for a press for baling hay and cotton.

applied to them), but their smaller size facilitated packing for railroad shipment. This was particularly important, as up to this point hay was sold locally; railroads had previously not accepted hay for transport in its loose form.

Advertisements from the period advised farmers that by baling their hay they could seek a better price than that paid locally. Similarly, this advantage was promoted in a 1917 International Harvester almanac, accompanied by an illustration of the hay press in action with the caption "Good Baled Hay tops the market." The text below the title "A Good Hay Press Means Added Profits" reads:

Neat bales of bright, clean hay bring good prices—enough to allow a good profit on the baling. The owner of a good hay press not only makes profit on his own baling, but can do the work for others, and thus establish a fine "side line." When hay is in the bale, the markets can be watched closely and shipments quickly made when prices are higher, and then, too, less storage space is required and bales are easily and quickly handled. The hay keeps better, there being a larger proportion of the nutrients retained than where the hay is loose.

It is interesting to note that the change in bale scale led to commercial success, as the smaller bales could be more easily manipulated by a single person. This same manipulability is also significant in the appeal of straw bales as an empowering building material in the contemporary straw-bale building movement. Historically, the larger social context is telling. The late 1800s were the period of the Industrial Revolution. Counterintuitively, this actually contributed to the increased use of horses for all kinds of transport and horse power in metropolitan centers. The demand for hay for feed and straw for bedding was at an all time high at the very period when crop yields in the eastern United States were declining due to disease and lack of crop rotation, providing a ready market for grain-belt hay. Dederick's advertisements illustrate his patented baler, which folded the grain sheaf, compressing it into separate grain-bundle flakes within the larger bale—a technique still used in the baling technology of today. The 1881 Dederick catalogue further points out the convenience of the flaked bale for stable feeding—small flakes can be removed one at a time from the bale as needed for feed. Competitors were quick to enter the market, developing both wooden and steel balers, powered by horse or by steam. Hay presses came in all shapes and sizes, to fit every pocketbook. By this time, balers powered by one, two, or more horses existed side by side with belt-driven balers

The Perpetual or Continuous Baling Press

THE PERPETUAL OR CONTINUOUS BALING PRESS was invented by the patentee, of the best of the old style presses, and who, for more than a third of a century, has made three-quarters of all such old-style machines. The Perpetual was a new departure in the art of baling loose material at the time of its introduction into use, and its construction, method of operation, and bale production, all patented, and more than one hundred United States and foreign patents were granted its inventor, P. K. Dederick, on the invention and various modifications of the same.

It was the first successful horizontal baling press ever invented or introduced, and the first to pack and complete a bale in charges, with a reciprocating traverser — the only successful method ever discovered for horizontal use, or making small bales economically; hence alone in the field with imperfect imitations.

It should be observed that, prior to this invention, the only successful method was to tramp or beat the hay into a upright box, after which it was pressed into a bulky bundle.

To bale compactly thus, requiring heavy, cumbersome and costly machinery, and, in consequence, used in but a few sections, so that baling and shipping hay was a new art and industry, in the most of the country, when Dederick introduced his new method or process, whereby hay is baled to load grain cars to their capacity in weight, with a small inexpensive machine, as portable as a wagon, and to this invention is due the universal movement of the surplus hay crop of the country, the commerce and trade resulting therefrom, and a great benefit to the hay interests of the country generally, from the farmer to the consumer.

The Press, Process and Bale.

The Perpetual Press is continuous in operation. The hay being pitched in one end of the press in successive charges, and the finished bale ejected at the other end in proportion as the forming bale is com-

pleted, the bales being divided by slotted partitions through which they are bound as they pass through the machine, so that there is no stop in the operation for any purpose whatever. The bale is in sections, and of oblong quadrilateral form bound lengthways.

Following we present some plain cuts, illustrating the manner in which the Perpetual Press folds and presses the hay into sections:

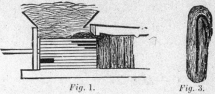

Fig. 1.   Fig. 3.

Fig. 1 illustrates the hopper filled with hay, which is pushed down at the center, when the traverser is withdrawn, as shown in Fig. 2, and then pressed against the solid hay and forced into the chamber, as shown in Fig. 1.

Fig. 2.

The hay overlapping the traverser, as shown in Fig. 1, is folded down as shown in Fig. 2, thus securing a smooth and even surface to the bale. Fig. 3 illustrates a section removed from the bale, end view.

Fig. 4.

Fig. 4 illustrates the neat and convenient manner of handling hay thus put up, and its advantage in feeding out.

Patents were granted Dederick on this press, and also on the continuous process and the bale itself.

United States Court Decision.

That the plaintiff's (Dederick's) press is the result of a new plan and a new combination, that invention was necessary to produce it, and that great benefit has resulted from its production and use, we cannot doubt.

* * * Such a press was never before constructed, and such a result never before obtained.

Hay Market Evidence.

Investigate any of the principal hay markets of the country and when neat, square ended, evenly pressed bales, with large uniform, nicely folded sections that fairly show the quality of the hay are found — trace them back to the press that made them, and in every instance you will find a Dederick press, as also you will find a swindling copy of a Dederick press if you trace back the soiled, crooked, chaffed sided, slant ended, unevenly pressed bales of varying sections, with u certain and bulging folds. Try it.

FIGURE 3

Dederick's continuous baling press. (Smithsonian Institution)

powered by steam traction machines. It was shortly after this introduction of hay-baling technology that was capable of producing uniform, brick-like bales that the first straw-bale houses were constructed in the Sand Hills of Nebraska. Though the terms "hay" and "straw" were defined less precisely during this period than at present, before the introduction of the hay baler only hay was of any value. Straw, the left-over stalk with no grain head, was usually burned as a waste product.

THE EARLY HISTORY OF STRAW-BALE BUILDING IN NEBRASKA

The Sand Hills of Nebraska, a region that produced impressive stands of meadow hay, is where the first straw-bale buildings were constructed. Nebraska pioneers of the late 1800s were poor people. Their plains environment was short on timber, and if they had used much of the sod of the western Sand Hills to build sod buildings they would have used up their topsoil. Building with straw bales was easier than building with sod because their larger size facilitated more rapid wall construction and larger structures. Moreover, the folklorist Roger Welsch proposes that by this time the specialized plows for cutting sod and the oxen needed to pull them efficiently were no longer in use, the tools and the skills both being rusty.[10] Though pioneers sometimes built small sod structures to establish their land claims, sometimes they also built larger straw-bale houses as inexpensive second dwellings. Though cereal grains were not yet being cultivated in the 1880s, natural meadow grass was available. Oral tradition suggests that a hay-bale home was constructed near Lincoln as early as 1889; however, the earliest actual documentation for a straw-bale structure is a 1902 report by Nebraska State Superintendent of Schools William K. Fowler of a one-room schoolhouse built near Minitare (then called Tabor):

Some five or six years ago in district No. 5 of Scotts Bluff County there was erected a temple of learning, the walls of which were of baled straw, the floor was the primitive mother earth and the roof above presented a face of earth to the heavens. This roof was made of poles laid across from side to side and covered with sod. The building was sixteen feet long, twelve feet wide, and seven feet high. There was a window in each side and a door in one end. The bales of straw were laid in mud instead of mortar, and with some half bales the joints were broken the same way that bricks are laid. . . . The building was used but two years as a schoolhouse. . . . Like the temple of ancient Jerusalem, of that schoolhouse there is not left one stone or bale of straw upon another, as the cattle were allowed to range around it.[11]

Evidence for actually placing specific brands and models of hay presses in Nebraska during this period can be found in the advertising testimonials of the hay press manufacturers. While testimonials, like any advertisement, are doubtful in their claims for the performance of the equipment, they do indicate where it was being purchased. The earliest Nebraska-based testimonial I have found in the Warshaw agricultural trade literature collection at the Smithsonian National Museum of American History's Archives Center is quoted in a one-page flier for the Scott Hay Press Company's horse-powered O.K. Hay Press. The testimonial, placed among others from states across the grain belt, reads as follows:

Germantown, Neb., Nov. 26, 1892
Gentlemen:—The O.K. Press we think the best in the market.
R. F. Seeman

Similar testimonials for the steam- or horse-powered Southwick Press in an 1895 Sandwich Manufacturing Company catalog show that the press was sold in Nebraska:

Revenna, Neb., Feb 14, 1895
Sandwich Manfg. Co.:
Gentlemen:—Last Fall I bought one of your "Southwick" Baling Presses of J. R. Patterson, of this place, and took it into the hay country about Whitman, Neb., and have baled about 1,200 tons of hay with it. Have laid out but $1.25 for repairs. I saw nearly all kinds of Presses working in that locality and will say the "Southwick" beats them all for capacity and durability and ease of operation. I am well pleased with it.
Yours truly, Jos. Simon.

Friend, Neb., March 4, 1895.
Sandwich Manfg. Co.:
Gentlemen:—In regard to the "Southwick" Baling Press, size 18×22, bought of you through Metz & Agee, of Friend, Neb., last October, must say that it has given me the best of satisfaction. Notwithstanding jobs were very small the past season, I have baled about 700 tons. Have baled ten tons in four hours and a half. Baler worked very smoothly and has cost nothing for repairs. Customers tell me they are better pleased with it than with any baler they have ever used, as the work is better and speedier.
Respectfully, J. W. Middleton.

FIGURE 4

"Lightning Hay Press" manufactured by Kansas City Hay Press Company, c. 1910. (Smithsonian Institution)

An undated catalogue for the Collins Plow Company from around 1918 includes a testimonial for the Model 112 Eli Baling Press, which could be run on steam, gasoline, or horse power. However, by far the best evidence placing the equipment of a particular company in Nebraska during the period in which the earliest straw-bale homes were built is a photograph taken by Solomon Butcher in 1904 in Dawson County. This photo, show-ing a horse-powered baler, two horses, a stack of baled hay, and a baling crew, closely matches illustrations in the Kansas City Lightning Hay Press catalog from around the turn of the century. Not only is the characteristic C shape of the pitman end of the press obvious in both photos; in addition, the shape of the loading chute, the horse lead, the baling shaft, the config-uration of the tension springs, and even the position of the horses and the automatic whip mechanism are identical. It is irrefutable that the Lightning model of the two-horse-powered Kansas City Hay Press was in use in Nebraska around the turn of the century and perhaps earlier. (The Kansas City Hay Press Company, founded in 1885, was one of the first baler man-ufacturers to be established west of the Mississippi. A geographically closer manufacturer would be a factor in more economical shipping costs for cash-poor farmers. Hay press manufacturers were well aware of the diver-sity of economic factors in their clientele and continued to manufacture a broad array of hay presses ranging from one- and two-horse-powered presses to steam and gasoline belt-powered presses from the 1880s through World War II to cater to everyone from the small farmer with a few acres to the traveling thresher men and the large landowners. An International Harvester Machines catalogue from the early 1940s includes baler configurations from single-horse-powered to the then-state-of-the-art pick-up baler, which baled in the field, picking up the hay directly from the windrows into which it had been raked.) People and things interacted in networks, and straw-bale builders had to deal with changes in baling technologies.[12]

The oldest straw-bale building still standing in Nebraska is the 1903 Burke homestead, covered with a shake-shingled hip roof. Located just west of Alliance, it continued in use as a home until it was abandoned in 1956. Because bale structures went up quickly, some families in immediate need of housing turned to bales as the quickest way to create shelter. In many cases these structures were regarded as temporary, to be used until a "proper home" of wood could be built. However, discovering that such homes were both durable and comfortable in the weather extremes of Nebraska winters and summers, people plastered over the straw walls and made them permanent homes. Such was the case of the Burke homestead, the exterior walls of which remained unplastered for 10 years after it was built. While it currently is showing deterioration because it has not been maintained and the roof is in poor repair, it has withstood more than 90 years of wide temperature swings and blizzard winds in relatively good condition.

Other historic bale buildings are the headquarters and bunkhouse at the Fawn Lake Ranch near Hyannis. Both structures were erected between 1900 and 1914 and are still in use. According to owner Mike Milligan, when the walls were opened to add an addition to one of the buildings in the 1950s, some of the original bales were set aside near a corral. The bales were so well preserved that horses in the corral reached over the fence and ate the sweet meadow hay in the 50-year-old bales.[13]

"Nebraska style" bale construction, used in all the examples discussed so far, is load bearing, meaning that the weight of the roof rests directly on the stacked bales. The 900-square-foot Scott house, built of wheat straw by Leonard and Tom Scott between 1935 and 1938 near Gordon, is an example of the technique at its best. The current resident, Lois Scott, has resided there since childhood and remembers her mother's opposition to building her home with straw bales. This unconventional building medium was considered to be low in status even by many of those who employed it of necessity. However, Lois Scott reports that her mother, despite her earlier

FIGURE 5
Catalog of International Harvester Company hay presses. (Smithsonian Institution)

opposition, lived happily in the house for 25 years until her death. Scott also reports that she has noticeably lower utility bills than her neighbors.

Welsch estimated that more than 60 straw-bale buildings were put up in the Sand Hills region, but a more recent inquiry by Matts Myhrman and Judy Knox turned up only 28 straw-bale buildings. Exactly how many straw-bale buildings may have been built in Nebraska between the 1890s and the 1930s is not known.

The first Nebraska post-and-beam straw-bale building still standing in good repair today was completed in 1938. In this technique, posts carry the weight of the roof and straw bales are used for infill walls.

The late 1960s and the early 1970s saw an increase in both counterculture and professional interest in innovative and vernacular building, often using recycled or waste materials.

In 1974 Roger Welsch published a one-page illustrated article titled "Baled Hay" in a book of essays titled *Shelter*. The lore of straw-bale building credits this one short article, which was based on Welsch's research on folk and vernacular Nebraska architecture, with doing more to launch the straw-bale building revival than any other single factor.

THE EARLY DAYS OF THE STRAW-BALE RENAISSANCE

The 1970s was a period of questioning authority in all fields, including architecture. It was also the period in which ecological concerns were introduced to the public, and the period in which two groups that would be important to the straw-bale movement of the 1990s set off in two different directions. The rebellious counterculture experimented with communal living, with the growing of herbal remedies and organic foods, and with the use of inexpensive, recycled, and found materials. Not only was *Shelter* published during this period; so were Bernard Rudofsky's *Architecture without Architects* (which documented the beauty of indigenous peoples' building with natural materials) and *Handmade Houses* by Art Boerick and Barry Shapiro.[14] However, not all members of the baby-boom generation took the alternative path, and many who experimented with the counterculture in the 1960s and the 1970s returned to the mainstream to pursue lucrative careers. In the meantime, architectural innovators as well as those who simply wanted to build shelter as cheaply as possible began experimenting with straw bales as a building material, drawing on Welsh's article.

In 1978 Dan Huntington applied for a permit "to put a roof over" a straw-bale structure with a concrete post-and-beam frame near rainy Rockport, Washington. Issued the most basic permit available, Huntington and his wife lived in the structure for 2 years while building a conventional home. More influential was the straw-bale post-and-beam cottage built by the northern California architect Jon Hammond, because it was featured in *Fine Homebuilding.* These two examples are early hallmarks of the grassroots innovators and professional architects and builders who have contributed to the growth of the straw-bale movement. A December 1984 article in *Fine Homebuilding* inspired a number of future leaders of the contemporary straw-bale movement, such as David Bainbridge, co-author of *The Straw House,* and Steve and Nena MacDonald, who built their straw-bale home in Gila, New Mexico for $7.50 per square foot in 1987. By 1986 Bainbridge was giving talks and publishing articles on straw-bale housing. Attending the 1987 Permaculture Design Course sponsored by the Sonoran Permaculture Association, at which Bainbridge spoke, were individuals who went on to build their own straw-bale homes and to play major roles in the development of straw-bale building, such as Sue Mullen of Gila, New Mexico, who completed her house in 1988. A 1988 newsletter summarizing the 1987 course and including an article on straw bale by Bainbridge influenced yet another circle of future activists in the movement, including Matts Myhrman, who, with Judy Knox, founded the information and education service "Out on Bale (un Ltd.)," which publishes a quarterly journal titled *The Last Straw.* (Much of the history reported here is derived from this journal.) A 1989 straw-bale workshop in Oracle, Arizona, brought together David Bainbridge, Bill Steen, Matts Myhrman, and Pliny Fisk of the Center for Maximum Building Potential in Austin to work out consistent methods for building with bales. This was a crucial juncture in the exchange of tacit knowledge, the standardization of practice (to an extent), and the structuring of a network to disseminate it.

In the 1990s, straw-bale buildings of all shapes and sizes proliferated throughout the United States. The early 1990s was also a period of dramatic increase in newspaper and television attention to straw-bale building along with numerous hands-on workshops. It was also the period in which the first explicitly straw-bale bank loans and building permits were issued. The groundwork for straw-bale building permits was laid in Tucson through the activism of straw-bale advocates and a building code culture ready to receive them. Today a number of counties and cities in Arizona,

New Mexico, and Texas have straw-bale building codes, and a movement is afoot to create a national straw-bale building code. A certain amount of testing has been conducted to lend scientific measurement to the body of data supporting the claims of durability and high insulation value for straw-bale construction. Straw-bale web sites have proliferated to the point that one site has indexed them alphabetically. Straw bales have been employed in at least one commercial development in Arizona and a few contractors have built conventional-appearing homes on speculation using straw bales. All this suggests that straw-bale construction is moving toward the mainstream. Though history reveals some of the factors contributing to this, it is not the whole story. Ethnographic field work and participant observation with those building straw-bale homes begin to reveal the answers to the question "Why now?" What combination of technical, material, cultural, economic, and human factors create and are created by this network that is able to successfully build with a low-status technology from 100 years ago?

ON AND OFF THE GRID: ECONOMICS AND BABY BOOMERS, THEN AND NOW

who?
why?

The obvious place to start in linking this history to the contemporary straw-bale network of people, things, and intangibles is to ask "Who builds a straw-bale house?" The second question is "Why do they build with straw bale?" Interestingly, it is often members of the baby-boom generation who came to adulthood in the status-quo-questioning era of the 1960s and the 1970s who are building homes and businesses in this new medium. Moreover, straw bale is a technique that seemingly reunites members of that generation who chose different directions in their youth. Those who took the counterculture road may or may not have university educations but have experimented in various alternative approaches to building and living, developing skills and knowledge in using alternative materials and techniques that are workable and affordable. Those who used their educations to pursue careers in more traditional lifestyles are now in their financial maturity and able to afford to build in a manner less dependent on commercial financing. Some individuals are "crossovers" who developed sufficiently marketable skills in alternative environments. In most cases they are individuals, of any age, seeking either to build their own home or to build for others as an expression of conscientious values. That many are individual home owners as opposed to speculative builders is one of the factors which has allowed for experimental innovation and has been facilitated by the flexibility in philosophy, cost, and housing style of straw-bale building.

Though some of the earliest straw-bale homes were built for exceedingly low cost (as little as $7.50 per square foot), cost alone is not a true measure of the commitment to straw-bale building. The homes built that inexpensively were done by individuals who collected recycled materials well beforehand, contributed much of the labor (unskilled and skilled) themselves, and built small. More luxurious straw-bale homes have run as high as $200 per square foot, so cost saving obviously is not an independent factor.

individuality

The earliest straw-bale revival builders were innovators who were intentionally looking for different ways of building that expressed a form of individuality. They tended to be artists and musicians, alternative health practitioners, sometimes disenchanted architects and contractors, but generally speaking, counterculture in political, ecological, aesthetic, or other lifestyle orientation. What is interesting is that the act of building with straw bale is sufficiently flexible both in practical terms and as a symbolic expression that people can use it in a variety of ways. It can be an expression of ecological values as a model for others in ways to "live lightly on the land" by coupling it with other "green" building materials, a water catchment sys-

"green"

tem, a composting toilet, and other conservation innovations. The first person to obtain a straw-bale building permit in Austin states such values for choosing straw bale:

> We were looking for alternative ways of building a house, that is, alternative to conventional stick-frame and something more in the direction of ecologically sound building practices. So we looked at rubber tire houses. . . . We researched that pretty carefully and decided we do not want to live underground. My mother-in-law found a straw bale house book . . . in the Real Goods catalog out of California. She bought and read it and said this looks pretty interesting and sent it to us. And I read it and saw that the systems were worked out carefully, thought through enough, and saw ways that I thought I could apply it.

aesthetic

The choice to build with straw bale could equally be an innovative aesthetic statement, using the qualities of the bales themselves to create stepped walls, window seats, and patio benches echoing Southwestern and Mexican adobe architecture without architects. A retired couple who choose straw bale for these reasons had looked at the modest dwelling an ecology-oriented home owner had built and thought they could improve on the design:

When I went in there the warm-like feeling [was there]. I wasn't crazy with the styling, it was very basic, and I am sure it suited their needs very well. But it was the feeling I got, we're gonna take this and show that it can be used aesthetically, wonderful as an art form, rather than just [as] economical building. So that we can be challenged, you know, an artistic challenge and have both. And engineering, you know, 'cause the techniques are evolving, so there's . . . a lot of room for creativity.

It can, of course, be both. The story told by a gay couple who decided to build with straw bale captures nicely the tension and resolution between the two orientations. Bill was very enthusiastic to build with straw bale. He works in landscape architecture and is ecologically conscientious. He took his partner, Bob, who is much more aesthetically oriented, to slide shows at Straw Bale Association meetings and on local straw-bale home tours in and around Austin and the surrounding hill country. Bob was unconvinced. Retelling their house story, now that their straw-bale showplace is finished, Bob says he was less concerned about the ecological issues than Bill was, though he did not object to straw bale on those grounds. It was the overly modest size and the aesthetics. "I hated them. I didn't want to live in a shack. I wanted a nice place, a place where we could entertain our friends." Obviously, one person's ecological model is another person's shed. By serendipity, a new architect came to town on a large project using innovative building materials. He had extensive experience in building with straw bale and an innovative design vocabulary for the medium, drawing on the facility of straw bale to create curved walls more easily than conventional materials. The result was the personal expression Bob wanted with the high-insulation-value walls Bill wanted.

The tension between these two views is not always resolved so amicably. Some purists and social justice activists point out, correctly, that homes in the United States are overbuilt and overconsumptive by world standards, and that simply building a luxury home with straw bale does not address the problem. There is more than one perspective here. Activists are taking straw-bale construction into low-income neighborhoods and into under-developed countries. However, low-income families do not immediately embrace straw bale as an acceptable home-building technique. Unlike comfortable middle-class home builders and avant-garde designers, their outlook is closer to that of the cash-poor Nebraska pioneers. They see straw bale as an inferior and cheap mode of building, nothing to be proud of. Building showplaces with straw bale lends it the credibility that makes it attractive to people in all income ranges.

Many of the earliest straw-bale homes in the Austin area were built in locations that builders refer to as "off the grid." This meant they were outside city or town limits, often in unincorporated areas where building codes did not apply and city amenities such as power, sewer, and water lines were not available. For early innovation and experimentation, this was advantageous, as no building codes for straw-bale building existed and so the technique was not subject to codes designed for timber building. However, for straw-bale building to move into the mainstream, codes were necessary. Straw-bale activists realized this and started working on codes in the 1980s, not only to allow straw-bale building to take place but also for at least two other significant reasons. First, cognizant that the budding solar energy movement had been set back by unskilled and sometimes unscrupulous entrepreneurs who promoted inferior and sometimes non-functional solar-powered technology, and incorrect or misleading information, straw-bale activists wanted to ensure that those who employed straw-bale building technology were held to a standard that would ensure the success of the technique and hence its overall reputation. Second, bank loans and insurance availability are tied to building code requirements.

BUILDING CODES: A CULTURE OF THEIR OWN

Serving as guest editor for a 1996 issue of *The Last Straw* (the primary printed medium for exchange of information in the straw-bale building movement) that focused on moving straw-bale building techniques into the mainstream, David Bainbridge acknowledged "the wisdom of committing to a cooperative approach to dealing with building code issues." Bainbridge's technical training is in engineering. He often opens his presentations with slides of the steel-and-glass buildings he helped build in the past, stating "I have a lot of bad karma to make up for." His reference is to the wasteful building practices of the contemporary building industry and the energy-consuming character of their products. His reference to a "cooperative approach" is in regard to the manner in which building codes for straw-bale building in individual districts have each built on previous work done elsewhere as well as the manner in which builders have shared information with one another regarding successful and unsuccessful tactics in getting building permit approval in new locales. This sharing takes place in articles such as those which make up the special edition of *The Last Straw,* formal presentations and informal conversations at workshops and "green building" conferences, web-site postings, and personal email messages. Aware that "how research and testing are conducted will affect how building codes are written and consequently determine how accessible and affordable [straw-bale] building will be," Eisenberg, together with others, formed BRAN (Bale Research Advisory Network) "to coordinate and maximize the effectiveness of this work."[15]

Virginia Carabelli's house near Tesuque, New Mexico, built in 1991, was the first insured, bank-financed straw-bale structure and also the first built with a building permit. Having granted Carabelli's permit, the State of New Mexico Construction Industries Division issued ten experimental permits for non-load-bearing straw-bale structures. About the same time, the Straw Bale Construction Association was organized in Santa Fe, bringing together professionals committed to straw-bale design and building. This activist group successfully sponsored a small-scale fire test and transverse load test for straw-bale building and succeeded in getting straw-bale construction guidelines into New Mexico building codes. Such testing and cooperative work with building officials in New Mexico followed similar efforts in Tucson. There, Eisenberg, Myhrman, and Knox worked closely with local officials to coordinate a structural wall testing program conducted by an engineering graduate student named Ghailene Bou-Ali for his master's thesis in structural engineering at the University of Arizona.

Each building permit office has its own culture, depending on its geographical region and its local personality and leadership, as David A. Mann, Codes Administrator for the City of Tucson, pointed out when asked by Eisenberg about the impression of straw-bale building on building officials given its long history but limited amount of quantified testing data: "Every building department, like any other organization, has a culture. Some depend on facts and calculation, others are more interested in construction community input, and still others are interested in what other departments are doing. My advice would be to have information on all of the above: calculations, test results, local designer and builder interest, and local jurisdictions that have straw-bale experience."[16]

Fortunately, the broader building culture of Tucson, where building officials were already accustomed to adobe building as an alternative to construction industry practice, set a precedent of open-mindedness for innovation with highly placed building officials. Leroy Sayre, the Chief Building Official of Pima County, Arizona, has stated that, when Matts Myhrman came to his office to see if he would be receptive to the idea of allowing straw-bale buildings in the county, he had wondered "Is this guy really serious?" But after reflection, he began to see the similarity between bales and other building blocks such as brick, concrete, and cinder. The testing done by Bou-Ali, even though merely for a master's thesis and even though "relatively primitive" according to Sayre, was accepted as "showing that the material could withstand the wind loads, and to a certain degree the seismic loads anticipated by the building code" and "became the basis for the development of prescriptive standards in the building code for the use of straw bales as a construction material."[17] Even so, it took 3 years of innumerable meetings and multiple rewrites before the written standard was acceptable to the code review committee. The standard was approved and added to the 1994 Pima County/City of Tucson Uniform Building Code as Appendix Chapter 72. On January 2, 1996, Pima County's Board of Supervisors and Tucson's mayor and city council adopted the 1994 edition of the Uniform Building Code, with Appendix Chapter 72, allowing the use of straw bales as load-bearing walls for single-story buildings.

Illustrative of the differences in building office cultures and political climate, the Board of Supervisors of Napa County, California, adopted the standards from the Tucson/Pima County code as an amendment to their building code, though the Building Department staff report had recommended against it. The supervisors felt that the standards were sufficiently developed and that the history of straw-bale construction adequately ensured safe outcomes, so that if plans had an engineer's stamp of approval they were acceptable. Nolo and Glenn counties in California followed suit.

This is not to say that every straw-bale project has met with success in dealing with building officials. The straw-bale Montessori School wall raising planned for the Natural Building Colloquium East had to be canceled because building officials of Davidsonville, Maryland, refused to even look at the compilation of materials amassed to date and the school subsequently was forced to chose another building technique. While public-oriented structures such as schools, even private ones, may have more of a challenge, independent home builders have been paving the way in testing codes along with a few enterprising independent businesses.

Austin also used the Pima County/Tucson code as a model, but held meetings to address the variation in regional needs, particularly questions of moisture and insect protection. The first permit was issued to Norm Ballinger, who reports that getting through the process was made easier by the abundance of alternative construction information resources in the Austin area, and that other straw-bale homes had been built nearby, off the grid. He found his greatest asset to be perseverance, likening the experience to an advanced game of Dungeons and Dragons: "I simply found on several occasions that I hadn't picked up or was not given the proper document, or I hadn't spoken to the right department, or asked the right question at the right time. The dragon, if there was one, would have been my own impatience in the thick of it. I talked with several geezers who were large-scale developers and such and found that they were having a lot of the same problems. So it hasn't been so much a matter of experience as diligence. . . . My best resource seemed to be to try to find a human behind every bureaucratic face I encountered."[18]

Ballinger's perseverance in finding the human face behind the bureaucracy is particularly useful in Austin, a city with a higher-than-average number of highly educated and activist citizens concerned with the local ecology. A kind of haven of liberal thought in a state and region stereotyped for its conservative, "Bible Belt" culture and politics, Austin is known as "the Berkeley of the South." The University of Texas, the state capital, the geography of hills and springs in a flat and arid state, and eclectic music and art venues draw "progressively oriented" individuals from all over the American South and elsewhere.

THE LANGUAGE GAME

The "progressive" and "green" orientation of Austin means that sometimes, lurking under the bureaucratic structure of city, county, and private agencies are people "inside the system," willing to work creatively for change. A case in point is a building inspector who does evaluations for mortgage companies. When he fills out the forms for a straw-bale building, instead of listing straw bales as the building material, he writes "cellulose fiber infill." Loan officers never question it. Particularly with post-and-beam construction rather than a load-bearing building, it is hardly necessary to mention wall structure, as timber beams support the roof. Sometimes straw bales are not mentioned at all in bank and insurance documents.

TACIT KNOWLEDGE

By the time I attended my first straw-bale wall raising, I had a general familiarity with the technique and its history. I knew the bales were stacked like bricks with metal rebars pounded down through them for stability. But I didn't know, until I started lifting them, that there was a difference in straw bales, that they came tied with wire or polystyrene cord and were inconsistent in length and density. Availability was also an issue. Hay and straw are harvested in late May and early June in Texas. The reason no wall raisings had occurred when I first started my field work in January was because that is the rainy season, particularly bad for exposing bales to moisture, and any bales purchased during that time would be last year's crop, including storage costs. A pioneer straw-bale contractor related his first lessons about bales on his initial project:

The biggest thing was that when we went to gather straw, we've got two trailers, and our first trailer, we pulled up one side of the barn, and the farmer told us to pull it to the other side of the barn for the rest of it. We, I didn't realize it, but the . . . strength of the first load was much better than the second load. They gave us some really loose straw [bales]; so we had two trailers of straw, and uh, also I didn't know about starting in the four corners at the same time. . . . And so we just started at one side of the building and worked our way around. So, by the time we got over to the first side of the building, the straw, we got to the soft straw [bales], and it compressed much and it just didn't stack up as high either, it just compressed on its own, even without a roof load. So with the walls just dry stacking, with no load on it, we had a two-and-a-half inch discrepancy. So, we had to tear all the walls down and redistribute the bales, and that came out of a phone call to Steve Kimble [an experienced straw-bale builder in Arizona] in the middle of the night, I was just totally freaking out. And . . . he says, "I've always been afraid this was going to happen, and I don't know what else you can do but you know, just tear it down and reassemble it." So, we did that. . . . We tore it quite a ways down, . . . enough to get distribution. And that worked much better.

The same contractor also related that he had mitered the corner bales at 45° angles, since that is what he was familiar with, using wood, rather than butting and overlapping them like a brick corner configuration. He noted that part of his problem was that he just jumped in without doing his homework, having missed the local workshop because of other commitments. Nevertheless, he turned to the then-small straw-bale network for knowledge and salvaged the project.

When I arrived at my first workshop and wall raising (bringing, as directed, work gloves, a tape measure, a hammer, and a long-sleeve shirt), a small portion of wall had already been erected, to be used as a model. We sat on a semi-circle of bales while the architect/contractor demonstrated how the bales should be stacked, a whole bale always spanning the break beneath the two bales under it, even if it had to be custom sized, so that the 8-foot rebars hammered through the bales from the top and the all-thread pins coming up through the foundation, added onto with fasteners at each subsequent row of bales, would succeed in pinning the courses of bales to one another top to bottom and bottom to top. "This seems very straight-forward," I thought. "Where is the tacit knowledge?" The architect/con-tractor then showed us how to make custom bales, those smaller than regular size, needed to fit the last odd space in a bale course while retain-ing full bales at wall turns and corners. Customizing the bale involved push-ing a huge needle, 2 to 2½ feet long (made from a flat iron rod or rebar with a hole in the end of it) and threaded with binding cord, through the bale exactly where it was already bound, cutting the new twine, and tying off each end of the bale. This was done twice, at the locations of both original bindings. When the original binding was cut, two smaller cus-tomized bales fell out. Again, I thought this technique was quite obvious, involving little tacit knowledge. We were also shown how to form curved bales for use in circular or semi-circular walls by setting a bale lengthwise, with one end propped up on another bale, and kicking it in the middle; it looked simple. Ten minutes into the first day I knew why I needed the gloves and the shirt. I put on my work gloves to protect my hands from binding wire and my long-sleeve shirt to protect my arms from scratchy straw. Then I tried the techniques shown in the workshop. They were not so obvious as they seemed, just as watching someone do a calculus problem is not the same thing as solving it yourself. Making custom bales was fraught with situational issues. The bales at my first wall raising were from a farmer in the Texas panhandle who specialized in producing what he called "construction-grade bales." Somewhat like the bales produced by nineteenth-century hay presses, these bales are denser than those produced by growers for conventional farm use, where the density and uniformity of the bale is of no consequence. Moreover, despite efforts to protect them by stacking them under the roof that was already raised in the post-and-beam section of the house and wrapping them in plastic tarps, they had become slightly damp in a recent downpour. Our first act after the workshop was to move the wettest bales into the sun to dry. Getting the thick needle through such dense and slightly damp bales was a real challenge. Working two to a bale, we began developing techniques of pushing and bracing to get the needle through. Twisting seemed to help, but that produced a new problem. Twisting the needle while it was going through the bale meant that the cords could overlap inside the bale so that when the original binding was cut and the bale was expected to fall into two nicely tied smaller chunks the two were held together by overlapping ties. Even without twisting the needle this sometimes happened. One solution was a needle with two holes, one for each thread, but it did not always work either. Finally, the only way to ensure no overlap was to thread each tie band separately, but even that did not always work. But we did get better with practice. Another issue was the knot of the new tie band. When I found my knots slipping as I tried to keep the tension on the bale the same as that of the original band, my partner showed me a better way to tie a square knot that she had learned in Girl Scouts. I was pleased with this solution and adopted it. However, at the next wall raising I attended, and I was again on the custom bale crew for a while, the owner-builder preferred that we use a slip knot that he had learned to tie the smaller bales. Not only does the variation in these techniques show that they are experiential knowledge, learned in practice; it also illustrates the distinction between tacit knowledge and local knowledge, as each region or even each site may have its own way of solving the problem. Since many variations work, standardization is not an issue. I also found, working on the custom bale crew, the beauty of the room for slippage in straw-bale building. Though some practitioners prefer very smooth walls, and they can be achieved with much attention to alignment and trimming, others, including myself, find the wall variation part of the beauty of the technique, a reference to its vernacular roots. Consequently, when measuring the space to be filled by the last bale in a course, which is somewhere in the middle to maintain the integrity of the corners, the measurement cannot be exact, because it will differ if taken at the edge of the adjacent bales or at the binding. The same is true of the bale, custom tied to fit the space. If the space is measured at 29 inches, the bale to fit it can be anywhere from 28 to 30 inches and still fit just fine. Forcing the elasticity of the surrounding bales is just as acceptable as stuffing the gaps with loose straw. One learns that limits exist, but their edges are fuzzy. I also found I could not curve a tight bale by kicking it while wearing tennis shoes. Those who succeeded (women and men) were wearing cowboy boots.

One of the leading straw-bale contractors in the Austin area, a former English teacher, pointed out that explaining things in great detail in workshops is counterproductive, stating that people learn more by asking questions as needed to accomplish the task they have undertaken:

I do know some other workshop leaders and they have their theoretical things and they go into tremendous detail. I come from a background where I taught English before [and was exposed to] a lot of study with learning mechanism. And you know, the one that has stuck clearly is never try to teach anybody with three things, they'll never get it. . . . Kind of what we do is people asking for more, more, more information. And we just kind of make some assumptions about how much they can actually take in. . . . We don't just sit there and feed them, when we [work out] what's the reason, [it allows people] to grow. So at our workshops we do very short introductions, we tell everybody, "Hey, there's lots of this we need to get done, you know, don't worry, you know, you don't have to break your back or anything like that, there's lots of us, let's just have fun. And that's [how] as you go along . . . and stop [to explain] little itty bitty pieces, you know. Anyone with half a brain could understand if they paid attention in the right way, [but] not trying to learn it all at the same time.

This statement matches my own experience. By the time I finished two days on the first site, I knew that felt paper had to be put under all the bale walls and window frames to wick moisture away from the bales, and that this was particularly important under windows not only because they were a potential source of moisture but also because bales could better stand to get wet on their long-stalk sides (where moisture would just run off) than on their cut-stalk side (where moisture might trickle down the stalk to the interior of the bale and start decomposition). Since I had helped roll that black paper for every wall of the house at the first site, novice that I was, I was the one who noticed the omission of paper under the kitchen window frame at the second site. I learned that a bale that was unusually warm or discolored should be discarded because rot was already underway. Observing the architect and owner discuss window size and placement, I learned about bale aesthetics, that single- and multiple-bale course heights were used to set window placement, and that windows should be high and narrow rather than broad to preserve the integrity of a load-bearing wall. I became familiar with new tools and with old tools used in new ways. A cap placed over the rebar made it easier to hit with a hammer; a small metal sign made a "bale horn" to ease slipping a custom bale into a tight spot; a weed whip became a bale wall trimmer to even up the surface for plastering. I watched those with more advanced tacit knowledge skillfully use a chain saw to cut difficult angles on custom bales and to cut recessed arches into the straw wall to be plastered over, becoming niches, part of the southwest aesthetic. All this took place in the two weekends that I participated in two wall raisings.

Visits to subsequent sites and conversations with owners and builders informed me of techniques to contend with moisture protection in the humid and rainy climate of central Texas. A much-told story concerned the importance of breathability for finished straw-bale walls. A client of one of the local straw-bale builders had gone against his advice and painted his walls inside and out with Elastamerica paint, which completely seals a surface with a plastic-like coating. With the walls completely sealed, temperatures varied widely between the inside and the outside. In summer, when air conditioning was in use, condensation built up inside them, making the bales wet and contributing to bale rot. If breathable natural plasters are used, such condensation will not occur.

The wall raisings in which I participated in Texas did not reach the state of completion for plastering before I left for Washington, so I only knew about the techniques from the small introductory workshop at the first site. There I learned about "robert pins" (the "grown-up bobby pins" used to fasten wire lath to the bales for rounding out window and door frames, corners, and other architectural details), and I learned that plaster (in amounts ranging from one scratch coat to three coats) is used to seal a straw-bale structure from the elements, including pests. While I was attending the first Natural Building Colloquium East (held in Maryland), I finally had the opportunity to pick up a trowel, helping finish a straw-bale garden wall. Tacit knowledge here starts with understanding the use of half a dozen different kinds of trowels, developing a technique to work the cement plaster scratch coat, the first applied, into the straw instead of leaving an air gap where the wire lath may not be flush. This mattered only for aesthetics and durability for the garden wall, but on a house an air gap with metal present could significantly lower the insulation value of the wall. This is a lesson the Department of Energy learned when they first set out to test straw-bale walls and found them not up to claims because they had built the wall incorrectly. (Fortunately, they rebuilt it, paying attention to the air gap, and came up with numbers that matched those of the claims made by the straw-bale community.) The final tacit skill lesson, which I was not ready to learn until I had worked with the tools and the medium for a while, becoming at least able to manipulate them to some degree without dumping most of the plaster on the ground, was how to "float" the concrete plaster. This involved using the correct trowel, a small oval one rather than the big triangular one or the rake-like one used to texture the scratch coat so subsequent coats would stick to it. Using a light hand to force the plaster into the straw, this technique let the water in the plaster "float" to the top, creating a desirable smooth surface which could be extended to the entire area. Again, the tacit knowledge I acquired was only the beginning of a deeper folk skill and wisdom which is just now beginning to be appreciated. While standard plaster recipes come with the commercial products, there is another knowledge, almost lost, of which I heard snatches in conversations here and there—vernacular knowledge, now residing in non-industrial cultures, about whether the addition of pulverized prickly pear, cow manure, or horse urine creates more durable surfaces for different floor and wall uses. Yes, there is a romance about this kind of building, but it is not purely about nostalgia for the old ways, though that is present too. Its rhetoric is about respect for the Earth and creating human-scale dwellings from non-toxic products and producing community in the process of doing so.

THE ROMANCE OF STRAW: BUILDING WITH LEGO BLOCKS AND COMMUNITY

Norm Ballinger expresses some of this sentiment in part of his article about contending with building codes: "Years ago, I only dreamed of building a house—a dream grown dim in the barrage of wastefulness and careless workmanship that gets passed for normal. Now I can see how building one's own place conscientiously can also rebuild the world in some small way." Two themes can be found here. One is a restatement of the straw-bale philosophy that, in building with such a renewable resource, one is contributing to the health of the planet. The other is about empowerment: that building one's own home has a kind of satisfaction about it. This sense of satisfaction is also available to those who participate in wall raisings. It doesn't come right away, though the novel experience for many of us who work behind computer screens, seldom engaging in manual labor, does have a level of satisfaction. It comes at the end of a long day when you are one of the few who have stayed to finish the last course on that wall you started, or to get that last rebar pounded or window set straight. And you feel a closeness to the person you have been doing this with all day, who also stayed until it was finished. This is part of the shared community of the wall raising that is the hallmark of straw bale. But it is more, too. It is that quality that Marx points out is missing in the division of labor, why it is alienating instead of empowering: the sense of a job well done. It is also the shared practice with those who have helped, the solidarity of the work space. Here it is not collective class consciousness; after all, the people involved are mostly middle class, are building individual homes, and may or may not work together again. It is about shared practice, learning, work, and shared knowledge gained in a situated practice that you and another person have in common—the building block of community. You look forward to seeing that person again at the next straw-bale wall raising. You will both come again, not because you are planning to start your own home next month and payback is fair play (the rationalist approach) but because community is being built with shared values and shared labor.

Several other layers of meaning have been incorporated into straw-bale culture. One is a tradition of having all those who participated in the wall raising write a good wish for the inhabitants of the structure and tuck it in among the bales. Another is the tradition of the "truth window," a small area, with or without a door, that reveals that indeed the straw bales are there underneath the plaster. These range from the simplest of framing to very elaborate shrines, port holes, custom, antique, and ready-made shutters and doors.

Another aspect of the romance of straw is its apparent simplicity. The size of the bales gives the sense of building with giant Lego blocks. With a good-size crew, all the walls can be up in a day. This is slightly misleading, since the foundation, roof, and finishing take much longer, but the bales themselves are symbolic of the human-scale possibility and the empowerment to take your environment into your own hands and build what you want. The simplicity also references the supposed simplicity of farm life and mentality. When I told a friend of mine that I had started researching straw-bale building, he replied that it gave a whole new meaning to the expression "hayseed." The straw-bale movement has taken gleeful delight in using such stereotypes and turning them to their advantage in their choices of newsletter and web site titles: "The Last Straw," "Out on Bale," "Save the Bales," "The Baley Pulpit," "Around the World in 80 Bales," and so on. Many such expressions are due to Matts Myhrman's unique sense of humor, but my favorite is his statement which captures the cultural ethos of the whole movement in its caricature of the simplicity associated with straw and farms and the radical activist's disgust with the status quo establishment: "The moral of the Three Little Pigs in *not* that you shouldn't build your house out of straw, it's that you shouldn't have let a pig build your house."

The community ethos that surrounds straw-bale building in terms of the romance of the wall raising is one that is socially constructed as a part of the movement. It could be otherwise, and indeed it may become otherwise if the movement succeeds in moving completely into the mainstream. Such an ethos existed for timber houses before Levittown. They, too, were regarded as a mode of building capable of expressing individual empowerment in that anyone who could pick up a hammer and saw could build their own shelter. That changed as industrialization took up the laboring hours of the homeowner. So, too, did the organization of the building industry into the generator of mass-produced houses, even though they used the same technology that the builder of a single home had used. The message is not in the technique alone but in what people do with it. As codes make straw-bale building available to more mainstream builders and buildings, the potential for commodification of straw-bale building appears on the horizon. This perspective is also expressed by Myhrman, who depicts himself as a closet "Struddite" (a contraction of "straw" and "Luddite"):

The traditional load bearing technique was created primarily by cash-poor, unregulated owner-builders in an environment that offered few options for affordable building materials, and those in a limited variety of forms. Now, in a very different materials and regulatory environment, the techniques are being tested, engineered and modified by "professionals." As a potential benefit of this effort, they see being able to build *for others* with less regulatory hassle, with fewer "callbacks," in less time, and perhaps, with less materials.

Hidden within this potential benefit, however, lies a potential danger—the "commercialization" and "professionalization" of the technique in a way that discourages, marginalizes or "handcuffs" the prospective owner-builder (be they male or female, groups or individuals). . . . I want to make it clear that I see no evil conspiracy by the "professionals" within the straw-bale revival to "take it over." Like the rest of us in the revival, they gratefully acknowledge our debt to the Sand Hills homesteaders. They are innovating creatively within the load bearing realm, and will undoubtedly continue to make contributions that will benefit the revival in general. . . . I will probably use some of these innovations myself. But deep down inside I will continue to be greatly nourished by stories like that of the Scott family [who moved into their house as soon as the roof, doors, and windows were on and spent their first winter with exposed straw, inside and out, finishing their home as money and time allowed] and to feel the "rightness" of the intuitive, uncomplicated methods used in the Sand Hills of Nebraska, especially for cash-challenged people anywhere who have no decent shelter.[19]

The straw-bale community is concerned that the very factors that have contributed to the success of straw bale as a sustainable system have the potential to make it into a commodity. Some feel that commercial application of the techniques would be a welcome sustainable asset in contributing to less deforestation and energy consumption. The more cautious, such as Myrhman, worry that the tradeoffs may come at a great price, indeed at the price of the community building network that is part of what makes straw-bale building not just another building technique. In some parts of the country, straw bale has become so accepted that alternative building is regarded as "something other than straw bale," while some R&D officials at the Department of Energy feel that the method is still too labor intensive and fraught with moisture issues to be taken up by the commercial building industry, as entrenched as they are with the timber companies. Whichever path the future brings, it will probably not be a clear one in either direction, but pockets of one or the other, depending on local networks. This is a movement in which networks are everything, and the practitioners recognize it.

NETWORKS INSIDE NETWORKS INSIDE NETWORKS

I have traced some of the components of the straw-bale movement, the network of people, things, and intangibles that are created by and create a network. Even so, I have only scratched the surface. Basic components that helped initiate the contemporary revival were baling technology, with its own developmental history, conjoint history with historic straw-bale building, and more recent developments in agriculture that have kept the small square bales available.[20] There were also the folklorist's account of the Nebraska Sand Hills straw-bale homes, a one-page publication, an architect's experiment in another publication, and some people. That network alone did not launch the straw-bale movement. Even more crucial was the ethos of the time that had people primed with various economic, cultural, and political/ecological reasons to try something different—the elements, essentially, of the "green building" movement. That the first experimental straw-bale buildings of the revival period were built in New Mexico and Arizona, where a precedent of vernacular architecture without architects had been set with adobe, also was no accident. Moreover, owner-builders initially building "off the grid" allowed for crucial experimentation that spawned the movement. Owners' financial and skill resources and the flexible appeal of straw bale to both aesthetic and environmental values also were factors. While diverse, this is a very conscious movement, spiced with activists who may have been part of other politically and/or ecologically motivated movements and who recognize the need for communication networks. One farmer dedicated to raising straw for home building had been a member of Students for a Democratic Society and then a grower of organic food before moving to "growing houses," as he says. At the same time, networks in the building, finance, and insurance industries have also played parts. Recounting the problems some of his early clients had in obtaining financing for straw-bale structures, an Austin contractor pointed out that after the first person had obtained a loan in Arizona because she was an excellent risk and had put up most of the money herself, other loan officers became more willing to consider loans for straw-bale building. They trusted other loan officers and would consult with them, even if they were out of state and working in competitors' firms. The same was true of the insurance industry. Once the first policy was made available for a straw-bale structure, others followed suit. Hence, from the very beginning the straw-bale building network, starting with workshops and continuing with informal communications via all available media (telephone, the Internet, publications, word of mouth) could make not only technical information available; it could also make names of straw-bale-friendly loan and insurance officers available to other owners, builders, and owner-builders.

Code building worked in a similar way. As the above quotes have shown, political as well as technical support had to be marshaled to get the first codes passed in Arizona. Activists in California then used that accomplishment accompanied with historical information for building a political network sufficiently strong to override the building officials, while Austin used the Arizona experience. Yet even while they recognize this, movement activists are worried about conveying consistent standards in the codes so that those outside the network do not build in ways that could damage its reputation. If the future brings change for straw-bale building, it will be because the networks supporting it have changed. As an early activist and a current good-will ambassador for straw bale to the world at large,

Myhrman worries about new commercially oriented members. Tensions already exist in the network, such as those over using straw bale for "living lightly on the land" as a model for the future while its very flexibility allows luxury homes, speculation homes, and models of sustainability to employ it. This suggests a proliferation of more networks in the future, some more tightly linked than others. This does not mean the demise of the simple owner built straw-bale home. The network to support that kind of building has the potential to remain as long as there are a few with Myrhman's philosophy: ". . . whatever happens as the revival grows in scope and impact, I'll be somewhere in the stubble working to keep those uncomplicated methods available to those who would choose to use them."

NOTES

1. In addition to the specific works cited below, this paper draws on the following: Wiebe Bijker, *Of Bicycles, Bakelites, and Bulbs* (MIT Press, 1995); Louis L. Bucciarelli, *Designing Engineers* (MIT Press, 1994); Joan Fujimura, "The Molecular Biology Bandwagon in Cancer Research," *Social Problems* 35 (1988), 261–283; Bruno Latour, "Where are the Missing Masses? The Sociology of a Few Mundane Artifacts," in *Shaping Technology/Building Society,* ed. W. Bijker and J. Law (MIT Press, 1992).

2. John Law, "Technology and Heterogeneous Engineering: The Case of Portuguese Expansion," in *The Social Construction of Technological Systems,* ed. W. Bijker et al. (MIT Press, 1987); Michel Callon, "The Sociology of an Actor-Network: The Case of the Electric Vehicle," in *Mapping the Dynamics of Science and Technology* (Macmillan, 1986); Bruno Latour, *Science in Action: How to Follow Scientists and Engineers through Society* (Harvard University Press, 1987).

3. See all references listed in the preceding note.

4. Eugene Ferguson, "The Mind's Eye: Nonverbal Thought in Technology," *Science* 197 (1977), 827; Ferguson, *Engineering and the Mind's Eye* (MIT Press, 1992); Bruno Latour, "Visualization and Cognition: Thinking with Eyes and Hands," *Knowledge and Society: Studies in the Sociology of Culture Past and Present* 6 (1986), 1–40; Gordon Fyfe and John Law, eds., *Picturing Power: Visual Depictions and Social Relations* (Routledge, 1988); Kathryn Henderson, "Flexible Sketches and Inflexible Data Bases," *Science, Technology and Human Values* 16 (1991), 448–473; Henderson, "The Political Career of a Prototype," *Social Problems* 42 (1995), 274–299; Henderson, "The Visual Culture of Engineers," in *Cultures of Computing,* ed S. Star (Blackwell, 1995), 196–218; Henderson, *On Line and On Paper: Visual Representations, Visual Culture, and Computer Graphics in Design Engineering* (MIT Press, 1998).

5. Jean Lave, *Cognition in Practice* (Cambridge University Press, 1988); Lucy Suchman, *Plans and Situated Actions: The Problem of Human-Machine Communication* (Cambridge University Press, 1987).

6. The term "tacit knowledge" was coined by Michael Polanyi to describe the personal way of knowing that informs the explicit knowledge characteristic of science. Such knowledge is present in the creativity of laboratory practice and the passion of discovery. See Polanyi, *Personal Knowledge* (Routledge and Kegan Paul, 1958) and *The Tacit Dimension* (Routledge and Kegan Paul, 1967).

7. Harry M. Collins, "The TEA Set: Tacit Knowledge and Scientific Networks," *Science Studies* 4 (1974), 165–186.

8. Athena Steen, Bill Steen, David Bainbridge, and David Eisenberg, *The Straw Bale House* (Chelsea Green, 1994).

9. *House of Straw* (US Department of Energy, 1995).

10. Roger Welsch, "Baled Hay," in *Shelter* (Shelter, 1973).

11. Matts Myhrman and Judy Knox, "A Brief History of Hay and Straw as Building Materials," *The Last Straw* 4 (1993), fall, 1–8, 18.

12. Today scale is again the issue in baling hay. Throughout Texas and the grain belt, technology producing large round bales predominates over that producing rectangular bales suitable for building. Just as the demand of stables for the patented "flake" bale (shipped by rail during the Industrial Revolution) supported the marketing of the smaller baler in the 1880s, the continued demand for baled straw for animal bedding has kept up square-bale production when the large mechanized farms have turned to round bales and (in some cases) large square bales. The large scale that almost kept the baler from succeeding as a commercial enterprise is now the rule. Ironically, a search of patent documents reveals that round-bale technology was introduced in the late 1800s but did not succeed commercially in its small-scale form.

13. The only early straw-bale home outside Nebraska was built by Warren Withee near Alsen, South Dakota. Withee, who had only one arm, used flax as well as meadow hay bales and mortared between them. Still standing, it is currently being used for storage. In 1925, the Martin-Monhart house in Arthur, Nebraska was constructed; it was originally plastered with mud, but re-stuccoed with cement by new owners in 1930. It is currently a museum. Arthur is also the site of a Pilgrim Holiness Church built in 1925. The bales used to build it were pressed at a homestead nearby using a stationary baler, hand tied, and carted to town by mule. The church, 28 by 50 feet in size, including four rooms used for living quarters, saw active use for 35 years and is now a museum.

14. Bernard Rudofsky, *Architecture without Architects* (Doubleday, 1964); Art Boericke and Barry Shapiro, *Handmade Houses: A Guide to the Woodbutcher's Art* (A&W Visual Library, 1973).

15. Steen et al., *The Straw Bale House*.

16. David Eisenberg, "Code Officials Speak Their Minds," *The Last Straw* 14 (1996), spring.

17. Leroy Sayre, "One from the Top," *The Last Straw* 14 (1996), spring.

18. Norm Ballinger, "Permitted in Austin," 1996.

19. Matts Myhrman, "Ruminations of a Straw-Bale Romantic," *The Last Straw* 14 (1996), spring, 15.

20. They are potentially in danger of extinction, as large growers have moved to the large round bales and large square bales. These are not new options, but rejected options of the past before automation of agribusiness.

THE WIMBERLEY HOUSE OF HEALING

MARLEY PORTER

Feng shui (wind water). Chi (energy). These are words and concepts not often understood or practiced in Western design philosophy, yet these functional and meaningful realities are every bit as important to the building as nails and glue and bricks and mortar.

Just as the human brain attempts to maintain a healthy equilibrium balancing the right (creative) and left (practical) hemispheres of thought, the boxes we live in should be no less balanced. Good design weighs equally the physical and the metaphysical, exposing an architecture that transcends the box. Great architecture may be found in the celebration of the marriage of Eastern and Western ideologies, the living together of the rational (mind) and the emotional (heart).

Far too many homes and offices in North America lack meaning. They are empty of soul and void of divinity. America's preoccupation with speed, modularity, cost, and size has evolved a sea of insanely proportioned and grossly unsustainable "tract houses with thyroid problems."

Far too many buildings have been "facadomized"—decorated with stylized facades in two-dimensional screaming matches with all the other billboard, paper architecture. Far too many buildings are built with toxic and unsafe materials, most utilizing dated construction techniques and systems that are costly to the environment and consumptive of natural resources. It is no wonder that our buildings contribute significantly to health problems and emotional unease.

The end result of any given project must transcend simple profit measures in order to get the biggest bang for the buck or to develop to the highest and best use possible. The Western fiscal invention of immediate gratification might be elevated to a more long reaching and ultimately more rewarding structure of balanced living; truly, a "higher" use.

FIGURE 1

Exterior of Wimberley House of Healing. (Living Architecture)

To create a living architecture there must first be holistic intent. The intent of the designer or architect, the owner, the builder, the subcontractors, the workers, and even the material suppliers engenders the architecture with meaning. The making of "conscious architecture" magnifies the intelligent intent of the makers.

Living architecture is about the joining of pieces, the making of magic through the marriage of the many into the *one*. This holistic approach to design unifies the distinct and individual energies and allows them to evolve into a synergistic whole.

Through the wisdom of its "owner," the openness of her architect, and the many sacred contributions of the many builders, craftspersons, and suppliers, the Wimberley House of Healing has evolved into a wonderful example of awake and conscious architecture.

It is more than a house. It is more than a building or structure or shelter. While the quiet retreat accepts its outward appearance as a work of art, an artistic interplay of light and motion and meaning, the Wimberley House of Healing is much more indeed. The home has risen to become "a

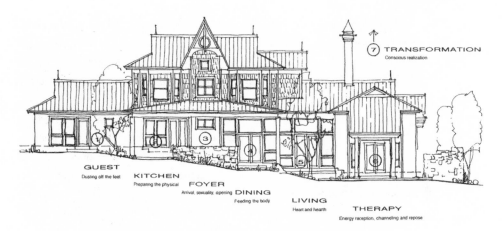

FIGURE 2
Elevation of Wimberley House of Healing. (Living Architecture)

living architecture," a breathing, moving, dancing thing that wraps and plays about its occupants like a mother or a lover, with caring open hands lifting through layer upon layer of overlapping spaces, fields of energy, and conscious thought.

All good architecture begins in the ground. The site in Wimberley, Texas slopes gently southwards to Lone Man Creek. At the water's edge, giant 20-ton limestone boulders cantilever over cool, green pools. The water falls from a strong, small dam and echoes in lapping ripples, slowly carving the stones' underbelly. This place is already and always a living architecture.

And so the site and the energy of the cedars and the oaks all lent their voices to the carving of the thing.

Living Architecture emerges from the mundane and the simple, the quiet things: straw and mud and spit and sweat. The Wimberley House of Healing is constructed from bales of straw gathered from fields not 3 hours distant from the site. Instead of being burned off, as with most wheat chaff, the straw is baled by a farmer who knows its intended use—the construction of a special home—and he thinks of this as his machine bales the straw.

And the bales of straw too seem to think for themselves. The bales and the height of two steps at a time, each and together being a lazy 15 inches tall, allow the foundation to cascade in perfect cadence with the slope. From the top then to the bottom of the home six levels emerge and the seventh rises up and out of the building; a conscious departure from the mundane into the divine.

When a human body is laid out over the floor plan of the home a surprising (not really!) coincidence is noted: the energy chakras of both body and home overlap. The feet are elevated, as we would recline after a brisk walk, into the guest room where the dust of the feet of our guests is wiped clean; the lower chakras.

The entry to the home, the entry to the body, the sexual opening faces east, faces rebirth, rejuvenation, the coming light and opens triumphantly, liberated into a three-story skylit foyer, a place of making decisions. "Ah now, which way will I go?" In this, the central chakra, yin and yang, female and male, curvilinear and angular meet and mingle and merge. Balance in all things is suggested as one enters, on the ground, over stone, humble and small but looking up and up through the glass to the light, to the sky, to the stars; infinite and finite at once.

The dining room is under the stomach where we are reminded to nourish the body and the soul, where we are fed real food of organic design, where we are designed of organic reality. The dining room opens westward to a sheltered patio and court. Set deeper yet than the entry, the dining room serves an even deeper holistic need, that of nourishment. Like King Arthur's court, the dining table is round, by choice, to not distinguish a head. All are one.

As the floor steps down toward the creek, the volumetric energies of each upper space flow down into and mix with the next. Like a brook with small waterfalls, the chi cascades down and eddies into each space, magnifying the cumulative whole.

Functioning under the same principals as a dam, a curved, twelve-foot tall limestone wall captures the accumulated energies and holds them back. The living room level is cradled against the stone wall as is the fireplace, the heart, the hearth of the home. Sitting comfortably in the light colored fabrics of the couches and chairs the energies are breathed in deeply, almost like being able to breathe under water.

A 4-inch slit of rose-colored stained glass, centered perfectly with the home's geometry, functions as a laser beam focusing the built up energies into the lowest space, the Transformation Room. The creek and ponds just feet away from this level serve as a deep gravity field, a well of energetic repose.

The Transformation Room itself is a chapel. Based on the Golden Section and natural geometric progressions, an inspiring harmony of perfect proportions maintains the high vibrational level of the space. Having a

FIGURE 3
Interior of Wimberley House of Healing. (Living Architecture)

ceiling height of almost 18 feet and a floor plan of about 12 by 16 feet, the space lifts one's vision and spirits up through five clerestory windows.

The proportions of the space are divided into sacred numerical divisions: five windows, seven spaces, a pair on the east wall depicting the dialectic separation of heart and mind, three windows on the south suggesting union of mind and heart into soul. Three niches just below the three windows hold pictures of Christ the Healer and flanking angles to assist in the work.

Throughout the home, elemental pairs are set up to demonstrate the concept of "The Space Between." Double columns flank each space and drop in the floor. Double steps on either side of the rooms allow movement against walls. Furniture, like the human bodies they support, float always away from the walls (and the sliding, slicing energy washing them) bathing in the eddies of space in the middle of the rooms.

The naturally finished, non-outgassing, plywood-paneled ceiling, with simple wood battens, maintains a level and horizontal plane, perhaps alluding to the even emotional plane from which one might realize a better perspective. The floor falls ever lower toward the creek, beneath the ceiling, from pool to pool, from space to space, again, always two steps at a time. Each room is elevated in hierarchical progression in space and volume and time.

Human beings live between the ceilings and the floors of our homes and our offices, between the walls, between the outside and the inside. We live always in the space in between. The best architecture is that which is carved from living space. What goes on a wall is never as important as what goes on between the walls.

The Wimberley House of Healing is owned by a clinical psychologist and energy therapist who uses music to open the chakras of her clients. Those in need of healing, adjustment, opening, containment, and enlightenment find themselves pried open by the vibrations of the music and the power of the energy of the place. Often, they find themselves between what appears to be and what really is; this, the real reason they are here, allows the self to heal itself.

A room of repose lies on the same level as the Transformation Room looking out under a covered patio directly in line with the waterfall. This suite functions as a decompression space encouraging connection to the landscape. Gravel paths lead from the porch to the water's edge. Limestone steps continue the cascading.

The Wimberley House of Healing uses natural and holistic building materials to augment the spirit of health and well-being. Two-foot-thick straw bale walls, with an R value around 50, allow the windows to be set deep, creating window seats and ledges. The walls are finished with a lime based natural stucco that allows them to breathe. This keeps the straw dry and encourages passive air exchanges, critical in maintaining healthy living spaces. The walls are also virtually sound proof and contribute to the feeling of retreat.

Stained concrete floors seem to be calm pools of water and do not outgas. The floors also function as thermal flywheels, storing kinetic energy to help keep the home at an always comfortable temperature.

Cedar beams, door and window trim, and stair rails are cut from trees that were removed from the site. The aromatic wood purifies the air and brings the outside inside. Again, the interplay of one plus one being three is manifest.

Rainfall is captured off the "Galvalume" metal roof and delivered underground from gutters and downspouts, integral to the architectural expression, to two 30,000-gallon cisterns. From these, the home receives the majority of its drinking water (after filtration and ultraviolet exposure for purification) and all of the water for landscaping.

Ultimately, the power of the place transcends even the understanding shared between owner and architect and embodies the positive spirit of intent of the many crafts-persons, the carpenters, the plumbers, the electricians and those supplying materials. A truly magical and peaceful energy reigns within the playful but elegant spaces and mystically, the energy focused in the Transformation Room encourages anyone so lucky as to be laying here, receiving the work, to be moved above and beyond the confines of architecture to the seventh chakra, the transcendental to contemplate one's connection to everything else and to the *one*.

PORTRAIT OF INNOVATION: DAVID HERTZ

MARTHA DAVIDSON

Concrete, used in ancient Rome, is one of the oldest building materials created by human invention. It is also one of the newest and most versatile, thanks to the ingenuity of David Hertz, an artist and architect whose work embodies his concerns for craftsmanship, the preservation of the natural environment, and social responsibility. Hertz is the president of a multidisciplinary design firm, Syndesis, and the inventor of Syndecrete, an advanced cement-based composite that incorporates post-consumer products destined for landfills—almost any kind of ground-up glass, metal, wood, or plastic components, from pencils to plumbing fixtures—to create customized surfacing materials of outstanding quality and beauty.

The fusion of the functional and the aesthetic grew out of Hertz's childhood experiences. Born in California, the elder of two brothers, he has spent nearly all of his life in Los Angeles. His father, an oral and maxillofacial surgeon, has always had a strong interest in sculpture and architecture, particularly the work of Frank Lloyd Wright. Hertz's mother is a painter and fine art photographer. "Architecture is in many ways a synthesis of both their professions, in that it's part science, part art," Hertz says. "Some of my earliest memories are of working with my father in the garage. Together we would repair things or work on projects. I remember pouring molten aluminum off the roof and into the swimming pool to create frozen shapes as sculptures. My mother says I was obsessed with disassembling and reassembling things, like the vacuum cleaner. I used to enjoy building models and coloring. I even won a national coloring contest sponsored by Kellogg's."

Hertz got his first hands-on experience in construction by assisting his father in ongoing renovations of their house. Then, as an adolescent, he worked as a construction laborer, an apprentice carpenter, and an artist's assistant. Through his enjoyment of surfing and backpacking, he was also developing a deep respect for the natural world. As Los Angeles grew, Hertz witnessed the impact of development: "I watched nature being overrun by urban sprawl, or by pollution and effluents from the city, and so I began to be involved with grass roots environmental organizations that fought to protect it." Care for the environment became a part of his philosophy.

While a teenager, Hertz had the opportunity to meet a disciple of Frank Lloyd Wright, the renowned California architect John Lautner, after watching a house by Lautner being constructed in his neighborhood. The clients, sensing Hertz's fascination with the building, invited him to lunch with the architect. "I had never been so impressed with anyone's work before," Hertz remembers. "Lautner, I think, was open to my enthusiasm, due to my youth. He told me that Frank Lloyd Wright liked to hire young people before they were spoiled by too much education about what architecture was supposed to be." Hertz began working for Lautner on a voluntary basis and later became a paid apprentice.

Lautner, who died in 1994, was known for designing houses with open, soaring spaces. He often used concrete because of its sculptural quality. Lautner emphasized the importance of a single, unifying concept as a basis for architectural design. "Working in his small office was indeed a privilege," Hertz says, "and an amazing opportunity to work closely with a genius."

It was fine art, however, that interested Hertz most deeply. After high school, he pursued this interest in a combined program of studies at the UCLA Extension School and at Santa Monica College. But gradually his focus shifted:

I'd always aspired to work as an artist, which means, from my perspective, a pure expression of creativity applied to some form. At the same time I've always been interested in the tactile and in creative problem solving. I entered into working with my hands through the construction end, so I began also be fascinated with the built environment and how it connects to the natural environment. I thought there was a real need to reconcile those two in a responsible way.

Deciding to study architecture, he chose an institution that, although relatively new, had already established a reputation for design innovation. The Southern California Institute of Architecture (SCI-Arc), located in Los Angeles, was not bound by tradition. "It doesn't really look toward the past for its sources of inspiration," Hertz says, "as much as allowing for creativity and creative problem solving. It doesn't paint narrow definitions of what the role of an architect is." That creative freedom appealed to him. He began to experiment with materials and grew captivated by concrete. "In architecture school I started to use concrete in an unusual way, to make models," he recalls. "That was looked at in a somewhat negative light at that point." Concrete became his medium of choice for projects including sculpture and furniture:

I started to play with concrete, based on the experience in construction. I started to make some furniture pieces. The first piece that I made using concrete was [an adaptation of] a coffee table I had inherited. It had a really nice glass top, but an ugly wood base. So I took some scrap wood and built a little form, and then I got a four-dollar bag of ready-mix concrete and poured it in my driveway. I stuck the glass into one end of the concrete, which then allowed for the glass to cantilever quite a striking distance.

Subsequently, I started to exhibit the furniture, really as functional art. That was in the early 1980s, when Memphis, the design studio in Milan, Italy, was getting a lot of attention, and the concept of art as doing functional things and architects doing art became possible.

While continuing his apprenticeship with Lautner, Hertz became associated with the Whiteley Gallery in Los Angeles, participating in several group shows. His first solo exhibition of furniture and drawings was installed in the historic Schindler House. It was an appropriate setting: the 1922 home of the Los Angeles architect Rudolph Schindler (who also had worked with Frank Lloyd Wright) was constructed of Schindler's preferred building material, concrete.

Hertz's training at SCI-Arc included a 5-month program of travel and study in Europe in 1982. There he developed an appreciation for classical and Gothic architecture. He also visited every building by Le Corbusier and studied with the Italian architect Mario Botta.

Returning to the United States for his fifth year of training, Hertz expanded his experience through an internship with another internationally acclaimed California architect, Frank Gehry. Winner of the 1989 Pritzker Architecture Prize, one of the profession's highest awards, Gehry is probably best known today for his design of the Guggenheim Museum in Bilbao, Spain. His projects defy traditional forms and materials in creations that are both functional and witty. Gehry's fundamental belief is that architecture is art, and he approaches it with the eye of a painter or sculptor. That approach resonated with Hertz, who worked on several Gehry projects in 1982 and 1983. Hertz also worked with a few fine artists, including Charles Arnoldi and Jean-Michel Basquiat.

Hertz completed his studies with a Bachelor of Architecture degree in 1983, obtained certification, and promptly started his own practice, which was incorporated in 1984. He named his firm Syndesis, a word meaning to link or connect. He was thinking not only about the physical connections of buildings but also about the various disciplines he wanted to pursue— architecture, furniture, and the development of materials and products. In his practice, they were all connected, and all based on the idea of sustainability. "Sustainability," Hertz explains, "means providing for the needs of the present without detracting from our ability to fulfill the needs of the future."

For the first 3 years, Hertz shared a small storefront studio space with his father, who, although still practicing medicine, was also engaged in sculptural projects. Hertz began working on commissioned pieces and public sculptures, occasionally collaborating with his father. He also continued his experimentation with concrete furniture:

There was a really good forum and a good economy for me to do commissioned furniture. So the furniture became little exercises, just like architecture. They had structural issues, clients, and budgets, they were little design problems, and they offered me the immediacy that architecture didn't afford me at that point.

While he relished the creative possibilities of concrete, he also felt its limitations. He was using commercial reinforced concrete. "As my furniture projects grew," he recalls, "and I started to do countertops and other things, I ultimately felt limited with the material, just wrestling with its sheer weight and fragility and the difficulty in tooling it, especially on site. So I began to think, if only I could make some of these pieces off site, in a controlled environment, and make them light enough that I could carry them, that should be the goal. That goal then led to, well, what can I do to lighten up the material?"

Concrete is a mixture of cement, water, and aggregates (which can be anything from gravel and sand to the kinds of recycled consumer products that Hertz uses today). The earliest evidence of concrete used in architecture dates back nearly 2,000 years, to Roman ruins such as the villa of Emperor Hadrian, built in the second century A.D. Roman concrete was made with a cement mixture of lime and pozzolana, a volcanic ash. After the fall of Rome, knowledge of concrete construction was apparently lost until late in the eighteenth century. In 1824, an English mason, Joseph Aspen, patented Portland cement, named after a fine building stone quarried on the Isle of Portland, off the coast of England. This cement was made from limestone, chalk, and shale, which were combined at extremely high temperatures in an industrial kiln.

Modern commercial concrete is made from Portland cement, water, and aggregates. To generate the 2,700°F temperatures needed to form Portland cement, high-energy wastes such as scrap tires and waste petroleum products (used motor oil, printing inks, solvents, and paints) can be burned instead of new fossil fuel. The intense heat entirely consumes the toxic substances. Thus the very manufacture of cement offers a means of reducing hazardous waste and landfill. Moreover, concrete normally is made from locally available natural substances on an as-needed basis and does not have to be transported to the building site.

Concrete also offers other advantages for building. It is naturally waterproof and fire resistant. It has great compression strength. Because it can store and re-radiate heat, it requires 70 percent less energy than wood to maintain comfortable indoor temperatures in a completed building. Chemically inert, it does not require any sealers, coatings, or preservatives. It can be colored in its raw state by the addition of natural pigments or other agents. It can also be molded into nearly any desired form, and the addition of aggregates offers an enormous range of textural possibilities.

In his experimentation with concrete, Hertz was in distinguished company. The use of concrete in architecture and even in furniture had been explored early in the twentieth century by the great inventor Thomas Edison as well as by Schindler and other architects. Edison, in fact, established a company to produce Portland cement, devised a new kind of industrial kiln, and made other improvements in the production process. Believing that low-cost cement housing offered a technical solution to problems of poverty, Edison patented a system for casting concrete houses from molds, but his plan for a city of such houses was never carried out.

As a material for furniture, however, Hertz found concrete too heavy. Searching for lighter alternatives, he began to devise his own mixtures. This was "almost more alchemy than chemistry at the beginning," he recalls. "I didn't want to work with synthetics to make the material so different from the qualities that I liked, the kind of humble qualities of the concrete. So I did a lot of research, meeting with sales people and basically doing empirical testing with different types of lightweight aggregates, mainly volcanic materials, hollow glass spheres, and other things to displace the material. Then I started to look at it as a composite, using a high percentage of fiber as a reinforcement, and kept making samples and then using them on limited projects. At that point, it was just me doing all the mixes and building the molds, pouring them, finishing them, and installing them. This was in 1983 to 1986." Eventually he hit on a formula that uses 40 percent recycled products, does not require much energy to manufacture, and results in a strong, light fiber-reinforced material that can be shaped with woodworking tools. Called Syndecrete (a registered trademark), it has a longer service life than wood, brick, or steel when used in construction, and it offers an alternative to nonrenewable materials such as stone, renewable resources such as wood, or synthetic solids made from petrochemicals. When its service life is over, Syndecrete can be crushed and recycled as an aggregate for new concrete or as fill for road beds. These qualities attracted the interest of other architects and builders, but what most appealed to clients was Syndecrete's terrazzo-like appearance.

Hertz's experimentation combined his environmental concerns with his search for pleasing textures and colors. He found that, rather than use sand or pebbles as aggregate, he could use a vast array of "the fossils of our landfill"—nuts and bolts, golf tees, plastic strawberry baskets, guitar picks, even polystyrene packaging. When broken up and mixed into Syndecrete, these familiar objects created striking patterns and textures.

Syndecrete, Hertz quickly discovered, filled a void in the market for architectural surfacing materials. The fact that it could be customized in form, color, and texture gave it wide appeal, as did its low environmental impact and its innate integrity. The Syndecrete furniture Hertz exhibited drew attention, and word of it was spread by articles and reviews in design and architecture magazines. Hertz began collaborating with other architects, some of them quite well known, who wanted to use Syndecrete in their buildings.

Struggling with the tension between art and commerce, Hertz began offering Syndecrete as a product on a custom-order basis. He found that product development afforded Syndesis a creative freedom that the firm might not otherwise have achieved. "The product idea," he explains, "became the thing that would allow us to practice architecture more selectively. Many architects have to run their practice as a business in the hopes that maybe they get one or two projects where they can really be creative. They end up participating in buildings that probably don't really deserve to be built. I felt I needed to keep architecture, for me, more sacred. But I wasn't really happy teaching full time, which a lot of architects do when they're more academic or more theoretical and idealistic. So the product became market led."

Hertz also struggled with the problem of protecting his innovation. Patent lawyers advised him that, although there were aspects of Syndecrete that were eligible for process or utility patents, such patents would be very expensive to enforce and would be susceptible to reverse engineering, particularly in countries where US patent law is not respected. Instead, Syndesis chose a strategic path toward trade secrecy and made Syndecrete a proprietary product.

As Syndesis grew, Hertz moved the business to a much larger space in Santa Monica, near SCI-Arc. But he found it difficult to manage both the creative and the administrative aspects of the firm. Fortunately, a friend in San Francisco had given Hertz's name to Stacy Fong, a recent graduate of the University of California School of Architecture at Berkeley, who was moving to Los Angeles. Fong did not contact Hertz for a long time. She had read about his work in an architectural magazine and thought that anyone that young and successful would likely also be very arrogant. But when they did eventually meet, Hertz says, "I knew that she was either the ideal employee or a great friend or a mate. And I got all three." A strong friendship grew between them when she joined Syndesis in 1989, and they were married in October of 1990, on Hertz's thirtieth birthday. Today they have three young children. Fong, who is a partner in Syndesis, handles all the administrative responsibilities, leaving Hertz free to pursue the creative end.

The Syndesis office occupies a 25,000-square-foot compound in Santa Monica, the site of a former foundry. The firm has fifteen employees, including two other architects, a research and development associate, five craftsmen, a shop foreman, and marketing and administrative personnel. Additional temporary employees are hired as needed. Syndesis employees do all the fabrication themselves, and the market has always been much larger than their ability to fill it. "I haven't been interested in being a big manufacturer," Hertz says. "I'm much more interested in invention and creativity. Every job is like reinventing the wheel. That's where the fun is, and the challenge."

Syndesis does not have a sales force. However, because Syndecrete is topical and appealing, the firm has received extensive editorial and national television coverage. People often contact the firm, wanting to know more about its products and services. A number of its clients have been major corporations, including Microsoft, Sony, and Rhino Records. Some projects have afforded Hertz an opportunity to put his commitments to environmental responsibility and to social responsibility into practice.

One Syndesis client, the author and entrepreneur Paul Hawken, co-founder of Smith & Hawken, inspired Hertz to view business as a vehicle for fundamental social and economic change. Hertz joined a group called Businesses for Social Responsibility, of which Rhino Records was also a member. Syndesis was commissioned to create flooring for Rhino's new headquarters and was able to use Rhino's connections with inner-city gang intervention programs to hire members of two gangs as an essential part of the project work force. As an aggregate in the Syndecrete flooring, Hertz planned to use about 15,000 recycled or unsold records, CDs, and audio and video cassettes. These products had to be pounded into small bits before they could be incorporated into the composite. It was shortly after the Rodney King riots in Los Angeles, Hertz reports, "so we actually got two rival gangs to work together to take their aggressions out on the Monkees and Partridge Family and other records . . . and that is evident in the final product." He reflects that "it was really a case of doing good by incorporating the company's waste stream in an environmental way and also in a social way, trying to solve problems."

Creative problem solving is what most appeals to Hertz, but it is not easy:

It's been a constant uphill battle. I think we've chosen one of the most difficult paths, the path of innovating something and then learning how to manufacture it where there is not really existing technology. Having to invest lots of time and energy in adapting equipment and manufacturing processes to suit our need, that was certainly an initial hurdle. Understanding the fundamentals of business and pricing was another one. If you were just making a repetitive product, the cost accounting would be much easier. I think we've reconciled those first two obstacles. Ultimately, although there was a tremendous initial market enthusiasm, I had to make sure we did all the proper independent third-party testing. If a large corporate architect was going to specify our material, and it was new, they would want to see a proven track record of installed jobs, which we could not yet provide. I think we've overcome that, too, but it was another hurdle.

We lost the opportunity for some very large contracts because we were just too new.

Hertz is developing a concrete product that makes use of some of the 4 billion pounds of carpet that is dumped in landfills each year. It would be a mass-produced product on a very large scale, and he has been discussing a strategic alliance to manufacture and distribute it with a large public company that has a strong environmental ethic and an international reach, both in production and marketing. Hertz envisions using community-based groups in emerging countries in the production process. An alliance with a large company would free Hertz to concentrate on innovation and initial product and process development, to be selective in his projects, and to retain his high standards of craftsmanship.

The Syndesis web site (www.syndesisinc.com), which includes an extensive bibliography and lists of clients, exhibitions, and awards, provides abundant evidence of how much interest Syndecrete has generated. Hertz is proud that his work has been recognized by the Museum of Modern Art in New York, which included Syndecrete in a 1995 exhibition titled Mutant Materials in Contemporary Design. But his creative endeavors have extended far beyond this one product. His designs range from household items (e.g., flower vases made from inverted, refinished road cones) to houses, including his own, that integrate the thermal properties of Syndecrete with passive solar heating.

Hertz and his staff know Syndecrete's capabilities so well that they do not have to offer standard items. "If we're going to do a bathroom," Hertz explains, "we don't have to pick a bathtub from a catalog, we can actually design it and fabricate it ourselves." This results in buildings and components that are not only rational, functional, and beautiful, but truly unique.

Hertz is also active in professional and environmental organizations, lectures widely, and serves on the faculties of the UCLA Extension School and SCI-Arc. At SCI-Arc he is involved in a new academic-industrial partnership for developing low-environmental-impact materials. In all that he does, Hertz is mindful of Albert Einstein's observation that "we cannot solve today's problems with the same level of thinking that created them." In a 1997 address at the Barnstall Municipal Art Gallery in Los Angeles, Hertz challenged other designers, architects, and inventors:

Those of us interested in the making of things must help shape the future. Products either should be designed for longevity or for ease of disassembly, re-manufacturing and recycling and biodegradability. . . . Products including buildings should be designed to allow for expandability, adaptability and reuse, with an appreciation for the long view. Buildings are not "finished" when constructed, they are just starting a new life.

Hertz, in that address and through his own multi-faceted work, has called us to "add to the three R's of Reduce, Reuse, and Recycle . . . Restore, Reinterpret, and Reincarnate."

Whether we live in the city, the country, or somewhere in between, the
quality of our water and air are critical to our existence. The historian Martin Melosi points out that one of the primary ways in which cities have
addressed this concern is by taking direct responsibility for it through the
construction of systems. Even before the true cause of disease was understood, he points out, people made a connection between filth, or "miasmas," and epidemics. That connection led to schemes for the delivery of
clean water and the elimination of wastes. Melosi focuses on the supply of
clean water to city dwellers through the construction of municipal waterworks and water filtration systems. Today's water and wastewater infrastructure, he argues, is the legacy of the "Age of Miasmas."

But what if one doesn't live in a city or town where that infrastructure
is already in place? This is the concern of the inventor Ashok Gadgil, whose
childhood in India provided opportunities to witness the effects of unclean
water on people's health. Though trained as a physicist, Gadgil nurtured a
growing interest in the problem of how to disinfect drinking water in poor
communities. As a result, he created the UV Waterworks, a portable device
that uses ultraviolet radiation from fluorescent light to rid water of disease-
causing bacteria and viruses. The UV Waterworks is designed to be inexpensive, to require little maintenance, and to be made primarily of
off-the-shelf components.

The epidemiologist Devra Davis also understands the costs of pollution
to personal health. She was a child in the Pennsylvania coal-mining town
of Donora when a "killer smog" struck there in 1948, causing illness in
about one-third of the town's residents. Davis's later academic interests in
environmental policy, toxicology, and epidemiology came together in her
ground-breaking work on identifying environmental causes of cancer. Her

studies of increased rates of breast cancer among women living on Long Island helped demonstrate the connection between disease and chemicals in the environment.

HOW BAD THEORY CAN LEAD TO GOOD TECHNOLOGY: WATER SUPPLY AND SEWERAGE IN THE AGE OF MIASMAS

MARTIN V. MELOSI

[handwritten margin note: "service delivery" "hidden function" of gov't]

"Delivering services is the primary function of municipal government. It occupies the vast bulk of the time and effort of most city employees, is the source of most contacts that citizens have with local government, occasionally becomes the subject of heated controversy, and is often surrounded by myth and misinformation. Yet, service delivery remains the 'hidden function' of local government."[1] So stated the political scientist Bryan D. Jones and his collaborators in a 1980 book. Service delivery is a "hidden function" largely because it often blends so invisibly into the urban landscape; it is part of what we expect a city to be. While economic forces are essential to the formation of cities in the United States, urban growth depends heavily on service systems which shape the infrastructure and define the quality of life.

In my book *The Sanitary City* (Johns Hopkins University Press, 1999), from which this essay is derived, I focus on sanitary services—water supply, wastewater, and solid waste systems—because they have been and remain indispensable for the functioning and growth of cities. For the purposes of this essay, I will concentrate on water supply, and to a lesser extent, wastewater. Sanitary services are not organic entities; they are specialized technical systems—technologies of sanitation—that help to shape the apparatus of modern cities.[2] The development of technical networks in the nineteenth century was a prime characteristic of the modern city. Whereas industrialization remained local or regional for many years, new technological innovations were quickly diffused nationally. This suggests that, although American cities did not uniformly benefit (or suffer) from the direct economic effects of the Industrial Revolution, they were physically modernized as a result of new technical systems developing in the era. By the late nineteenth century, many cities in the United States had entered a period of dynamic system building in a number of areas, including energy, communications, transportation, and sanitation.

In the mid to late nineteenth century, when sanitary services became essentially municipal responsibilities, the decision to choose between available technologies was informed by the prevailing environmental theory of the day. Before the twentieth century, when the initial technologies of sanitation were implemented, the miasmatic (or filth) theory of disease strongly influenced choice. From the 1880s through the end of World War II, choices were informed by bacteriological theory. Sometime after the war, new theories of ecology broadened the perspective of sanitary services beyond the health outlook. These health and environmental theories were sufficiently widespread to constitute environmental paradigms.

The "Age of Miasmas" began in the seventeenth-century, when American cities faced poor sanitary conditions and suffered the crippling effects of epidemic diseases with only a vague understanding of their cause. By the nineteenth century, a few larger cities had developed community-wide water-supply systems with rudimentary distribution networks, but continued to regard waste disposal as an individual responsibility. The powerful worldwide influence of the English "sanitary idea" in the middle of the nineteenth century, however, linked filth with disease, and provided a clearer rationale and newer strategies for improving sanitary services in England and beyond. In his *Report on the Sanitary Condition of the Labouring Population of Great Britain* (1842), the English lawyer and sanitarian Sir Edwin Chadwick took a bold stand on the need for an arterial system of pressurized water that would combine house drainage, main drainage, paving, and street cleaning into a single sanitary process. Although this remarkable hydraulic system was never implemented, nineteenth-century English sanitarians and engineers became the leaders in setting standards for water and wastewater systems throughout Europe and North America.

English theories of sanitation helped to provide the environmental context for augmenting new technical systems in the United States and elsewhere. The development of North American water and wastewater systems in the mid to late nineteenth century depended heavily on the expertise of English civil engineers and English public health leaders, on the implementation or adaption of English sanitation technology, and on the absorption of English environmental values.

Rapid population growth and the proliferation of cities produced great potential breeding grounds for disease and increased the need for improved health and sanitation measures. Though no American city grew as rapidly as London, English sanitary reform attracted a receptive audience in the United States. The sanitary idea was persuasive because it became easier to compare urban problems after 1830 than it had been in a previous era of limited urban growth.

Cholera

Charles E. Rosenberg, in his classic study *The Cholera Years,* acknowledged the transformation in American thinking about disease that occurred in the nineteenth century. Focusing on New York's cholera epidemics of 1832, 1849, and 1866, he observed: "Cholera in 1866 was a social problem; in 1832, it had still been, to many Americans, a primarily moral dilemma. Disease had become a consequence of man's interaction with his environment; it was no longer an incident in a drama of moral choice and spiritual salvation."[3] The change in mindset was gradual and not entirely conscious, but the days of viewing disease as God's wrath were passing.

Throughout much of the nineteenth century, however, it remained simplest to blame the poor, the infirm, or members of non-white races for the scourge of epidemic disease. Rosenberg further noted that "when in the spring of 1832 Americans awaited cholera, they reassured themselves that this new pestilence attacked only the filthy, the hungry, the ignorant."[4] Newly arriving immigrants raised the greatest fears, especially when they were crammed into grimy and dilapidated housing. Ironically, cholera— "the poor man's plague"—made victims of the very people accused of breeding the disease. In New York, blacks and Irish immigrants were the most frequent casualties. Persons born in Ireland accounted for more than 40 percent of the deaths in New York. In Philadelphia, the case rate was nearly twice as great among blacks as among whites.[5]

In southern cities, cholera also was considered a class disease, and most especially a race disease. As Howard N. Rabinowitz noted, "although poor whites could be found near industrial and other unpleasant sites, the alleys and rear dwellings of the cities were almost entirely the province of the blacks." These were areas where cholera lurked, and poor housing often meant high mortality rates. In Richmond, Nashville, Atlanta, and other southern cities, cholera appeared first in black neighborhoods.[6] Local governments in Charleston and other southern cities emphasized social cohesion as a major objective. Thus, poor health conditions threatened all citizens, and public and private funds were intermingled in an attempt to develop an effective health-care system. Some historians have even argued that health and disease-control facilities were generally more advanced in southern cities. Disease was "a constant companion" there, since freezing temperatures that killed bacteria and viruses arrived so late in the autumn. Nevertheless, fighting epidemics often was not successful, especially since an understanding of contagion was so primitive.[7]

Yellow fever, unlike cholera, spared blacks more than whites. People of West African extraction suffered least. In the great yellow fever epidemic of 1878, only 183 of 4,046 victims in New Orleans were black, even though one-third of the city's population was black. In Memphis, where at least 14,000 of the 20,000 people remaining in the city through the epidemic were black, only 946 of the more than 5,000 yellow fever deaths in the city came from the "colored population."[8]

Anti-contagionism was eventually discredited, but its widespread adoption in the nineteenth century was a victory for empiricism and rationalism over sermonizing and moral outrage. Environmental sanitation appealed to simple logic and to the senses, offering a way for people to participate directly in cleaning the cities, and ostensibly to eradicate disease. That it misrepresented the root cause of disease was a serious (sometimes fatal) flaw, but its call for the removal and disposal of waste materials was a worthy objective.

In the wake of England's sanitary idea and Chadwick's report, the miasmatic theory of disease "emerged into practical vitality" during the 1850s. Based on empirical observation of the relation between filth and disease, the theory at first was somewhat crude, inferring that organic decomposition *per se* caused disease—i.e., that no specific relationship existed between a particular kind of decomposing substance and a particular infirmity. Eventually filth was recognized as the medium for transmitting disease instead of the primary source of contagion. This perspective provided a bridge to the eventual acceptance of the germ, or bacteriological, theory.[9]

Although the germ theory was not firmly established until after 1880, in one form or another the idea of contagion had been circulating since the sixteenth century or earlier. It was, however, incorrect in detail until Louis Pasteur and Robert Koch clearly linked a specific organism with a specific disease. In 1871, an advocate of the germ theory was severely criticized in *Scientific American* for postulating that yellow fever was caused by a living organism. Only a few years later, during the epidemics of 1878 and 1879, the belief that a germ caused yellow fever was more widely accepted.[10]

Controversy over contagionism in the middle of the nineteenth century, especially the competition among theories dealing with the generation of

living particles, made it easier for anti-contagionism to find a cordial reception.[11] The first two of four major cholera pandemics in the nineteenth
century won converts to the anti-contagionist cause. Erwin H. Ackerknecht
perceptively argued that anti-contagionists were "motivated by the new
critical scientific spirit of the time," while contagionists supported old theories that seemed never to have been carefully examined.[12]

environment over
personal
hygiene)
The emphasis on environment over personal habits of hygiene in Chadwick's report set the tone for the strategies to be employed in combating
disease and improving sanitation for much of the remaining century. Perhaps the earliest graphic example of Chadwick's influence in the United
States was *The Sanitary Condition of the Laboring Population of New York,* an
1845 study published by New York City Inspector John H. Griscom. *The
Sanitary Condition* began as an addendum to Dr. Griscom's first report as
City Inspector. That 1842 document, titled A Brief View of the Sanitary
Condition of the City, included a commentary with particular emphasis on
the state of the poor. New York's aldermen were so displeased with
Griscom's characterization of the city's sanitary condition and with his call
for preventive action that they chose not to reappoint him. Undaunted,
Griscom expanded the commentary into a small book with a title reminis-
cent of Chadwick's own report.[13]

The Sanitary Condition—the first in-depth study of health problems in
New York—ranged over many topics. It called for the expansion of the
public water supply, the construction of an underground sewer system, and
the development of a program of street cleaning and refuse removal. The
study was infused with the environmental view of disease and was based on
what the historian John Duffy called "the general acceptance of the symbi-
otic relationship between physical and moral health." Griscom, like Chadwick, decried the unnecessary physical evils that were responsible for
widespread sickness and premature deaths among the poorer classes and
that subsequently inspired moral decay. Prevention of disease was his greatest hope for reversing these ominous trends.[14]

Despite the dramatic disclosures, and the reformist momentum provided
by Chadwick and other European sanitarians, little immediately came of
Griscom's efforts. New York was infamous as an extraordinarily unhealthy
city with problems so complex and political webs so entangled that one
report could not change entrenched practices or reverse long-standing
policies of neglect.[15]

The earliest technologies of sanitation nevertheless began to spread to several American cities in an era of rapid urban growth after 1830, and especially in the wake of the English "sanitary idea" in the 1840s. Primary attention was given to water supply and, to a lesser extent, sewerage. Before 1800, most cities and towns had depended on a combination of water carriers, wells, and cisterns to meet their needs. Even during the first several decades of the nineteenth century, several larger cities and many smaller towns continued to rely on local sources of supply. Unless they hired water peddlers, each citizen used no more than 5 gallons per day.

water supply

Community-wide water-supply systems developed slowly in American cities. In 1801, Philadelphia had become the first to complete a waterworks and municipal distribution system, sophisticated even by European standards. The necessary health, economic, and technical factors converged to produce what became a model for future systems. Yet the Philadelphia waterworks also was something of an anomaly, since it did not spark an immediate nationwide trend. Concern for the health of the citizenry prompted the campaign for a waterworks in Philadelphia. Despite imprecision in determining disease causation in the late eighteenth century, the correlation between pure water and good health was nevertheless an early driving force in dealing with epidemics. *Scott's Geographical Dictionary* described the water in the densest areas of the city as having become "so corrupt by the multitude of sinks and other receptacles of impurity, as to be almost unfit to be drank."[16]

Phila 1801

Unsettled by ravaging yellow fever attacks in 1793 and 1798, political and business leaders formed a committee of the Common Council to deal with epidemics. The consensus view was that polluted water from wells and cisterns caused the fever, and that the city's private wells should be replaced by a community-wide system. Not only would the waterworks eradicate the disease; it also could be used to clean the streets, to provide fresh water for drinking and bathing, and to enhance the beauty of the city by supplying water to public fountains.[17] Attention to finding a new source of water had arisen before the Common Council's action. In 1789, after a yellow fever epidemic struck Philadelphia, Benjamin Franklin suggested that it was necessary to go beyond the city limits to find a pure water supply. In fact, in 1792 he had added a codicil to his will leaving money to the city to finance a central water system using Wissahickon Creek as a supply. Until the inception of water filtration and treatment, and methods derived from the study of bacteriology, seeking new supplies (as opposed to purifying the old ones) was the only alternative to tainted sources.[18]

After examining various options, the committee accepted the proposal of Benjamin Henry Latrobe. The English-born engineer recommended building a steam-powered pumping plant—soon to be called the Centre Square Waterworks—that would distribute water to the city from the still-protected Schuylkill River, located more than one mile away. Latrobe began the task in 1799 and completed it in 1801. In 1811 the city's Watering Committee replaced the Centre Square Waterworks with a larger plant. The engineer Frederick Graff, Latrobe's former assistant, called for pumping water to a reservoir and then releasing the water by gravity to the city. The Fairmount/Centre Square Waterworks served Philadelphia until 1911.[19]

Though not without its flaws, Philadelphia's waterworks was considered by many to be the most advanced engineering project of its time. Especially after its construction, Philadelphia had a system with a much greater capacity than existing demand (until the 1870s at least), unlike comparable cities such as New York, Boston, and Baltimore.[20] To promote its use, citizens were initially offered free water for several years. Despite the fear of epidemics, many citizens had not been completely convinced to give up "their cold well water for the tepid Schuylkill water." But 2,850 dwellings were receiving water from the new system by 1814.[21]

well >
river

The construction of a major waterworks in Philadelphia was widely publicized, but a national trend of adoption was not evident until late in the century. Inexperience in dealing with such a major project, in part at least, helps to explain why urban population growth exceeded construction for so many years. Despite the limitations of the new water systems, the few American cities that turned to community-wide approaches set patterns for the modern sanitary services of the near future. Proto-systems offering rudimentary distribution networks, pumping facilities, and new sources of supply were precursors to more elaborate centralized city-wide systems adopted by many cities and towns by the late nineteenth century. As in England, the application of these new technologies ran ahead of an effective understanding of the causes of disease and pollution; nevertheless, they attempted to enhance the healthfulness of the city and provide better protection against fire.

national pattern

Early in the nineteenth century, a few water-supply proto-systems began to appear in major American cities.[22] However, the number of waterworks multiplied at an increasingly accelerated rate from 1830 to 1880. At first, they almost doubled in 20 years. In the 1870s alone, they more than doubled. By 1880, some water supplies were evolving into modern city-wide systems. Not only did they deliver greater quantities of water over a larger area; they also included rudimentary safeguards to ensure purity. A growing preoccupation with water quality (a direct result of the sanitary movement) was bringing attention to filtration techniques and new methods of water treatment. City leaders and sanitarians were demanding more from their water-supply service than convenience at the tap.

By the 1870s, the trend toward more public water supplies was evident. The crucialness of adequate supplies of water to meet the needs of citizens, commercial establishments, and industry—and the emerging mandate of cities to protect the public health—meant that authorities in the largest urban areas at least wanted centralized systems under their direct control. Boosterism was an additional motivation, since an effective water system was a powerful promotional tool for city leaders seeking to enhance a city's economic base. Though many water companies had been very profitable, capital investment in the more modern systems (with reservoirs, pumps, and elaborate distribution networks) was steep, and operating costs were on the rise. Private service, therefore, was gradually phasing out in several communities. In addition, public control of the water supply enhanced the authority of city government relative to that of the legislature or that of rival cities, thus private owners often were under pressure to sell out, particularly through less lucrative franchises.

The desire of city leaders to convert private systems into public ones, or to build new public systems, rested on more than the will to do so. The central issue was the ability of cities to incur debt to fund major projects and to sustain the high costs of operating the new technologies of sanitation. As the nineteenth century unfolded, city finances underwent changes in scope and complexity that ultimately made the development of public sanitary systems achievable.[23] In most cases, a combination of local circumstances and the experience of other cities influenced the shift from private to public.[24]

Boston suffered under many years of water-supply politics before it developed a system in the 1840s. In 1796, Governor Samuel Adams approved a General Court act creating the Aqueduct Corporation, which constructed a line from Jamaica Pond in Roxbury to the city. The system was extended in 1803 with the addition of new mains and fire plugs. There was no further attempt to improve the system until 1825. Between 1825 (a year in which the city suffered a great fire) and 1846 (a period in which several epidemics rocked the city), civic leaders were embroiled in debate over the water supply.

Those in favor of a municipal supply were persistent, and the issue remained alive in the 1840s. A water referendum in December of 1844 resulted in a major victory for proponents of a municipal system. Long Pond (later renamed Lake Cochituate) was selected as the city's source of supply, to be purchased at city expense. But who should hold general powers over the water-supply system was not resolved. The city did not gain the legal authority to establish a municipal supply until 1846.[25] In that year the General Court passed Boston's Water Act, which provided for the development of Long Pond under the direction of a Water Board. The city was authorized to finance construction with up to $3 million in municipal bonds. (The entire Cochituate system cost almost $4 million.)   *Cachicluate*

In October of 1848, the Cochituate Aqueduct opened, ushering in an era of municipal control of water supply in Boston. The switch from private to public management was not the simple reason for the significant improvement. As the historian Fern L. Nesson argued, the completion of the Cochituate system also changed the focus of water-supply debates. It placed control and monitoring of the water system in the hands of experts, under whose influence Boston was favored by avoiding future water shortages and escaping the devastation of epidemic disease. "What had been a popular, political issue," Nesson stated, "became a technical issue initiated by an administrative request to the General Court for permission to add to the water supply system."[26] The success of the Boston system, therefore, elevated the stature of technical experts and also reinforced the faith in environmental sanitation.

Outside of a core of emerging major cities in the industrial East and a few others sprinkled throughout the country, the transition from private companies to municipal service more typically occurred in the late 1860s and after. For example, Buffalo did not establish its municipal system (by acquiring a privately held plant) until 1868. The Jubilee Spring Water Company had distributed spring water through log pipes as early as 1826, followed by the Buffalo Water Company (1849), which tapped the Niagara River. The latter's facilities provided the nucleus for the public system.[27]

The first waterworks in Chicago was not established until 1840, under the auspices of the Chicago City Hydraulic Company. Lake Michigan was used as the primary source of water, and the company built the city's first pumping station and reservoir. The distribution lines only reached a small portion of the southern and western divisions, while four-fifths of the city continued to obtain its water from the polluted Chicago River or from water carriers. In the wake of a cholera epidemic in 1852, city officials exercised their option and assumed control of the system. Since the epidemic was believed to have originated with the wells, the Lake Michigan supply gained greater significance in the years after 1852.[28]

Milwaukee

Milwaukee's first recognized waterworks was built in 1840 for the United States Hotel, while most citizens received their water from local springs and wells. In response to citizen pressure, the common council passed an ordinance in 1857 to authorize the issuance of bonds to finance the Milwaukee Hydraulic Company. It also gave the company some property for a water tower. The project was never completed, and a second attempt by Hubbard and Converse of Boston (1859) was sidetracked by the Civil War. In March of 1861, the state legislature passed a bill preventing the issuance of new city bonds. Serious progress on a waterworks did not commence until 1868.[29]

St Louis

The St. Louis Water Works was built in the 1830s. As early as 1821, however, general concern about fire hazards led to demands for a better water supply. In 1823 the mayor began promoting the idea for a city-wide system, and finally in 1829 the city council offered a $500 prize for the best plan. A committee also made inquiries in other cities, especially Philadelphia and New Orleans.[30] Within a short time, St. Louis officials signed a contract with Wilson and Company. The work began in 1830, but, as the historian Richard Wade noted, "no water moved through the pipes until the next decade."[31]

In the South, municipal water systems were rare in this period. In Reconstruction-era Atlanta, plans moved ahead for a new waterworks, but its primary thrust was to serve business and industrial needs and fire protection, not to supply potable supplies. Without a creditable municipal water supply, the more affluent turned to purchasing spring water or other pure sources, or depended on their own wells. In black neighborhoods, however, drainage was poor, sewer outfalls often dumped wastes there, and wells were badly polluted. Likewise in Memphis, little attention was given to residential water service.[32]

As changes in the administration of waterworks were slowly evolving after the middle of the nineteenth century, far less subtle changes were taking place in water-supply technology. New sources of supply became necessary when old sources could no longer meet growing demand or became severely polluted; questions of quantity *and* quality. The only viable alternatives were abandoning older sources, then digging new wells; pumping water from nearby lakes, rivers, and streams; seeking more distant sources; and filtration (but not until the 1870s). Good location was a significant advantage for cities forced to change or augment their water supplies. Filtration (and treatment), however, eventually helped to defy the limits of location, especially as the miasmatic theory was challenged by those who found dangers in polluted water beyond smells and discoloration.

St Louis

In some cases, cities took only modest precautions to ensure a good supply of water. St. Louis, for example, used the same system of pumping muddy river water into a single 330,000-gallon reservoir until 1871. The reservoir also served as a sedimentation basin. Pumping, with two rotary pumps originally bought for use on fire engines, occurred only during daylight hours, while the reservoir was used to supply demand at night.[33]

Chicago

For Chicago, location offered new sources in close proximity to population centers. When the town was founded in 1833, the water of the Chicago River—a relatively sluggish stream with two branches which divided the city—was considered pure, with some variation in quality from season to season. Water also was drawn from shallow wells, since the site of the town was only a few feet above the level of Lake Michigan. In the 1850s, especially as the Chicago River became more of an open sewer, the public water supply was pumped from an inlet basin on Lake Michigan near Chicago Avenue (a distance of 3,000 feet from the mouth of the Chicago River). Lake Michigan offered a magnificent alternative as a water source, extending over 22,400 square miles and with a watershed of 69,000 square miles.

As the city grew (more than 100,000 people in 1860), and as the lake water close to shore became increasingly polluted, the intake pipe was moved further out and deeper into the lake. In 1863, the Common Council approved a plan of the Board of Public Works to construct a two-mile tunnel burrowed under the lake bottom connected to a new intake. This first lake tunnel was completed in 1866 at an estimated cost of $600,000. The project proved to be a much more difficult and complex engineering task than anyone imagined. Duel and Gowan, the Harrisburg, Pennsylvania engineering firm that won the contract, was faced with several problems in tunneling below the lake bed. Most difficult was connecting the shore and lake points on a straight line. Despite the arduousness of the task, this first lake tunnel only supplied the needs of the city until 1871. After the Chicago fire of that year, a new tunnel and pumping station were built for the west side of the city.[34]

Most major cities that were growing as rapidly as Chicago did not have the advantage of such a convenient water source to meet the increasing demand. More consideration had to be given to distant sources. But these cities would need to confront two of the same problems Chicago faced in developing effective city-wide service: skyrocketing capital costs and the sheer scale and complexity of the engineering task required to develop a new supply and distribution system.

New York

The Old Croton Aqueduct (1842) is regarded as a sublime engineering feat, and as a symbol of the conquest of nature in service to the urban population explosion. The Croton Aqueduct project is also an important example of changes in the scale and complexity of modern water-supply systems. Several attempts to solve New York's water problem failed in the early part of the nineteenth century. The Manhattan Company's willingness to build an aqueduct from the Bronx River to Manhattan Island never materialized. Efforts to revive the plan in the 1820s likewise fizzled. In 1835, the fortunes of the city changed. Citizens, frustrated by the poor state of well water and frightened by the most recent serious outbreak of cholera, were ready to support a new plan. In a rare moment of political harmony, the voters, the state legislature, and the New York Common Council agreed to construct an aqueduct from the Croton River in Westchester County, running 41 miles to New York.

The Croton project won out over the Bronx project because the source was much larger—estimated at 40 million gallons per day—and could be delivered to the city without pumps. Even with the savings in the construction and maintenance of such machinery, the aqueduct cost approximately $9 million–$10 million.[35] The task of building the aqueduct was first entrusted to Major David Bates Douglass. Douglass, however, lacked experience with large public works, especially one that required building a variety of structures: a dam, an enclosed masonry conduit, bridges and embankments, and a huge reservoir. A good surveyor, he had consulted on railroad and canal projects, and taught civil engineering at West Point and New York University, but he had never carried out such a large project. In addition, he had a major personality clash with the Chairman of the Water Commissioners, Stephen Allen. Douglass was fired in October of 1836; at least he had routed most of the aqueduct before his departure.

The Commissioners replaced Douglass with John B. Jervis. A self-trained engineer with vast work experience, Jervis served as a supervising engineer on a portion of the Erie Canal (1823) and as Chief Engineer of the Delaware and Hudson Canal (1827). In both cases, he had worked his way up through the ranks of the projects. Based on his canal experience, he was in demand for many other similar ventures.[36] In regard to the task facing Jervis on the Croton project, the historian Larry Lankton noted: "This was no ordinary engineering work, no mundane railroad or canal. It was literally to become a lifeline to Manhattan, sustaining hundreds of thousands of lives. It had to be exceptionally dependable and durable. It had to work and it had to last."[37]

Jervis confronted the task of building what was to be the largest modern aqueduct in the world with great aplomb. Innovative design techniques had to be employed to make sure that the aqueduct could remain operational across a variety of terrains and could withstand the intense winter cold. To maintain a uniform grade, the aqueduct ran through tunnels dug into hills and was carried by bridges constructed over ravines and streams. When the Croton Aqueduct was opened on July 4, 1842, it safely carried 75 million gallons per day, 15 million more than Jervis originally calculated. At the time, it seemed that the project would meet the demands of the city for years into the future. But by 1860 the Croton Aqueduct was delivering its maximum, and it was pushed to provide as much as 105 million gallons per day before a new line was built.[38]

New York's need to supplement water supplied by the Croton Aqueduct became apparent in the 1870s. During droughts and in cold winter months, more water was consumed than was received, requiring the drawing of extra water from other sources in the city. Refilling the reservoirs took a great deal of time because of the limited capacity of the aqueduct, while millions of gallons of water ran over the Croton Dam simply unavailable for use.[39]

Despite its great overall success, the Croton system had not been built without technical difficulties, input from several special interests, contract irregularities, and discrimination in service delivery. The insurance industry, for example, wielded substantial influence in seeing that the aqueduct was completed quickly in order to reduce fire damage claims. The industrial community likewise was anxious for a pure and abundant supply of water. Construction work however, was sometimes haphazard, as when water mains were run through sewer lines. And despite the fact that the aggregate water production increased dramatically with the aqueduct, availability of supplies tended to favor the middle class over the poor. Affluent lower wards, for example, received more pipes than poorer upper wards.[40]

Major projects in a few other cities followed the construction of the Croton Aqueduct. Boston completed its aqueduct (half the length of the Croton) in 1848. Another major engineering feat, much of the Cochituate Aqueduct ran through deep trenches covered with dirt. It terminated at Brookline in a 20-acre reservoir, where the water was then moved along large mains to two distributing reservoirs. The cost of the project was approximately $4 million.[41] The Washington Aqueduct, built to supply the nation's capital with water from the Great Falls of the Potomac (14 miles from the city), began construction in 1857 and was completed in December of 1863.[42]

Distant sources of supply received great attention from major cities because they offered large and dependable quantities of water, but also because they provided alternatives to polluted or infected sources in the local area. Smaller communities, without a sufficient tax base or other financial resources, were hard pressed to seek distant sources. For many communities the lack of viable options for dealing with polluted water supplies was the weakest link in the early systems. The transformation of proto-systems into modern waterworks required methods for ensuring—or at least improving—water quality. The gradual introduction of filtration and new techniques for water distribution held some promise for accomplishing that goal.

Complicating the search for pure water was the fact that determining what constituted a tainted supply was little understood in the Age of Miasmas. Taste and smell substituted for scientific testing in most assessments of water quality. Some physicians warned patients not to drink hard water or water with vegetable and animal matter in it, fearing that it would harm the kidneys or produce stomach and intestinal maladies. In 1873, the president of the New York Board of Health, a chemistry professor at Columbia University, advocated the consumption of lake or river water stating that "although rivers are the great natural sewers, and receive the drainage of towns and cities, the natural process of purification, in most cases, destroys the offensive bodies derived from sewage, and renders them harmless."[43]

It was the British physician John Snow's research on cholera in the 1850s that established a clear link between epidemic disease and polluted water that wasn't based simply on the test of the senses. Snow's work on waterborne transmission of disease inspired Dr. William Budd's studies of typhoid fever. Like Snow, Budd determined that typhoid was spread through water supplies contaminated with human feces.[44] Of the possible water-borne diseases that threatened American cities, typhoid fever was the worst. "The disease means little to us today since it is no longer a threat to modern cities," stated the historian Michael P. McCarthy, "but it frightened the urbanizing world of the late nineteenth century."[45]

typhoid

The typhoid bacillus could be contracted by direct contact with a "carrier" or through contaminated food such as milk, raw fruits and vegetables fertilzed with night soil, and shellfish found in polluted waters. Most often it was spread when excreta from a victim entered the water supply directly or as untreated sewage. Detecting the disease posed something of a problem because the incubation period was approximately 14 days. Typhoid fever produced vomiting and diarrhea leading to dehydration, and was accompanied by high fevers. Children, in particular, were most susceptible to the disease.[46] Not only was the disease a threat to human life, but it could severely damage the reputation of a city trying to attract new citizens and new business enterprise.[47] It was a scourge to be avoided if at all possible.

By the turn of the century, various approaches to ensuring a pure water supply and limiting water-borne disease emerged from the bacteriological and chemical laboratories supported by agencies such as the Massachusetts Board of Health. But the first means of water purification readily available to cities in the late nineteenth century—one that fit within the context of the filth theory—was filtration through sand or gravel to improve the clarity, odor, and color of the water.

sand or gravel filtration

As early as the ninth century, Venetians filtered water from cisterns through beds of sand. The first filtering system for a public water supply, as stated earlier, was likely established in Paisley, Scotland in 1804. The Chelsea Water Works in London (1827) employed a "slow sand" (or English) filter, which was the archetype for later models and which eventually found its way to the United States. Berlin's water was filtered in 1856, and by 1865 several European cities followed its example.[48]

In 1835, a noted American engineer, Samuel Storrow, was the first to recommend the use of filtration in the United States. "If the supply be originally taken from a river," Storrow wrote in his book on water works, "it will be liable at some seasons of the year to be very much loaded with impurities, and it is important to have, in connection with the reservoirs, filtering beds, by means of which it may be cleansed of them, before it is introduced into the distributing pipes."[49]

Albert Stein, the designer of Richmond's waterworks, was the first to attempt to filter a public water supply in the United States in 1832. Pumping water from the James River, Stein prepared a sand filter in the reservoir, but he could not get it to operate effectively. During the next 40 years, several major cities, including Boston, Cincinnati, and Philadelphia, considered installing sand filters, but they were too expensive at the time.[50]

A major step forward, but not recognized immediately as such, was James P. Kirkwood's "Report on the Filtration of River Water, for the Supply in Europe, Made to the Board of Water Commissioners of the City of St. Louis" (1869). In 1865, Kirkwood (an engineer) recommended that Cincinnati and St. Louis employ filters in their water systems. At the time, Kirkwood was engaged by a joint committee of the city council and the waterworks trustees of Cincinnati to report on a new water source. He suggested that one or more of the committee visit Europe to examine the filters in person, but nothing came of it.

Kirkwood was then engaged by the City of St. Louis to survey locations for supply works along the Mississippi River. Upon recommendation of a plan which included filtration, the water commissioners instructed him to travel to Europe "and there inform himself in regard to the best process in use for clarifying river waters used for the supply of cities, whether by deposition alone, or deposition and filtration combined." While he was gone, however, opposition to Kirkwood's plan had led to a clean sweep of the commission and replacement with members unwilling to underwrite the cost of a filtration system. The city would not even publish his report as a city document, and did not filter its water until 50 years later.[51]

Kirkwood's report ultimately became a bible of sorts for those cities interested in copying the European experiments. For several years after its completion, however, little additional firsthand knowledge was gathered about the various European systems.[52] By the early 1870s, a few cities began to recognize the value of filtering water. Poughkeepsie, New York, built the first American slow sand filter in 1870–1872; it was based on the designs in Kirkwood's report. This decision was particularly noteworthy because the filter was to be used on Hudson River water, a notoriously polluted supply.[53]

By 1880, there were only three slow sand filters in the United States and none in Canada. But experimentation continued even if adoptions were sluggish. The Europeans, to the contrary, forged ahead with several slow sand filters, and in Buenos Aires in the 1880s experiments were conducted on the suitability of various filtering materials. Information about the experiments in filtration, as well as other crucial information about water supplies, was disseminated more effectively because of the organization of the American Water Works Association (1881) and the New England Water Works Association (1882), both of which published their proceedings. Other engineering societies and public health organizations added to the rich body of data making its way even to the smallest town in the 1880s and after.

Several Americans were pioneers in disseminating information about the quality of water and the value of filtration in this period. Professor William Ripley Nichols of the Massachusetts Institute of Technology, a leading authority on water quality, had argued that sand was the only practical medium for large filtration operations, but that evidence was poor on whether sand filtration would purify polluted water efficiently. Colonel John T. Fanning, a well-respected hydraulic engineer, published a major treatise on water works in 1877, in which he discussed the water purification methods used in Europe.[54]

pumping & piping

Along with experiments in filtration, a variety of pumping techniques and changes in pipe technology helped to transform older proto-systems into modern, centralized waterworks. Aside from gravity systems, steam pumps were increasingly employed directly at the source (especially in the 1870s), and as a way of moving water to reservoirs, tanks, and standpipes.[55] Wooden pipe was adequate in low-pressure gravity systems, but could not withstand the action of high-pressure pumping engines. Wooden mains were sometimes modified by strapping iron bands around the staves. By 1850, iron pipe was coming into wider use in the United States, especially in high-pressure systems. The transition from wood to iron, however, was not uniform throughout the country. Some water utilities continued to use wooden pipe until the 1930s. In the West, especially, wood was used for large aqueducts, irrigation, hydroelectric plants, and hydraulic mining.[56]

distribution disparity

For all the improvements begun by private companies through municipal franchises, accessibility to water supply was still largely linked to class. Affluent neighborhoods and the central business district received the lion's share of water, while the working class districts often relied on polluted wells and other potentially unhealthy local sources. As Sam Bass Warner astutely observed about a later period, "The mode of construction of water-supply and sewerage systems divides the responsibility between municipal capital on the one side and the individual installations of middle-class homeowners and home builders for the middle-class market on the other."[57] This observation applies to the pre-1880 period of private water companies as well, insofar as those outside of the middle and upper classes were unable to tap into the water supply which at that point was available only to a limited market.[58]

[handwritten margin note: "modern" systems by 1880]

[handwritten margin note: still "miasma" theory]

The basic form and function of modern waterworks were established by 1880. Major cities began to devise financial plans based on enhanced revenue generation and long-term debt to plan construction and maintenance of new systems or to secure old systems from private companies. Sources of supply were no longer limited to local wells, ponds, and streams. Distribution extended over wider areas, owing (in part, at least) to the use of iron pipe and a variety of pumping techniques. And a concern for water quality led to research on, and in some cases implementation of, filters. All these changes occurred in the Age of Miasmas, which placed a pure and plentiful water supply squarely at the heart of environmental sanitation.

By the late nineteenth century, waterworks were generally regarded as a public enterprise, justified as such because of the need to protect the public health and to supply water on a city-wide basis. As with wastewater systems and solid waste collection and disposal in later years, modern city-wide water-supply systems were conceived in an Age of Miasmas. The structure of those systems and their functions were linked inextricably to the goals of environmental sanitation, that is, to utilize the prevailing sensory tests of purity to deliver a product that would not only be free of disease but also would be utilized to mitigate against disease. Plentiful supplies of water, for example, could help to flush away stench-ridden wastes.

Despite the fact that massive increases in piped-in water proved to be a major reason for wastewater systems, water supply and sewerage were addressed separately in the nineteenth century. However, the justification for wastewater systems was also graphically linked to the precepts of environmental sanitation. The leap of logic from liquid wastes to solid wastes was made possible by the adherence to environmental sanitation, to a point of view which almost etched in stone the primary directive that wastes had to be removed from the presence of humans as quickly as possible to preserve the public health. In essence, sanitary services in this period were little more than elaborate transportation networks (or water and waste redirection systems).

sewer
landfill
incinerator

The underground sewer was a logical embodiment of the goals of environmental sanitation, as was the sanitary landfill and the incinerator. These technologies were meant to distance humans from their wastes and discards—materials which presumably had imbedded in them the threat of disease.

Whole systems were designed, therefore, to meet the ends of environmental sanitation. Water-supply systems began with a protected source—or one that could be purified through filtration or treatment. The new distribution system of pipes and pumps removed from the individual responsibility for filling containers at a public well or local watercourse, and made the waterworks—private and public—responsible for bringing water directly to each consumer. Implicit in this system was a guarantee that the supply met the prevailing standards of purity. With respect to wastewater systems, citizen responsibility also was at a minimum. Combined or separate sewer pipes whisked effluent from homes and businesses, and placed responsibility for disposal in the hands of the city. The objectives of environmental sanitation had been met because human contact with the waste—at least at the source—was dramatically reduced. In the case of refuse, on-site pickup served the same purpose as water or sewer pipes.

Within the context of nineteenth-century environmental sanitation, sanitary services were at their best in the areas of collection and delivery of water, and collection of effluent and solid wastes. Insofar as pure water was delivered efficiently and initial human contact with wastes was minimized, these services fulfilled their major objective. A change in environmental paradigm—from miasmas to bacteria—did not disrupt these methods of collection, nor influence major changes in "front-of-the-pipe" components in these technologies of sanitation. The new age of bacteriology, however, did help to identify and ultimately confront "end-of-the-pipe" problems, namely pollution. The major weakness of environmental sanitation as a concept was its limited attention to disposal of effluent and refuse after they had been directed away from homes and businesses.

With the onset of the Bacteriological Revolution, water pollution, in particular, received greater and more pointed attention. Scientists, physicians, and engineers now had a much better idea of what they were looking for in the fight against communicable disease, and gave them a clearer idea on how to combat biological pollutants. Older methods, such as dilution of wastes in running water, had value under proper circumstances, but the primary focus centered on treatment. The goals of environmental sanitation were no longer regarded as sufficiently broad to deal with the several confounding problems of waste disposal—or to ensure the purity of a water supply. Without attempting to change the basic structure of the technologies of sanitation, experts introduced new methods of water testing and focused increasing attention on refining treatment technologies. Never in question, however, was the basic precept of designing permanent, city-wide sanitary systems; the basic designs originating in the Age of Miasmas were not substantively changed in the twentieth century. How extraordinary it was, however, that such significant technical systems as water and wastewater infrastructure depended so heavily on what proved to be flawed science.

NOTES

1. Bryan D. Jones, *Service Delivery in the City: Citizen Demand and Bureaucratic Rules* (Longman, 1980), 2.

2. Joel A. Tarr and Gabriel Dupuy, *Technology and the Rise of the Networked City in Europe and America* (Temple University Press, 1988), xiii.

3. Charles E. Rosenberg, *The Cholera Years: The United States in 1832, 1849, and 1866* (University of Chicago Press, 1987; orig. pub. 1962), 228.

4. Ibid., 7.

5. Ibid., 37, 38, 55–57, 59–62, 135–137.

6. Howard N. Rabinowitz, *Race Relations in the Urban South, 1865–1890* (University of Illinois Press, 1980), 114–121.

7. David R. Goldfield, *Urban Growth in the Age of Sectionalism: Virginia, 1847–1861* (Louisiana State University Press, 1977), 152, 153, 160; William H. Pease and Jane H. Pease, *The Web of Progress: Private Values and Public Styles in Boston and Charleston, 1828–1843* (Oxford University Press, 1985), 90, 93, 99.

8. Khaled J. Bloom, *The Mississippi Valley's Great Yellow Fever Epidemic of 1878* (Louisiana State University Press, 1993), 10, 11.

9. Charles-Edward Amory Winslow, *The Conquest of Epidemic Disease* (reprint: Hafner, 1967), 266.

10. Howard D. Kramer, "The Germ Theory and the Public Health Program in the United States," *Bulletin of the History of Medicine* 22 (1948), May-June, 234, 235.

11. Several theories of disease transmission vied for acceptance in the middle of the nineteenth century. See J. K. Crellin, "The Dawn of the Germ Theory: Particles, Infection and Biology," in *Medicine and Science in the 1860s,* ed. F. Poynter (Wellcome Institute of the History of Medicine, 1968), 57–67, 71–74; J. K. Crellin, "Airborne Particles and the Germ Theory: 1860–1880," *Annals of Science* 22 (1966), March: 49, 52, 56, 57; Mazyck Ravenel, ed., *A Half Century of Public Health* (American Public Health Association, 1921), 66, 67.

12. Erwin H. Ackernecht, "Anticontagionism between 1821 and 1867," *Bulletin of the History of Medicine* 22 (1948), September-October, 567.

13. Stanley K. Schultz, *Constructing Urban Culture: American Cities and City Planning, 1800–1920* (Temple University Press, 1989), 132, 133.

14. Introduction by C. E. A. Winslow to Lemuel Shattuck, *Report of the Sanitary Commission of Massachusetts, 1850* (Harvard University Press, 1948), 237; John Duffy, *The Sanitarians* (University of Illinois Press, 1990), 96, 97, 137; Ellis, *Yellow Fever and Public Health in the New South,* 7.

15. See Charles E. Rosenberg, *Explaining Epidemics and Other Studies in the History of Medicine* (Cambridge University Press, 1992), 126, 127.

16. John C. Trautwine Jr., "A Glance at the Water Supply of Philadelphia," *Journal of the New England Water Works Association* 22 (1908), December, 421.

17. See Michal McMahon, "Fairmount," *American Heritage* 30 (1979), April-May, 100, 101; Donald C. Jackson, "'The Fairmount Waterworks, 1812–1911,' At the Philadelphia Museum of Art," *Technology and Culture* 30 (1989), July, 635.

18. City of Philadelphia, Department of Public Works, Bureau of Water, *Description of the Filtration Works and Pumping Stations, Also Brief Historical Review of the Water Supply, 1789–1900* (1909), 57–59; Michal McMahon, "Makeshift Technology: Water and Politics in 19th-Century Philadelphia," *Environmental Review* 12 (1988), winter, 24.

19. Jackson, "'The Fairmount Waterworks, 1812–1911,'" 635; McMahon, "Makeshift Technology," 25, 26.

20. Joel A. Tarr, "The Evolution of the Urban Infrastructure in the Nineteenth and Twentieth Centuries," in *Perspectives on Urban Infrastructure,* ed. R. Hanson (National

Aacdemy Press, 1984), 19; "Golden Decade for Philadelphia Water," *Engineering News-Record* 159 (1957), September 19, 37.

21. The financial security of the waterworks took longer than popular acceptance of the water supply. See Martin J. McLaughlin, "Philadelphia's Water Works from 1798 to 1944," *American City* 59 (1944), October, 86, 87.

22. See Charles Jacobson, Steven Klepper, and Joel A. Tarr, "Water, Electricity, and Cable Televison: A Study of Contrasting Historical Patterns of Ownership and Regulation," *Technology and the Future of Our Cities* 3 (1985), fall, 9.

23. Letty Donaldson Anderson, The Diffusion of Technology in the Nineteenth Century American City: Municipal Water Supply Investments (Ph.D. dissertation, Northwestern University, 1980), 102–104, 117; Letty Anderson, "Hard Choices: Supplying Water to New England," *Journal of Interdisciplinary History* 15 (1984), autumn, 218; Joel A. Tarr, "The Evolution of the Urban Infrastructure in the Nineteenth and Twentieth Centuries," in *Perspective on Urban Infrastructure,* ed. R. Hanson (National Academy Press, 1984), 30, 31; Ernest S. Griffith and Charles R. Adrian, *A History of American City Government, 1775–1870: The Formation of Traditions* (University Press of America, 1983), 198–217. In an important case study concerning the financing of physical improvements in Chicago, the historian Robin L. Einhorn makes an impressive alternative case for privatization in this period. Utilizing special assessments, the city government operated through what she labeled the "segmented system," especially between 1845 and 1865. Einhorn argued that "what made this system 'segmented' rather than simply elitist was that a voice in decision making . . . required the ownership of property that was 'chargeable' or 'interested' in the decision at hand. An owner had to show that he owned a lot that was liable to a particular special assessment before he could participate in the decision to levy that assessment." The segmented system, Einhorn concluded, did not disfranchise the propertyless, but was a way to distribute costs and decision-making power among the propertied class. Einhorn's analysis is especially persuasive with respect to public works projects such as street paving and bridge building. However, as she suggests, water and sewer systems could not be segmented "with the rigor of street projects or fire limit rules" because they required central planning and large initial investments. See Robin L. Einhorn, *Property Rules: Political Economy in Chicago, 1833–1872* (University of Chicago Press, 1991), 16–19, 133.

24. See Anderson, "The Diffusion of Technology in the Nineteenth Century American City," 108.

25. The water commission discovered that the Indians called Long Pond "Cochituate." The mayor proposed the name change to Lake Cochituate—thus the name of the aqueduct. For information on Boston's water supply, see Nelson Manfred Blake, *Water for the Cities: A History of the Urban Water Supply Problem in the United*

States (Syracuse University Press, 1956), 172–198; LaNier, "Historical Development of Municipal Water Systems in the United States," 174; John B. Blake, "Lemuel Shattuck and the Boston Water Supply," *Bulletin of the History of Medicine* 29 (1955), 554–562.

26. Fern L. Nesson, *Great Waters: A History of Boston's Water Supply* (University Press of New England, 1983), 6–12.

27. George C. Andrews, "The Buffalo Water Works," *Journal of the American Water Works Association* 17 (1927), March, 280; "History of the Buffalo Water Works," *Engineering Record* 38 (1898), September 24, 363, 364.

28. James C. O'Connell, "Chicago's Quest for Pure Water," *Essays in Public Works History* 1 (Public Works Historical Society, 1976), 1–3; W. W. DeBerard, "Expansion of the Chicago, Ill., Water Supply," *Transactions of the American Society of Civil Engineers CT* (1953), 588–593; LaNier, "Historical Development of Municipal Water Systems in the United States," 176.

29. Bruce Jordan, "Origins of the Milwaukee Water Works," *Milwaukee History* 9 (spring 1986): 2–5; Elmer W. Becker, *A Century of Milwaukee Water* (Milwaukee Water Works, 1974), 1–3.

30. Despite its pioneering effort, Philadelphia's water supply system deteriorated in the middle of the nineteenth century. Shortages struck the system and pollution infested the Schuylkill and Delaware rivers, once sources of pure supplies. See Sam Bass Warner Jr., *The Private City: Philadelphia in Three Periods of Its Growth* (University of Pennsylvania Press, 1987), 108, 109.

31. Richard Wade, *The Urban Frontier* (Harvard University Press, 1959), 297; LaNier, "Historical Development of Municipal Water Systems in the United States," 176; Gurdon G. Black, "The Construction and Reconstruction of Compton Hill Reservoir," *Journal of the Engineers' Club of St. Louis* 2 (January 2, 1917), 4–8.

32. John Ellis and Stuart Galishoff, "Atlanta's Water Supply, 1865–1918," *Maryland Historian* 8 (1977), spring, 6, 7; John H. Ellis, *Yellow Fever and Public Health in the New South* (University Press of Kentucky, 1992), 29, 142.

33. Black, "The Construction and Reconstruction of Compton Hill Reservoir," 4.

34. The cost of the tunnel would have been less had not the Civil War been raging, since some of the materials used were in great demand. Louis Cain, *Sanitation Strategy for a Lakefront Metropolis: The Case of Chicago* (Northern Illinois University

Press, 1978), 37–51; DeBerard, "Expansion of the Chicago, Ill., Water Supply," 593–597; Frank J. Piehl, "Chicago's Early Fight to 'Save Our Lake,'" *Chicago History* 5 (1976–77), winter, 223, 224; Samuel N. Karrick, "Protecting Chicago's Water Supply," *Civil Engineering* 9 (1939), September, 547, 548; John Ericson, *The Water Supply System of Chicago* (Chicago: Bureau of Engineering, 1924), 11–13.

35. Voters approved the project by a 3–to-1 margin, except in the sparsely populated northern part of the city which had good well water. See Larry D. Lankton, "1842: Old Croton Aqueduct Brings Water, Rescues Manhattan from Fire, Disease," *Civil Engineering* 47 (1977), October, 93.

36. Lankton, "1842: Old Croton Aqueduct Brings Water, Rescues Manhattan from Fire, Disease," 94.

37. Ibid.

38. Ibid., 95, 96; Stuart Galishoff, "Triumph and Failure: The American Response to the Urban Water Supply Problem, 1860–1923," in *Pollution and Reform in American Cities, 1870–1930,* ed. M. Melosi (University of Texas Press, 1980), 36.

39. Ibid., 90.

40. Eugene Moehring, *Public Works and the Patterns of Urban Real Estate Growth in Manhattan, 1835–1894* (Arno, 1981), 31, 32, 44–47, 50.

41. Blake, *Water for the Cities,* 199–218; Nesson, *Great Waters,* 11, 12. In 1878 the Sudbury system, drawing water from the Sudbury River, complemented the Cochituate.

42. William R. Hutton, "The Washington Aqueduct, 1853–1898," *Engineering Record* 40 (1899), July 29, 190–193.

43. Cited in John B. Blake, "The Origins of Public Health in the United States," *American Journal of Public Health* 38 (1948), November, 1541.

44. Galishoff, "Triumph and Failure," 37, 38.

45. Michael McCarthy, *Typhoid and the Politics of Public Health in Nineteenth-Century Philadelphia* (American Philosophical Society, 1987), 1.

46. Ibid., 1.

47. Galishoff, "Triumph and Failure," 37, 38.

48. The first efforts in water purification probably occurred in China and India several thousand years ago. It was common in China and Egypt to put alum in water to clarify it. Sir Francis Bacon wrote about water purification experiments, which were published one year after his death in 1627. The first known illustrated description of sand filters was published by the Italian physician Luc Antonio Porzio in 1685. For filtration history, see "Community Water Supply," in *History of Public Works in the United States, 1776–1976,* ed. E. Armstrong et al. (American Public Works Association, 1976), 235, 236; M. N. Baker, "Sketch of the History of Water Treatment," *Journal of the American Water Works Association* 26 (1934), July, 904, 905; Harold E. Babbitt and James J. Doland, *Water Supply Engineering* (McGraw-Hill, 1949), 4, 5; John W. Clark and Warren Viessman Jr., *Water Supply and Pollution Control* (International Textbook Co., 1965), 2–4; George W. Fuller, "Progress in Water Purification," *Journal of the American Water Works Association* 25 (1933), October, 1566.

49. Cited in Baker, "Sketch of the History of Water Treatment," 905.

50. Stein designed the first settling basin in the United States for Lynchburg in 1829. See "Community Water Supply," 236; Baker, "Sketch of the History of Water Treatment," 906–908; George E. Symons, "History of Water Supply 1850 to Present," *Water and Sewage Works* 100 (1953), May, 191; M. N. Baker, *The Quest for Pure Water vol. I The History of Water Purification from the Earliest Records to the Twentieth Century* (American Water Works Association, 1981; orig. pub. 1948), 127.

51. Baker, "Sketch of the History of Water Treatment," 908–910. See also "Water Purification—A Century of Progress," 83; Baker, *Quest for Pure Water,* 133, 135; City of Cincinnati, Water Commission, *Report of the Commission to Take into Consideration the Best Method of Obtaining an Abundant Supply of Pure Water* (1865), 3–9.

52. In 1878 Professor William Ripley Nichols studied water purification in Europe for the Massachusetts State Board of Health. He published his findings on filtration and related matters in a state report, and 5 years later expanded his observations in a book, *Water Supply.* See George C. Whipple, "Fifty Years of Water Purification," in Mazyck Ravenel, *A Half Century of Public Health* (American Public Health Association, 1921), 163.

53. See Baker, *The Quest for Pure Water,* 148.

54. Baker, "Sketch of the History of Water Treatment," 912–914; Baker, *The Quest for Pure Water,* 136–138.

55. See J. Leland FitzGerald, "Comparison of Water Supply Systems from a Financial Point of View," *Transactions of the American Society of Civil Engineers* 24 (1891), April: 252–256.

56. Ellis et al., *History of Public Works in the United States,* 232, 233; Frederic Stearns, "The Development of Water Supplies and Water-Supply Engineering," *Transactions of the American Society of Civil Engineers* 56 (1906), June, 455; Turneaure and Russell, *Public Water-Supplies,* 7, 8; Jean-Pierre Goubert, *Conquest of Water: The Advent of Health in the Industrial Age* (Princeton University Press, 1986), 56–58; Anderson, "The Diffusion of Technology in the Nineteenth Century American City," 10–14; Allen Hazen, "Public Water Supplies," *Engineering News-Record* 92 (1924), April 17, 696; John W. Alvord, "Recent Progress and Tendencies in Municipal Water Supply in the United States," *Journal of the American Water Works Association* 4 (1917), September, 291, 292.

57. Sam Bass Warner, *The Urban Wilderness: A History of the American City* (Harper and Row, 1972), 202.

58. Wade, *The Urban Frontier,* 294, 295; O'Connell, "Chicago's Quest for Pure Water," 3; Tarr, "The Evolution of the Urban Infrastructure in the Nineteenth and Twentieth Centuries," 14; Joel A. Tarr, James McCurley, and Terry F. Yosie, "The Development and Impact of Urban Wastewater Technology: Changing Concepts of Water Quality Control, 1850–1930," in *Pollution and Reform in American Cities,* ed. Melosi, 60.

CLEAN WATER FOR THE WORLD

ASHOK GADGIL

I am trained as a physicist. I conduct research at Lawrence Berkeley National Laboratory, applying physics to problems of the indoor environment, such as indoor radon, mostly in the industrial countries, specifically the United States. However, this essay discusses a specific invention that I developed over the period 1993–1995, prodded by an outbreak of a mutant strain of cholera in India, in the state of Bengal.

Even if we know exactly how to engineer a vaccine and manufacture it on a large scale, it takes about 2 years between the outbreak of a mutant strain as an epidemic and the market availability of relevant vaccines. Thus, for 2 years the population is a sitting duck. For example, in May of 1993, in just one state (Bihar, which borders on Bengal), 2,000 people died from this mutant strain. First found in Bengal (hence the name "Bengal cholera"), it quickly spread into adjacent states along India's east coast, into Bangladesh, then into Thailand. That is about the time I started getting uneasy that we were not doing anything about it, particularly because I had been interested in disinfecting drinking water for poor communities in poor countries for some time.

I had been sending references and literature on the disinfection of drinking water to my colleagues in India, but their responses had been that they had other things to do. I was saying the same. We had proposals to write and papers to finish, our bosses had other agendas, we had other agendas, and nobody had the spare time and energy to tinker around with ideas like these—particularly without funding.

In the summer of 1993, by a fortunate coincidence, I found a graduate student who agreed to work with me on this topic. I worked on my own time and paid the student. We began investigating the disinfection of drinking water.

About 2 billion people do not have tap water supplied to their home. About 4 million children, mostly under the age of 5, die each year from drinking biologically contaminated water (that is, water contaminated with human and animal wastes containing various pathogens). A single person sick with a diarrheal disease may have initially ingested a few hundred pathogens; however, with each diarrheal stool he or she passes, 10 billion pathogens enter the environment. That is an enormous multiplier. If waste is not treated adequately, the pathogens get in the water supply and we have an outbreak. Then there is the problem of stunted children. Children who do survive diarrheal diseases have reduced absorption of nutrients in their system for periods that are up to four times longer than the actual diarrheal episode itself. That results in a child whose growth is stunted. In addition, billions of work hours are lost to diarrheal diseases each year.

In downtown Mumbai (Bombay), there are slum areas right next to high-rise apartments. One often sees women who live in the slums fetching water from a collective tap. These women probably work nearby as housemaids; many of the men living in the slums work as office boys or in factories.

The common methods of treating drinking water are chlorination and boiling. Chlorination requires maintaining a supply chain of necessary chemicals and a trained operator. Water overdosed with chlorine becomes unpalatable. Boiling, of course, requires a huge amount of biomass, or a lot of kerosene fuel, which is very expensive. A family of five requires at least 10 liters, or 10 kilograms, of drinking water per day. To boil 10 liters of water requires about 3 kilograms of fuel wood per day, and that is beyond their means. Three kilograms is more than their current daily consumption of fuel wood just for cooking food, so doubling that amount is not something that people can manage or that the environment can sustain.

What we were looking for in the summer of 1993 was something that would be energy efficient and low in cost, would treat unpressurized water, and would be able to disinfect water that people carry in pots and buckets and cans. We wanted rapid throughput. People do not want to invest much more time than they already do for collecting water. For good field performance, we wanted a device with low maintenance and high reliability, and we wanted it to be based on mature components—preferably off-the-shelf components, so we would not have to worry about technical failures or immature manufacturing processes.

Our device does not appear to have been difficult to design; it is just that no one seems to have tried to design it before. Something like this could have been done 10 years ago by someone else with far less technical knowledge and support than our team had. It did not require a national laboratory and all the high-tech machinery we ended up bringing to bear on the problem.

The use of a low-pressure mercury plasma, as in kitchen fluorescent lamps, is a mature technology; it has been in use for half a century. Some 95 percent of the radiation from that plasma actually comes out as "UV-C," at a wavelength to which DNA is extremely susceptible. There is no light of that wavelength available in the solar spectrum on the Earth's surface. The ozone layer absorbs it all. Maybe that is how the DNA on the planet evolved to remain susceptible to it, since it can receive no damaging UV-C from sunlight. It turns out that a standard fluorescent lamp produces lots of UV-C light of this wavelength. But the inside of the tube is coated with a fluorescent powder, which absorbs the UV-C and emits visible light. If you do not have that powder and if your tube is made out of quartz rather than glass, so all the UV will come out, then you already have a mature technology that the factory perfected in some sense, mass produced, robust, like your kitchen fluorescent lights again. And any DNA that is exposed to this UV light gets damaged severely; adjacent base pairs in the DNA double helix get covalently bonded. You can think of them as fused together. It's like having a zipper jammed so that when the bacterium of the virus tries to unzip the DNA for replication, the two adjacent base pairs are fused together, and the DNA cannot be replicated, so the bacterium or the virus is unable to reproduce. If the UV dose to the DNA is high enough, you will fuse a huge number of adjacent base pairs in this DNA, essentially making that DNA useless for the organism, and the organism dies. It cannot infect, it cannot replicate.

We designed and built the unit, we went through the first level of field testing in India, and based on the feedback from the field we redesigned the unit. The final design is compact, simple, lightweight, and delivers more than 120 milliwatt-seconds of UV-C energy per square centimeter of water surface, in 12 seconds. We tested the unit's performance in reducing the concentration of E. coli introduced in the inlet water. E. coli is the standard test organism used by World Health Organization and the US Environmental Protection Agency to test the effectiveness of various disinfecting methods against waterborne pathogens. We found that our apparatus, which we named, UVWaterworks, reduces *E. coli* concentrations by a little more than a factor of a million, and it reduces virus concentrations by a little more than 10,000.

The viruses against which we tested this were polio virus and rotavirus, the most common pathogenic waterborne viruses. The rotavirus gives diarrheal dysentery and is the most resistant to UV disinfection among all diarrheal pathogens. We also tested the technology against waterborne concentrations of real pathogens clinically isolated from hospital patients in Bombay, including cholera, shigella, and typhoid pathogens, and found the technology to be very effective.

With UVWaterworks, 5 cents' worth of grid electricity—about half a kilowatt-hour—suffices to disinfect a day's drinking water for 1,000–2,000 people. So it seemed pretty cheap. It is also extremely simple, embarrassingly simple. As I said earlier, nobody appears to have bothered to do it before us, but it was there for anyone to build. And the reason I think nobody bothered to do it is because, on the one hand, these people's lives are cheap. You cannot make too much money off somebody who dies of cholera in Bangladesh. These people do not have votes, they do not have the technical know-how to do it themselves.

Inside the UVWaterworks, you have an aluminum reflector. Below it is a UV lamp suspended in air above the water's surface. All parts that contact the water are made with stainless steel so there is no corrosion to worry about. Below the UV lamp, water flows in an open channel under gravity in the stainless pan and comes out at the other end of the device. That is essentially what is inside the device.

By the way, disinfecting with UV-C light is not a new idea. It has been known for almost a century. However, all other water disinfection devices using UV light require the UV lamp to be encased inside a quartz tube, and the whole thing placed coaxially inside a stainless steel cylinder and immersed in axially flowing water. This design requires pressurized water to drive the flow through this narrow gap between the quartz tube and the stainless chamber, which means that the 2 billion people who fetch water in buckets cannot use it. So, it is extremely surprising and slightly irritating that a simple device such as this has not been developed for all these years.

The second innovation was, of course, ensuring correct hydrodynamics for the flow in this pan. We wanted to ensure that all the water moves almost as a block rather than various parts of the water moving at different speeds in what mechanical engineers would call a fully developed channel flow. In a normal channel flow, the top surface and center of the water channel has higher speed than that of water near the edges and the bottom. That would make the machine inefficient.

In 1995 this box, UV Waterworks, was handbuilt by research technicians at a national laboratory. It cost $1,000 to build, which means it would have to be sold for about $5,000 if somebody were to take it on as a business. At this point there was a socially responsible group of investors who came together wanting to start a business and licensed this technology and the design from Lawrence Berkeley National Laboratory. They promised to raise the money to redesign this product from a research prototype that cost $1,000 to build into a factory-manufacturable product which we can mass produce cheaply. It took at least one day for our lab technicians to build one research prototype. On the other hand, the box made by WaterHealth International is built in 12 minutes. And they can sell it for about $700 with a decent profit margin. Urminus Industries, the water-treatment company that partially funded our first field trials in India, negotiated and obtained the rights for the Indian market.

The design process took 2 years. We tested the early prototypes in India. The final prototype has been tested in eleven different laboratories in five different countries. We have ongoing field testing in South Africa, funded by the US Department of Energy and by Lawrence Berkeley Laboratory. Commercial production started in 1998.

Now, nothing works perfectly. Even this box does not work perfectly— it is not a silver bullet to eliminate diarrheal waterborne diseases. In addition to making it available and affordable, you need to convince people that they need to drink only safe water in order to avoid diarrheal diseases. They need to scrub their hands after they to go to the toilet. They need to dispose of human waste and animal waste in an appropriate manner. All of these other actions and activities are needed so that they all work together to eliminate diarrheal disease and save lives. Nevertheless, safe or clean drinking water is a necessary condition for getting rid of diarrheal deaths. Furthermore, UV Waterworks does not remove toxic chemicals or *Giardia* spores, and we still do not claim that it disables *Cryptosporidium,* or amoebae. If these larger protozoan parasites are present in the water, they must be removed by either a sand filter or a roughing filter. On the other hand, if the raw water supply is obtained from a hand pump, then you do not need to worry about *Giardia* and *Cryptosporidium* because the layers of soil and rock between the surface water sources and the deep aquifer filter out the larger pathogens. Then all you need to worry about is viruses and bacteria. And then this unit can handle it well.

Finally, UV transmittance in water is hindered by high turbidity and certain dissolved salts. However, we estimate that most waters are adequately transparent to UV-C to enable this technology to disinfect the water. We know that we can disinfect water up to 20 Nephalometric Turbidity Units of turbidity. Raw water with this turbidity is 20 times more turbid than what the US EPA and the WHO recommend for drinking water. So, if the raw water has higher turbidity, one must filter it to reduce the turbidity anyway.

Those who travel to developing countries are familiar with the conditions in which much of the world lives. In any given village, the creek running through the slum collects all the wastewater. This is the stream along which children play, and this is also the water for these people to wash their clothes. In a community, for example a typical village in Orissa (a poor state in India), one UVWaterworks unit would save about one diarrheal death per year in a community of 1,000 people. Presently, there are about three diarrheal deaths per year of children below age 5 in a typical village of 1,000 in Orissa. We can save (and take credit for) only one; the other two deaths will have to be avoided through public hygiene, public education, and adequate treatment of human and animal waste. If all of that is done, you would avoid all three deaths. So, over the 15 year estimated life of a UVWaterworks unit, it would certainly avoid 15 deaths by our estimate. It would also avoid the stunting of 150 children growing up in this community over these 15 years. UVWaterworks uses 6,000 times less energy than is needed for disinfection by boiling the same amount of water over biomass cook stoves. Including grid electricity, consumables, and amortization of capital, the annual cost of disinfection is less than 14 cents per person.

The units have undergone limited field trials in India and in South Africa. In Manila, a flourishing business has been established by the distributor of these devices. The distributor sets up water vending kiosks built around pumps, filters, stainless steel water storage tanks, and one unit of UVWaterworks. As feedstock, he uses the available city water supply (which is of uncertain quality), or he obtains water from wells. The raw water is filtered, disinfected with UVWaterworks, and sold at very low cost to people from low-income or slum communities. But he recovers his investment very quickly because the device is so efficient. The kiosks also provide employment to one person, who manages the whole enterprise.

Starting from scratch in the beginning of 1998, this business has grown to serving about 20,000 people daily who buy their water from UV Waterworks in and around Manila. It is expected to expand quite rapidly, so maybe this is an effective business solution. No individual has to invest; instead, consumers pay as they go.

PORTRAIT OF INNOVATION: DEVRA LEE DAVIS

MARTHA DAVIDSON

The sooty gray skies and coal-paved alleys of Donora, Pennsylvania, are among the earliest memories of Devra Lee Davis, a leading epidemiologist and former director of the Program in Health, Environment and Development at the World Resources Institute in Washington. But the path from Donora (a steel- and zinc-mill town that was the site of the first "killer smog" in the United States) to Davis's pioneering studies of environmental risk factors for cancer and other chronic diseases was not a direct one.

Davis was born in 1946, when her parents were stationed at a military base in Virginia, but most of her childhood was spent in Donora, south of Pittsburgh in the Monongahela Valley, where her grandparents had settled when they emigrated from Eastern Europe. Her grandfather earned a living by scavenging and reselling good bits of steel from the slag heaps of the mills. Her father, after working as a chemist and machinist in the mills and serving in the Army, expanded the family business to resales of office and industrial equipment. Feeling that, as Jews, they owed a debt to the United States for defeating Hitler, her father remained an officer in the Army Reserve for 30 years. During the Korean War he was called up as a drill sergeant, and the family lived for a while at Army bases in Indiana and Texas. With hindsight, Davis is glad that at least a few of her early years were spent away from the polluted air of Donora.

As a child, however, she took Donora's gray skies, dazzling sunsets, and sulfuric air for granted. She and her three siblings had fun sliding down the town's slippery, barren hills. Nor did she find her grandmother Pearl's heart disease at all unusual. "I only realized that not all blue-haired grandmothers stayed in bed tethered to oxygen tanks when I met someone else's granny who actually walked around," she recalls. Many years later, all of Pearl's five children, including Devra's mother, developed heart problems.

When Devra was 2 years old, Donora made national headlines. In October of 1948, a smog of coal and coke fumes, mixed with toxic fluoride and cadmium gases from the mills, became trapped in the valley by an atmospheric inversion. It hung there for nearly a week, so dark and thick that cars could not navigate the streets even with headlights on in daytime. A third of the town's residents fell ill within a few days, and the death rate for that month was 50 people more than average, 20 of those "extra" deaths occurring in the first 4 days of smog. The event was never fully investigated, although a few chemists and physicians believed that the fluoride and cadmium gases from the zinc mill were connected with those deaths. "The idea that air pollution played any role in when and how many people die had never been imagined," Davis explains. Nonetheless, the incident in Donora, along with similar lethal smog in Belgium in 1930 and England in 1952, raised public and professional concern about air pollution. Donora's killer smog ultimately contributed to passage of the federal Clean Air Act of 1956. It may also have contributed to the heart problems in Davis's family. Fluoride gas leaves no traces on the lungs, but it directly weakens the heart.

That early experience with pollution did not inspire Davis to pursue a career in public health. Rather, her passionate interests were in science, religion, and music. One of the fundamental precepts of Judaism is to study the Torah, the five books of the Old Testament. As a child, Davis loved to hear the stories of the rabbis. Her devotion to Torah study carried over into other areas of learning as well, and has remained central to her life. She also spent hours each day playing the cello, eventually becoming the first-chair cellist of the Pennsylvania High School State Orchestra. But it was science, and questions about the realms of science and religion, that led her to pursue higher education.

Although no one in Davis's working-class family had a college education (her mother had dropped out to get married, but completed her degree later in life, after raising four children), Davis was recruited by the University of Chicago when she was 15 years old. Her parents would not let her go away to college at that age, but Davis, already tall and mature, felt more comfortable on a college campus than with high school peers. Her family had moved to Pittsburgh, and she began taking classes at the University of Pittsburgh while still in high school. Her most exciting course was graduate-level statistics. "The teacher was wonderful," she says. "I was like a sponge!" (She still has strong feelings about statistics, always conscious of the people behind the numbers. She sometimes describes statistical data as "human beings with the tears removed.")

As a teenager, she was also involved in the civil rights movement, serving as a local organizer for marches in Washington, D.C. and in Selma, Alabama. Although she was not aware of it at the time, she found out many years later that her father had been active in integrating the armed forces.

After graduation from high school, Davis enrolled as a junior at the University of Pittsburgh. She majored in sociology with a minor in biology, and gained additional science training by working in the labs and typing dissertations. She graduated in 3 years with a B.S. in physiological psychology and an M.A. in sociology. A Danforth fellowship—one of the few given to female students in 1966—allowed her to pursue a doctorate at the University of Chicago.

Davis completed requirements for a Ph.D. in science studies (an interdisciplinary program in the history of culture) in 2 years. She passed her exam 2 years later, in 1972, after a brief marriage and divorce. By that time, she was already an assistant professor of sociology and director of interdisciplinary studies at Queens College of the City University of New York. There she continued her social activism in the antiwar movement and met another young faculty activist, an economist named Richard Morgenstern, who was studying the disparity in salaries paid to male and female academics. They married in 1975 and had two children, a son, Aaron, born in 1976, and a daughter, Lea, born in 1979. In the years that followed, Davis and Morgenstern often found their work dovetailing, as both were drawn into the world of environmental policy.

To have more time for her children while they were very young, Davis left teaching and accepted three postdoctoral fellowships between 1971 and 1982. The first was in neurology and neurotoxicology and included a study of acupuncture. The second was in neurotoxicology and toxicology with Larry Ng of the National Institutes of Health and resulted in their book *Strategies for Public Health: Promoting Health and Preventing Disease,* published in 1981. Davis had by then moved to Washington to work as an advisor to the US Environmental Protection Agency. Her husband was also at the EPA, as chief economist, running major environmental programs. Davis's third fellowship was with Abraham Lilienfeld, a renowned epidemiologist at Johns Hopkins University. It was Lilienfeld who awakened her interest in epidemiology and propelled her toward a career that from its inception pitted her against established authorities in cancer research, most notably Sir Richard Doll.

Doll, a physician at Oxford University, had worked with A. Bradford Hill in the 1950s to prove the link between smoking and lung cancer. In 1981, Doll published another treatise on the causes of cancer, this time with Richard Peto, which dismissed data on rising rates of cancer in people over the age of 64 as being an "artifact" of better diagnosis. Lilienfeld disagreed with those findings, and Davis assisted him in researching and writing an article challenging Doll's conclusions. Lilienfeld died, suddenly, just before the article was published. Davis was left, a postdoctoral fellow in her early thirties, to carry on the debate. "Can you imagine the chutzpah of it?" she asks.

Davis had an opportunity to meet with Doll at the International Agency for the Treatment of Cancer. She admired his pioneering work and felt very flattered that he would give her so much attention. During the meeting he proceeded to explain why she was wrong, seeking to convince her of her errors. After that meeting, she spent nearly 2 years checking, point by point, what he had said, going directly to official death and population records to compute the rates of cancer among older people from specified and unspecified causes. If Doll were right and the overall rate of cancer was not increasing, better diagnoses would mean that increases in cases of specified cancers would be offset by a decline in the number of cancers from unspecified causes. But the evidence seemed to contradict Doll's conclusions. Statistics indicated an increase in all kinds of cancer, both specified and unspecified. Davis's determination to find answers was intensified by news of her own father's bone cancer, multiple myeloma, which had been diagnosed in 1979, when he was 53 years old.

In 1983, after getting a master's in public health from Johns Hopkins and working at the Environmental Law Institute, Davis was recruited by the National Academy of Sciences to direct the Board on Environmental Studies and Toxicology. There she headed a large staff and coordinated the work of hundreds of experts throughout the country in studies such as the one that led to the ban on smoking on most domestic airline flights. When her father died, in 1984, she decided she needed to do more on the issue of cancer in the older population, this time comparing evidence from a number of industrialized countries. She wrote to Alice Whittemore, a highly respected professor of epidemiology at Stanford University, who told Davis that the work she proposed was very important and had to be done.

Davis then requested a leave from the National Academy of Sciences and was instead given a 16-month sabbatical. Colleagues—particularly David Rall, director of the National Institute of Environmental Health Sciences— helped her contact authorities in Germany, France, Italy, Sweden, Denmark, and Great Britain. Davis spent two months visiting those countries, taking her 9-year-old daughter with her and lugging an early-generation laptop computer, with which she "vacuumed up" data from computerized government records. David Hoel, associate director of NIEHS and a collaborator in the project, obtained statistics from Japan. Returning to the United States, Davis divided up the data among an international team of epidemiologists who proceeded to analyze it. Their findings were presented at a workshop at the Collegium Ramazzini in Italy in 1989 and published in both the British medical journal *Lancet* and the *Annals of the New York Academy of Sciences* in 1990.

Basically, the report presented strong evidence that the incidence of many forms of cancer was increasing in all the countries studied, particularly among people over 55, and that brain cancer was increasing in those under 45. Davis and Hoel concluded that the rise was too great and too widespread to be simply an artifact of improved diagnoses. They urged further research on possible causes of these cancers and emphasized the need for prevention.

The publication brought her greater attention and gained her more critics. Doll and others dismissed Davis's work. They said she was seeking publicity. By generating new hypotheses and raising issues about possible causes of cancer, she made many people uncomfortable. But she had her supporters as well. Vilma Hunt, an Australian-born epidemiologist who had done important work a few decades earlier, advised Davis that she would be taken more seriously when she was older. In 1992, after an article appeared in the *New York Times Magazine* discussing the controversy about Davis's research, Hunt wrote a letter to the editor stating that "Davis's professional experience had a unique impact on the very small group of female scientists from earlier generations." The letter continued: "We read familiar words to describe Davis and her work— 'out for publicity,' 'she's not that reliable,' 'unoriginal,' 'wrong.' They were such devastating words for us thirty or forty years ago. They seem so trite and inconsequential today, now that we know what those words really mean."

Other supporters applauded Davis for finding new ways of synthesizing information and interpreting data. Gradually, as concerns about the environment, public health, and global changes became more widespread, her once-radical ideas filtered into the mainstream. By now, even Doll has published papers reporting unexplained increases in certain cancers.

Davis, an avid skier, likens the experience of coping with attacks from colleagues to knowing how to take a fall: "You have to learn to fall, and relax in your fall, and get back up again. I think that applies to professional life as well, because one always learns from it. If you are putting things out there that people aren't going to like to hear, you have to be prepared for the fact that not everybody's going to be happy." She notes that she has had many fruitful critical exchanges with people with whom she doesn't always agree, but from whom she always learns something.

When her sabbatical concluded, Davis returned to the National Academy of Sciences as a scholar in residence. At the Collegium Ramazzini workshop in 1989, Davis had founded the International Breast Cancer Prevention Collaborative Research Group. One day in 1992 she received a phone call from Representative Bella Abzug, whom she had never met, telling Davis to appear at City Hall to testify on breast cancer at a special hearing. Women on Long Island had been developing breast cancer at a high rate. Those who had lived there more than 40 years developed 4 times more breast cancer than those who had lived there less than 4 years. A study conducted by the Centers for Disease Control concluded that these statistics could be explained by the standard risk factors, but Long Island women's organizations had mobilized to demand better answers. Davis testified, and breast cancer became a major focus of her work, particularly after she joined the World Resources Institute, a nonprofit environmental research group, in 1995. At WRI she studied links between the environment and health and examined effective public policy options. "Spurred by Bella's demands," Davis says, "my colleagues and I began to look again at what was known about breast cancer and to consider whether avoidable environmental agents could be involved."

In countless lectures and publications, on the WRI web site, in a video titled *Exposure,* and in a CD-ROM (requested by women in developing countries), Davis has raised questions about possible links between cancer and chemicals in the environment. She explains that fewer than 10 percent of breast cancers develop from inherited genetic factors. Known risk factors, such as early menarche or late menopause, account for a relatively small portion of other breast cancers. In the *World Resources 1998–99* report, she wrote: ". . . a growing and complex array of evidence suggests that the general external environment—including behavior, diet, and physical and chemical exposures—plays a major role in fostering breast cancer. . . . Environmental exposures may damage genes directly or they may affect the overall production of growth-regulating hormones, such as estrogen, progesterone, and other such naturally produced substances."

Davis suggests that hormone-mimicking substances, which she calls "xenoestrogens," can be produced from some plastics, pesticides, fuels, and pharmaceuticals. Some xenoestrogens may be responsible for triggering breast cancer or sexual aberrations, such as a preponderance of female births or reproductive disturbances in men. There is growing evidence from laboratory experiments and studies of wildlife that supports this theory. Human evidence is more difficult to obtain, owing to the complex nature of cancer (which is rarely due to one single cause) and the near impossibility of maintaining strict experimental controls with human subjects. "The debate isn't whether the environment causes 5 percent of cancer or 10 percent of cancer," Davis emphasizes. "The debate is that whatever about cancer is due to the environment, we can do something about it. It's avoidable. It's controllable. Unlike so much of cancer, which comes from who you were born to, or what you'd eaten three decades ago, or whether or not you had children, the things in the environment are something you can change, the government can change, the private sector can change. That is why we have to pay attention to these things: because it's something that can be controlled."

Called upon to serve as senior advisor to Assistant Secretary for Health Philip Lee in 1993, Davis also advised Surgeon General Joycelyn Elders on epidemiology. She worked on the National Action Plan on Breast Cancer and held several visiting professorships. In 1994 she was appointed by President Clinton to the National Chemical Safety and Hazard Investigation Board. Davis has received many awards and honors for her work, including the Breast Cancer Awareness Award of the American Cancer Society. She was designated a "Global Guru" by GLOBE, a European parliamentary and environmental organization.

While continuing her work on breast cancer, collaborating on research projects as well as a public awareness campaign called "Better Safe than Sorry," she has also engaged in studies of urban air pollution and its effects on children. "The air pollution angle," she says, "evolved out of a concern about the obvious issue of climate change, and the fact that people can't relate to what's going to happen fifty years from now. . . . I was concerned about this question of urban excess. Then I got to thinking about what creates air pollution. The big issue 50 years ago in Donora was the power plants and the factories. In the United States today, new plants and factories are pretty well controlled, although the older ones can still be problematic. The really big issue now is cars, trucks, and buses, especially those that use diesel fuel. . . . What we are doing to our health and to the planet by the use of cars now, and lack of use of public transport, is astonishing!"

Children, particularly in developing countries, are being exposed to heavily polluted air in unprecedented numbers. Their growing bodies are more susceptible to toxins than those of adults, particularly where children do not have adequate diets or medical care. A report issued by Davis's office concludes that investing in improved transport and energy technologies that reduce the use of fossil fuels could have an immediate, positive effect on the health of millions of the world's urban children. "I really think," Davis says optimistically, "that within my lifetime people are going to ask 'Remember when we used to drive everywhere, poisoning the air and weakening our bodies?'"

Davis believes that there will be support to move in this direction. "Resistance may come from some in industry," she says, "but not from all. I think there's generally a move now, as one of my colleagues, Wolfgang Sachs says, to make sure that we do the right things, and that we do things right. Those are two different things. . . . What are the most important things to do? Should we all be drinking designer water? I don't think so. I think we ought to get our water clean for everybody in the city. Should we all be buying organic produce? If you can afford it, I suppose it makes you feel better and it does taste better, but we ought to get the food supply cleaned up for everybody. Those are the kind of things we need to pay attention to."

One of Davis's aims is to generate greater public awareness about the issues that concern her. "When does innovation have an impact?" she muses. "It's a very interesting question to me. I think that the opportunity to share information is what distinguishes modern science from, if you will, ancient science." She is writing a book about environmental risk factors in public health, intended to share information she has gathered with a wide popular audience; the working title is *Uninformed Consent: How the Environment Shapes Sex, Life, and Death.*

Two near-death experiences—one from an adverse reaction to a drug that should not have been on the market (and which she later worked to get off the market) and the other from multiple bee stings—have deepened her religious feelings and intensified her commitment to her work. She used to insist that there was no real connection between her childhood in Donora and her crusade against environmental toxins. Now she is more reflective: "I think some things happen for a reason, and sometimes, if we're lucky, we figure it out."

Davis's former office at WRI (she left in 2000 to become a visiting professor at the Heinz School for Public Policy and Management at Carnegie Mellon University) overflowed with publications, papers, and family photos. There were snapshots of ski trips with her daughter, then a student at Oberlin College, a portrait of her son in uniform as a lance corporal in the in US Marine Corps, and pictures of Davis mountain climbing with her husband, who is now a senior economist and a chief negotiator on global climate issues at the US State Department. Among her books is a volume by Rashi, an eleventh-century French rabbi and scholar famous for his commentaries on the Torah and the Talmud, the collection of rabbinical writings essential to Judaic studies. Davis often draws on stories from the Talmud to communicate her ideas. One that she tells is about a group of farmers and workers who are complaining to a rabbi that they have been given too much to do. "We can't complete it. We'll never get the work done," they protest. "How can you possibly expect us to do this?" The rabbi says: "Look, it's not for you to finish the task. But you must begin it."

"That story really applies to the work we're trying to do now," Davis observed. "It's a very complex task. It involves rethinking how our lives are organized and how we can do things better, more efficiently. How can we de-materialize, do more with less, be smarter? Those are the challenges. We won't finish it. But we need to start." After a pause, she adds: "Sometimes people face a task that looks hopeless. But I think, being a Jew, that we always have hope. That's the lesson of the Bible, isn't it?"

HOW CAN INNOVATIONS IN ALTERNATIVE ENERGY SOURCES AFFECT THE ENVIRONMENT?

Oil shortages, nuclear waste, heating fuel prices, smog, blackouts and brown-outs—energy issues are part of everyday life. An inherent tension between increased demand for energy and heightened awareness of potential damage to the environment has spurred new generations of inventors to seek solutions. The historian Rudi Volti writes about one example: inventions in emissions control for automobiles. Volti argues that air quality in Southern California has improved as a direct result of innovations ranging from PCV valves to vapor recovery at gas stations to the catalytic converter. Is it enough, though, to reduce emissions?

Amory Lovins believes it is not. In his overview of energy use, he offers a number of alternative methods for getting the most from a variety of energy sources. One of these is the "Hypercar," a lightweight automobile prototype powered by a combination of gas, electricity, and/or fuel cells and capable of traveling nearly 200 miles per gallon of gas while producing zero emissions. Lovins also supports new methods of building construction, including superefficient windows and thick insulation, to cut energy use dramatically.

The inventor Subhendu Guha offers his own innovation in energy production: solar roofing shingles. Based on his research in amorphous silicon, Guha devised an alternative to heavy, rigid solar panels mounted on rooftops. His thin, flexible solar cells are built into shingles that are installed much like traditional roofing shingles. Backpack versions of the solar cells are used by campers, boaters, and the military to power equipment in remote areas.

The innovations highlighted in these essays combine increased technological efficiency with heightened protection of the environment. Since it appears unlikely that energy demand will decrease in the twenty-first century, devising alternative ways to create and conserve energy while minimizing adverse effects on the environment is critical.

REDUCING AUTOMOBILE EMISSIONS IN SOUTHERN CALIFORNIA: THE DANCE OF PUBLIC POLICIES AND TECHNOLOGICAL FIXES

RUDI VOLTI

For many years I have taught at a small liberal arts college in Southern California. When first-year students arrived at the college in the early 1970s, they settled into the usual things that occupy freshmen. A few weeks would go by, and then they would make a remarkable discovery: tall mountains would appear to the north as autumn weather dissipated the heavy blanket of smog that had obscured them. Today, the air is not perfectly clear in September, but students are aware of the mountains from the day they move into the dormitories. The region's partial victory over smog illustrates the successful use of technological fixes for a problem that was itself caused by technology. But it also shows that technological fixes have to be complemented by appropriate policies if they are to be successful. Moreover, these policies have to resonate with the political environment if they are to have a chance of success. This essay does not attempt to provide a comprehensive review of the war against smog. Rather, it presents a brief description of technologies used to reduce the emissions of cars and light trucks, followed by a summation of the government polices that have motivated the development and use of these technologies. These provide a background for the final section of the essay, which notes some of the circumstances under which technological advances can be stimulated by appropriate public policies.

HOW CARS MAKE SMOG

In the 1940s, residents of Southern California began to notice an atmospheric condition that obscured their vision, irritated their eyes, and hindered their breathing. It was dubbed "smog," an etymological mixture of "smoke" and "fog," although its actual constitution was far more complicated.

impact of cars & trucks

A major contributor to air pollution in Southern California was the exhaust from cars and trucks, a fact vehemently denied by the automobile industry until it was irrefutably proven by Arie Haagen-Smit at the California Institute of Technology in the early 1950s. Today, cars and light trucks account for about 60 percent of smog-creating emissions in the region, so any successful effort to reduce air pollution has to take full account of the emissions produced by the region's large vehicle population.

Motor vehicle emissions are converted to smog through a series of chemical reactions that occur in the presence of sunlight.[1] Uncontrolled vehicles produce the constituents of smog in a number of ways: through the venting of vaporized gasoline, the emission of gases from the engine's crankcase, and most important through the combustion process that converts gasoline into the power that propels them. When a charge of air and vaporized fuel is compressed and then ignited in an engine's combustion chamber, not all of the fuel is completely combusted; some unburned hydrocarbons are emitted. At the same time, high temperatures and pressures within the combustion chamber convert atmospheric nitrogen into various oxides of nitrogen (NOx for short). The exhaust gases are then released into the atmosphere, where the ultraviolet portion of sunlight breaks down NO_2, one of the oxides of nitrogen, into NO. The liberated oxygen atoms then combine with atmospheric oxygen (O_2) to produce one of the major constituents of photochemical smog: ozone (O_3), a major irritant to the respiratory system. At the same time, other oxides of nitrogen are converted into a variety of compounds, notably the peroxyacyl nitrates that contribute to the eye-burning effects of smog. Residual NO_2 adds to the general nastiness by obscuring vision with a brown haze.

chemistry of photochemical smog

other byproducts

Combustion of gasoline in an engine also produces carbon monoxide (CO), carbon dioxide (CO_2), water vapor, sulfur dioxide, and particulates. Strictly speaking, these are not constituents of photochemical smog. They are still a significant problem, however. Recent years have seen a growing concern about the emission of CO_2 into the atmosphere because it may contribute to a "greenhouse effect" and consequent global warming. Solid proof of this phenomenon remains elusive, but the increasing likelihood that today's cars and trucks are contributing to global warming may necessitate the eventual supplantation of fossil-fuel burning internal-combustion engines by other sources of power; no matter how clean it is in other respects, an internal-combustion engine powered by a carbon-based fuel will always produce CO_2.

INDICATIONS OF IMPROVED AIR QUALITY

Although no technology for the control of CO_2 exists, substantial progress has been made in the reduction of other emissions, and the skies over Southern California are much cleaner as a result. This accomplishment has been significant, for Southern California is the ideal location for the production of photochemical smog: it has a huge car population, valleys that trap stagnant air, and a frequent inversion layer that prevents emissions from dissipating into the upper atmosphere. For many years it seemed as though the region would be perpetually blighted, but technological advances have produced a noticeable improvement in air quality through a reduction of vehicular emissions and other measures. By the 1990s, tailpipe emissions of carbon monoxide and hydrocarbons had been reduced by 96 percent compared to cars built in the 1960s, while oxides of nitrogen have been reduced by 76 percent. The impressive progress that has been made in reducing air pollution can be seen in the dramatic reduction of one index of smog severity, the con-

Specific situation in So. CA

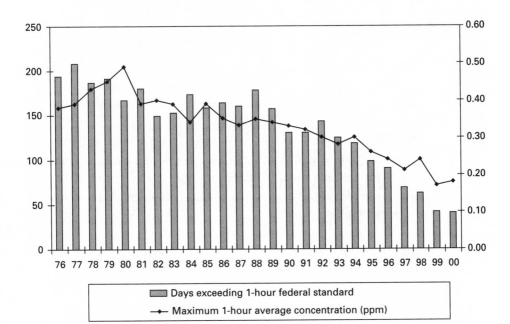

FIGURE I

Ozone air quality trends, South Coast Air Basin (California), 1976–2000. Left vertical axis: days exceeding 1-hour federal standard. Right vertical axis: maximum 1-hour average concentration (ppm). (South Coast Air Quality Management District)

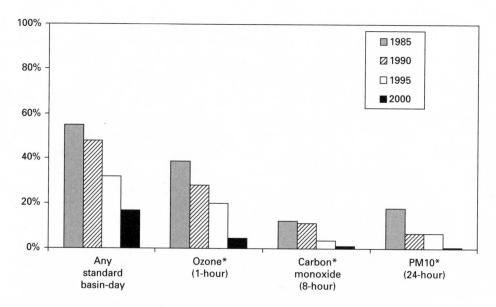

FIGURE 2

Percentage of days exceeding federal health standards, South Coast Air Basin (California), 1985–2000. *: percentage of days exceeding at site with highest exceedences. (South Coast Air Quality Management District)

centration of ozone in Southern California's skies. As can be seen in figure 1, the number of days when one-hour ozone concentrations exceeded the federal standard declined from more than 200 in 1977 to fewer than 50 in 2000. Encouraging gains have also occurred in the reduction of other pollutants. As figure 2 indicates, along with experiencing a decrease in ozone, Southern California has seen substantial progress in the reduction of carbon monoxide and PM10 particulates (particles with diameters of 10 microns or less).

PUBLIC POLICY AND THE REDUCTION OF AUTOMOTIVE EMISSIONS

A substantial portion of these gains can be attributed to a set of technological fixes that have dramatically lowered automotive emissions. But these fixes were implemented because the federal government and the government of California made clean air a major policy objective.[2] Government initiatives were essential for addressing the problem of poor air quality because individual efforts will never produce cleaner skies. Air pollution is the classic example of a "negative externality," that is, a market transaction's negative effects on parties not involved in the transaction. When an individual buys a car, he or she gains the benefits of car ownership; the seller benefits from the money earned through the sale of the car. But that is not the end of the matter; another car is now on the road, and its emissions contribute to the poor air quality that affects everyone in the region. Although they both suffer from poor air quality, the buyer and the seller of the car have no stake in addressing the problem by themselves. A vehicle equipped with pollution-control technology costs more, yet it provides no additional benefit to its purchaser. Even a buyer with a yearning for better air quality will not pay extra for a car or truck with reduced emissions, as an improvement made to a single vehicle results in an infinitesimal gain in air quality. Air pollution can be successfully addressed only when all or most vehicles have cleaner exhausts, and this requires a collective effort of some sort.

One possible way of producing a collective effort is to levy a pollution tax, perhaps coupling it with the establishment of a market for pollution credits.[3] People could drive dirty vehicles, but they would have to pay a tax tied to the amount of pollution they produce. Conversely, the operator of a vehicle that falls below some stipulated emission level might receive a credit that could be sold to the operator of a vehicle that failed to meet the standard. Programs of this sort have been used to reduce the emissions of stationary power sources, but the large number of motor vehicles requiring monitoring would make such a program very difficult to administer, at least with present information-gathering capabilities. Efforts to reduce vehicle emissions have instead been based on another kind of policy weapon, the setting and enforcement of emissions standards and the mandating of certain technologies for the achievement of these standards.

In the United States, the process of limiting emissions through regulation began in 1961, when the State of California began to require the installation of positive crankcase ventilation (PCV) valves on new cars beginning with the 1963 model year. That simple step reduced emissions of unburned hydrocarbons by about 20 percent. In 1966 the California legislature established the nation's first emissions standards for automobiles, which for a number of years forced manufacturers to build "California cars" that were cleaner than those destined for the other 49 states.

The first piece of federal regulation to directly target automotive emissions, the Motor Vehicle Air Pollution and Control Act of 1965, simply allowed the Secretary of Health, Education, and Welfare to set emissions standards, but later in the same year Congress passed legislation that required that Secretary do so. Beginning with the 1968 model year, federal standards set hydrocarbon emissions at no more than 275 parts per million (ppm) and put the acceptable level of carbon monoxide at 1.5 percent of total emissions. By the 1970 model year these were required to drop to 180 ppm and 1.0 percent, respectively. The federal Clean Air Act, passed in 1970, mandated a 90 percent reduction in emissions of nitrogen oxide by 1976. New standards for hydrocarbons and carbon monoxide were even more ambitious; amendments to the act passed in 1977 required a 96 percent reduction of these pollutants.

Another set of amendments passed in 1990 gave the automobile industry its current emissions standards, which are now defined in terms of pollutants per mile rather than as percentages of total emissions. Beginning with the 1994 model year, carbon monoxide was to be limited to 3.4 grams per mile, hydrocarbons to 0.25 gram per mile, and NOx to 0.4 gram per mile. There was also a requirement that emissions controls perform acceptably for 100,000 miles. Regions currently not in compliance with current air-quality standards were obliged to meet them according to a specific schedule or face a possible loss of federal highway construction and maintenance funds.

In 1990 the state of California went one step further. The California Air Resources Board (CARB) decreed that by 1998 2 percent of new cars sold had to be zero-emissions vehicles (ZEVs), with the ratio rising to 5 percent in 2000 and 10 percent in 2003. Faced with intense opposition from manufacturers and the absence of a receptive market, in 1996 CARB rescinded the 1998 and 2000 mandates while retaining the one for 2003. The program was further modified in early 2001, when CARB enacted a complicated schedule that granted manufacturers extra credits for such things as the early introduction of ZEVs and the sale of ZEVs with ranges beyond 50 miles, as well as an award of half a ZEV credit for "partial zero emission vehicles" such as gasoline-electric hybrids. The program also stipulated that the ZEV requirements would increase after 2008, rising to 16 percent of the light vehicle fleet by 2018.[4]

Whether these standards will be met is an open question. The only zero-emissions vehicles available for at least the next decade are battery-powered electrics. The high cost and limited range of these vehicles makes their widespread usage problematic, and up to now the few electrics that have been put on the market have met with a tepid consumer response at best.

hybrids

fuel cell

Hybrid vehicles have enjoyed greater acceptance, but they will not be a significant part of the vehicle fleet for many years to come.[5] At a considerable distance over the commercial horizon are vehicles powered by fuel cells that, theoretically at least, emit nothing but water vapor.[6] Although prototype fuel-cell vehicles are currently being tested by a number of manufacturers, they are not likely to be a commercial reality until well into the second decade of the twenty-first century, if even then.

TECHNOLOGIES FOR CLEANER AIR

With alternatives to the internal-combustion engine many years away from practical application, the motor vehicle industry has had to develop a number of technologies to reduce the tailpipe emissions produced by conventional engines. The centerpiece of these efforts has been the catalytic converter. California was the first place to require its use, mandating that all 1975 model cars be so equipped. A catalytic converter has an internal structure made of tiny ceramic pellets or a ceramic honeycomb that gives the interior of the converter a surface area the size of a football field. The converter's internal surfaces are coated with metals that catalyze chemical reactions: palladium, rhodium, and platinum. The first catalytic converters supported only an oxidation process that turned unburned hydrocarbons and carbon monoxide into water vapor and carbon dioxide. Within a few years cars began to be equipped with three-way catalytic converters that, in addition to the first two functions, support a reduction process that turns oxides of nitrogen into free nitrogen and oxygen.

A car produces the most emissions when a cold engine is started, as catalytic converters work effectively only at operating temperatures of 250–300°C (480–570°F). Tests conducted by the Environmental Protection Agency have shown that in the course of a 10-mile trip made by a catalytic converter-equipped car 80 percent of hydrocarbon emissions occur during a warm-up period of $2\frac{1}{2}$ minutes.[7] To counter this problem, it is likely that the next generation of catalytic converters will be kept at operating temperature by the car's electrical system.

Catalytic converters require just the right amount of oxygen admitted into an engine's combustion chamber. If there is too little, the unburned hydrocarbons and carbon monoxide are not oxidized; if there is too much, the NOx will not be reduced to free oxygen and nitrogen. Consequently, today's cars are equipped with oxygen sensors, fuel injection, and computerized engine management systems that keep the fuel-air ratio within precise limits. The reduction of automotive emissions also has required the use of other technologies such as exhaust gas recirculation (EGR) devices and vapor recovery systems for the vehicle's fuel tank. Finally, since catalytic converters are quickly destroyed by lead, that additive (which was used to allow engines to have higher compression ratios) has been removed from gasoline, which has had the additional benefit of reducing the presence of a highly toxic element. Today's engines are vastly more complicated and sophisticated than the ones found under the hoods of the cars of a few decades ago. Given the demands for cleaner air, the continued use of the internal-combustion engine, a nineteenth-century invention, now requires a host of technologies developed in the late twentieth century.

KEEPING CARS CLEAN

Emissions-control devices have made a major contribution to cleaner air, but their effectiveness depends on their working properly. As was noted above, federal law requires that anti-pollution devices have to maintain their effectiveness for 100,000 miles. But this mandate is of value only if there is some way to ensure that the vehicle population remains in compliance. Although emissions standards are set by the federal government, the enforcement of these standards is the responsibility of individual states. The federal government, however, is able to retain some control over the process by threatening to withhold highway funds from states deemed to have inadequate testing procedures.

The 1990 amendments to the Clean Air Act required that testing procedures had to simulate actual driving conditions. The place where the procedure was performed was left up to the individual states. The federal government has favored a network of facilities that do nothing but testing, but this has been strongly resisted by owners of gas stations and repair shops who perform emissions testing as an adjunct to their other services. The problem with this arrangement is that the presence of a large number of facilities makes it difficult to inspect the equipment and personnel performing the tests. There is also an inherent conflict of interest when enterprises are in the business of making repairs in addition to conducting emissions tests.

It is likely that the main value of smog inspection systems is that they allow the detection of the relatively small number of cars and trucks that are responsible for a disproportionate share of vehicular emissions. A study conducted by the National Research Council reported in 1991 that 50 percent of the ozone-forming emissions from mobile sources come from fewer than 10 percent of the vehicles in operation. Getting these "gross polluters" into compliance (or off the road, if this is not possible) may be the most cost-effective way of reducing automotive emissions.[8]

The issue of cost effectiveness must be faced squarely in any serious discussion of emissions control.[9] Government-mandated technological fixes have added hundreds of dollars to the cost of a car, while the costs of California's inspection program and required repairs come to about $500 million annually. In return, Southern California has benefited from cleaner air, but the region is still not in compliance with federal standards, and more drastic (and expensive) measures may be required in the years to come. We are well past the point where the installation of simple devices like PCV valves, and even complex ones like catalytic converters can effect substantial improvements. It is likely that diminishing returns have set in with regard to the benefits obtained from anti-pollution expenditures.

At the same time, there is no easy consensus in regard to additional expenses that should be borne in pursuit of cleaner skies. There is some evidence that air pollution in Southern California is associated with higher risks of bronchitis and asthma,[10] but it can always be argued that all good things, such as the personal mobility afforded by the automobile, will have some unfortunate consequences. Greater precision can be brought to the issue by conducting cost-benefit analyses that attempt to put a monetary value on the illnesses engendered by air pollution, but carrying out a precise epidemiological study would be a very difficult task, given the many factors involved in ill health, as well as the continual movement of people into and out of Southern California. And there is simply no way to put a dollar value on the ability to see the mountains or to play a game of tennis without feeling that one's lungs are being reamed out.

The complexities of cost-benefit analysis aside, there can be no question that the skies over Southern California are much better than they were before the effort to build cleaner cars was launched. Reduced levels of air pollution show that some problems can be successfully addressed through the application of one or more technological fixes. Smog pollution has been effectively addressed through the development of new technologies because it is a problem with a definite point of origin. A large portion of the emissions that cause smog can be traced to a single set of sources—cars and light trucks. Its unambiguous origins make smog a categorically different problem than, say, violent crime, which is the result of a vast number of things ranging from poor family environments to chronic unemployment to violent media programming. When the source of a problem can be definitively identified, a crucial step has been taken toward the development and application of successful remedial technologies.

The next step in the solution of a problem may be the pursuit of a technological fix through the invention of new devices like catalytic converters. But these devices will remain bottled up in research laboratories unless decisions are made to put them to use. This is an inherently political process, one in which priorities are set and resources are allocated. Technology can fix things only when we can collectively agree on what needs fixing and how much we are willing to pay for it. When such agreement exists, there is at least a reasonable chance that effective policies will emerge from the political arena. For decades, smog in Southern California was a problem that was never far from public consciousness, especially since much of the appeal of the region lay in its benevolent climate and the outdoors-oriented lifestyle that it fostered. Consequently, policies oriented toward the reduction of air pollution emerged in a receptive political environment where smog was producing universal discomfort. Everyone living in Southern California was affected by air pollution, irrespective of their social class, race, ethnicity, gender, age, or political affiliation. The reduction of air pollution has been an issue that has galvanized the citizenry as a whole, an issue that elected and appointed officials cannot easily disregard.

Political efforts to alleviate smog also benefited from the fact that, in contrast with what has happened in regard to other sources of health problems (cigarettes come to mind), there has been no organized group with a vested interest in the perpetuation of the problem. To be sure, car manufacturers and others implicated in the production of smog were at times unenthusiastic and even hostile to efforts to clean up the emissions generated by their products. Quite naturally, they feared that the cost of emissions equipment would raise the price of the vehicles to the point where sales and profits would be harmed. In fact, nothing of the sort has happened. The cost of vehicles has gone up as the negative externalities have been internalized, but cars appear to be relatively price inelastic, so higher prices have not significantly limited sales, and consumers have absorbed the costs of emissions-control equipment. Another cost of cleaner air that has been sloughed off onto the consumer has been the time and money expended in passing an emissions test every 2 years in order to renew a car or truck's registration. California's vehicle inspection system has not always been a model of either efficiency or consistency, but there has been no widespread revolt against it. Again, we see that elements of a technological fix are more likely to be implemented when they have at least tacit political support.

THE FUTURE OF SMOG IN SOUTHERN CALIFORNIA

As with most things, efforts to reduce vehicular emissions are subject to diminishing returns. Thanks to the improvements noted above, cars and light trucks are much cleaner than they were 20 years ago. Even so, the region still has the worst air in the United States. In 1989 the regional agency responsible for smog reduction, the South Coast Air Quality Management District, released an Air Quality Management Plan that would have required massive changes in ensuing years, such as having 40 percent of all passenger vehicles powered by methanol or electricity, while at the same time reducing the number of vehicle-miles traveled to the 1985 level. Loud protests accompanied the plan, which subsequently went through a number of revisions that diluted its proposed mandates. Yet if reducing smog in 1989 was problematic, it is even more so today and will be into the future. The population of California now numbers nearly 33 million. The state has nearly 17 million cars and light trucks, more than one for every two inhabitants.[11] According to recent projections, California will have up to 45 million people by 2020, and a large percentage of that increased population will be located in Southern California. Under these circumstances, just holding the line on air pollution will be difficult; making the skies cleaner will pose Herculean technical and political problems. Engineers, politicians, bureaucrats, and the citizenry will encounter many challenges down the road—or should I say the freeway?

NOTES

1. On the chemistry of smog formation, see Robert Jennings Heinsohn and Robert Lynn Kabel, *Sources and Control of Air Pollution* (Prentice-Hall, 1999), 281–313.

2. For a review of federal laws and regulations regarding air pollution, see Gary C. Bryner, *Blue Skies, Green Politics: The Clean Air Act of 1990 and Its Implementation* (CQ, 1993, 1995). For a brief history of efforts by the local and state governments in California, see Jeffrey Fawcett, *The Political Economy of Smog in Southern California* (Garland, 1990), 78–91.

3. Bryner, 256, 257.

4. California Air Resources Board, "Zero-Emission Vehicle Program Changes," accessed June 10, 2002 at http://www.arb.ca.gov/msprog/zevprog/factsheets/zevchanges.pdf.

5. Today's hybrid vehicles run cleaner because they consume less fuel than conventionally powered cars of equivalent performance. Future hybrids employing engines running at constant speed have the potential to be even cleaner because it is easier to control the emissions of engines that are not continually speeding up and slowing down. For a discussion of hybrid vehicles, see Victor Wouk, "Hybrid Electric Vehicles," *Scientific American* 277 (1997), no. 4, 70–74.

6. A. John Appleby, "The Electrochemical Engine for Vehicles," *Scientific American*, 281 (1999), no. 1, 74–79.

7. I. Gottberg, J. E. Rydquist, O. Backlund, S. Wallman, W. Maus, R. Bruck, and H. Swars, New Potential Exhaust after Treatment Technologies for Clean Air Legislation. SAE Technical Paper 10840, 1991, 3.

8. Lamont C. Hempel et al., Going After Gross Polluters: Remote Sensing of On-Road Vehicle Emissions. Center for Politics and Policy, Claremont Graduate School, 1992.

9. RAND Corporation, Rx for Urban Smog: Find and Fix Those 'Clunker' Cars. Accessed June 10, 2002 at http://www.rand.org/publications/randreview/issues/RRR.fall94.calif/smog.html. For an evaluation of the costs and benefits of pollution reduction, see J. Clarence Davies and Jan Mazurek, *Pollution Control in the United States: Evaluating the System* (Resources for the Future, 1998), 123–150 and 278–280.

10. James M. Lents and William J. Kelly, "Clearing the Air in Los Angeles," *Scientific American* 269 (1993), no. 4, 38.

11. State of California, Department of Finance, Table J-3, "Transportation and Public Utilities," accessed June 2002 at http://www.dof.ca.gov/html/FS_DATA/STAT-ABS/sec_J.htm.

NEGAWATTS, HYPERCARS, AND NATURAL CAPITALISM

AMORY LOVINS

This being a publication of the Smithsonian Institution, I will start with a little history. For many of us, a lot of the history of ideas about what to do about energy, not a new problem in society, starts rather recently. You could take it much further back. There are many examples of technical fixes for energy shortages, such as in Sweden in 1767, when they were running out of firewood and the king commissioned Baron Carl Johan Cronstedt to develop a double-efficiency stove. Baron Cronstedt, an architect, enlisted the help of Baron Fabian Wrede, who was experienced in stove design. Together they devised a double-efficiency stove, and the king then decreed that everybody would use it. That was the end of the firewood shortage.

But in the United States, we first had to get over some mistakes in thinking about what the problem is. I had an article in *Foreign Affairs* in 1976 that started with the conventional view of what the energy future of the United States would look like over the next 50 years (figure 1). According to that view, we would use depletable fuels faster and faster, convert them in ever larger, more complex, more centralized plants into premium forms, mainly electricity, and subsidize them enough that the high cost of those new supplies would not depress demand below the projected level. Now, there are a lot of reasons that future did not happen. It was too slow, too costly, and too disagreeable. But along the way we redefined the energy problem. My article suggested what is now called the end-use least-cost approach, which asks a different question and therefore gets a different answer.

The question had been assumed to be "Where can we get more energy, of any kind, from any source, at any price?" But it made more sense to start by asking "What do we want this stuff for in the first place?" People do not, in general, want lumps of coal or barrels of sticky black goo. They want instead the services that energy provides: hot showers, cold beer, mobility,

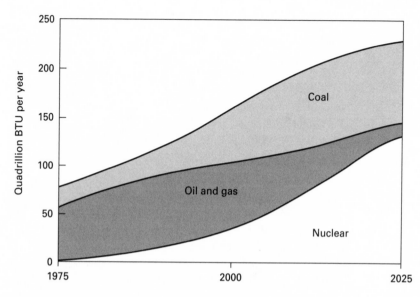

FIGURE I

"Schematic future for US gross primary energy use." (Rocky Mountain Institute)

comfort, illumination, the ability to bake bread and make steel. If those are the kinds of services we are after, then we ought to start by asking "How much energy, of what kind or quality, at what scale, from what source, will do each desired task in the cheapest way?"

If you ask that question, you get a very different answer, even when mainly assuming the same economic growth as before (figure 2). You end up with a stabilizing, and even declining, amount of total energy used as you wring out the losses in converting, distributing, and especially using it. So you use less and enjoy it more. At the same time as the depletable fuels become less available, or less pleasant, I thought they could be gradually displaced by appropriate renewable sources ("soft technologies"), a category which the Royal Dutch Shell group, planning in 1984, thought might plausibly be providing half the world's energy by 2050, which is a bit off the right end of the graph in figure 2. Now they say that is highly probable because the renewables are getting ever cheaper as we make more of them. A lot of funny things happened along the way.

We subsidized energy very heavily and still do, spending $20 billion–$30 billion a year to make it look cheaper than it really is. Most of the subsidies go to the least attractive and competitive kinds of energy, which happen to have the strongest lobbies. The same is true in R&D spending. In the crit-

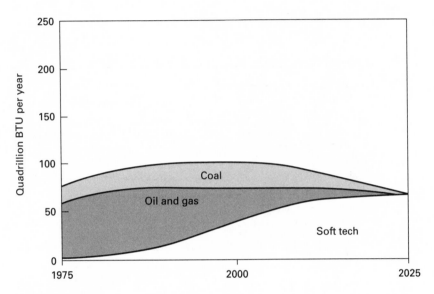

FIGURE 2

"Alternative illustrative future." (Rocky Mountain Institute)

ical period 1973–1989, for example, this is where our investments in energy research went, at a federal level, and they bear essentially no relation to where we got our new energy supply. In fact, from 1979 to 1986 we got about 4.5 times as much new energy from savings as from all increases in supply, and slightly over half of those increases in supply were from sun, wind, water, and wood, the minority from oil, gas, coal, and uranium. Not too bad for the poor cousins in the investment lottery.

It is kind of fun now to look back on what actually happened. The official forecasts of US energy use were all heading toward about 150 quadrillion BTUs per year, and I was very heavily criticized in 1976 for saying they might remain below 100 BTUs. But we have already cut $150 billion–$200 billion off the nation's annual energy bill by wringing more work out of the energy we use. Almost all of that happened before 1986, when efficiency had depressed demand so much that it crashed the price, which has been going down ever since. Today we have gasoline that is much cheaper than bottled water and cars that are cheaper per pound than Big Macs. It is not surprising that we have a whole lot of driving going on. Since 1986, there has been rather little energy saving in those sectors. In fact, we have even gone back slightly in "car" efficiency in recent years. (Many of today's "cars" are juggernauts.)

As a result, energy use has taken off again, slowly and pretty much in line with economic growth. Yet we are still wasting $300 billion a year worth of energy. We have barely scratched the surface of how much efficiency is available and worth buying, and that waste, that low-hanging fruit that is mushing up around our ankles, actually keeps accumulating, because we have learned new tricks about saving energy faster than we are using up the cheapest opportunity. This is not a technologically static field; quite the contrary. There are also stunning advances on the supply side. Oil is getting ever cheaper, even though we are using up more of it faster, but the cost of finding and lifting it is going down even faster.

In the electricity business, plants kept getting bigger until about 1970. Since then, the big power plant business has fallen off a cliff. The ordering rate of fossil-fueled steam power plants at the end of the twentieth century was back to Victorian levels. We returned to the size distribution of new plants that we had in the 1940s, and were very rapidly heading for the size distribution we had in the 1920s.

So history takes plenty of twists and turns along the way as technology changes, and that has enormous implications. I think the biggest surprise amidst all of the dramatic transitions of the past 20-odd years is not in economics, although there were certainly a lot of bad surprises in that sphere. We had a little bulge in investment among investor-run utilities in the 1970s and the 1980s, almost all of which was the nuclear experiment. By the late 1990s that nuclear experiment was delivering somewhat less energy than wood, but it cost about a trillion dollars and was at best a near-death experience for quite a lot of utilities that violated Miss Piggy's fourth law: Never eat more than you can lift. The biggest change is not in economics; it is in technology.

I think the biggest change of all is in design mentality rather than in a specific invention; in new ways to think about how we use technology to get, in this case, big energy savings that cost less than small energy savings. That is the big surprise, because we are all conditioned by neo-classical economics, and by a lot of everyday experience, to think that the more energy you save, the more and more steeply the cost of saving the next unit rises through diminishing returns until you hit the limit of cost effectiveness, and you have to stop. That is often true, up to a point. But there is often another part of the curve that the economic theorists don't tell you about and don't know about. It is something that you only discover from engineering practice. That is, if you go on further, and save even more, you can often make the cost come down again. From the physics analogy, we call this "tunneling through the cost barrier." You end up with big savings that cost less than small savings. Let me give some examples of how that works.

"creativity"

In the 1960s I worked for Edwin Land, the inventor of the instant camera, who I once heard say that invention is simply a sudden cessation of stupidity. (Someone else recently called it a "spasm of lucidity.") Land also said that people who seem to have had a new idea have often simply stopped having an old idea, which is quite true, and not easy to do. We are slowly getting better at it. Consider figure 3, which comes from the great engineer Paul MacCready, chairman of AeroVironment in Monrovia, California.

Many of you have probably seen textbooks on creative thinking, in which you are given this problem of finding the solution that connects these nine dots with only four lines without lifting your pen from the paper. You are supposed to think "Let's see, one, two, three, four, oops, five, that doesn't work. One, two, that doesn't work." Eventually you are supposed to think out of the box. One day, Professor Edward DeBono, the British guru of cognition, reportedly came into his class quite irked because one of his students had just solved the problem in three lines. If it makes you feel better, I didn't see this at first, either, but of course, these are not mathematical dots with zero diameter, and therefore, as long as your paper is wide enough, you can always draw the lines in that way.

Then the students started feeling a bit liberated, and they started solving the problem in one line. It turns out there are a lot of ways to do that. For example, there is the Origami method, and there is the geographic method, and there is the one where you get out the scissors. Or, there is the statistical method, in which you just crumple up the paper, and if you stab it enough times with a pencil, eventually it will go through all nine dots at once. The one I like best came from a 10-year-old girl who took a big wide thing, like a paint brush, and went "slush" through all nine dots and said

FIGURE 3

"Hey, it didn't say it had to be a skinny line!" Right, that was not part of the rules. I think the lesson MacCready tells us that we learn from this is that the original design assignment was misstated as "Find one solution with only four lines." If we had been asked instead to find as many solutions as possible with as few lines as possible by eight tomorrow morning, we would have been more creative. This tyranny of the word "the," the just-one-solution idea, put us back in a box.

I think we need to approach energy efficiency and resource efficiency generally in that spirit. Another thing we need to get over is the common notion that design is the art of compromise and tradeoff. Our friend J. Baldwin was being taught this one day in design school. He looked out the window and saw a pelican flying around catching fish and said: "Wait a minute. Out there is a 3.8 billion-year-old design lab that has a great deal of experience embodied in it. Whatever we see living today is what worked, whatever did not work got recalled by the manufacturer." As any naturalist can see just by looking at what is going on out there, nature does not compromise. Nature optimizes. "A pelican," Baldwin said, "is not a compromise between a duck and a crow. It is the best possible pelican." But, of course, it has to be designed in context. You could not design a pelican without knowing about fish, oceans, air, and so on, any more than you could design a good comfort or lighting system without a building, or a building without its landscape, climate, and culture. So the context is important, and where we draw the boundary determines the whole system.

That brings me then to a first example of tunneling through the cost barrier. My wife and partner Hunter and I live 7,100 feet up in the Rocky Mountains, where it gets as cold as −47°F on occasion. The growing season between hard frosts is from about June 26 to August 16, but you can get a frost any day of the year, and we have had as long as 39 days of continuous mid-winter clouds. Nonetheless, you can come into the atrium of the house out of a snowstorm, and you will find yourself in a semitropical jungle. We have harvested more than 26 banana crops in that atrium and you can get your advanced lizarding lesson there, too. If you qualify (the prerequisites are elementary hedonism and intermediate decadence), you can watch the little pygmy African hedgehog running around eating bugs and so on.

Then you realize that, other than the 50-watt portable supplementary unit (our dog), there isn't a heating system. Why not? It is cheaper not to have one. There are a couple of wood stoves, which provide about 1 percent of the heat, because we've got to burn the energy studies somehow, but there is no furnace. This is, of course, quite contrary to normal assumptions that if you have 87 100° days you need a furnace. The normal calculation engineers do about how thick your insulation should be compares the extra cost of thicker insulation with the present value of the heating fuel that you will save over time by having thicker insulation. It sounds reasonable, except that it leaves out something rather important. How about the capital cost of the heating system?

That isn't counted. It ought to be counted, though, because it turns out that if you make the insulation about twice as thick and put in proper thermal masses, uncoupling, air-to-air heat exchangers, and superwindows, guess what? You don't need a furnace, with all the duct work and control systems and fuel and power supplies, and so on, that go with it. The total capital cost of the building goes down a little bit when you do that, because you save more up-front getting rid of the heating system than you pay to do so. Then we took the saved money plus some more, totaling $1.50 a square foot after we paid for the superwindows and super insulation and so on, and we used it to pay for things like super-efficient appliances, a refrigerator and freezer using 8 percent and 15 percent of the normal amount of electricity. We could save two-thirds of what's left, but we haven't bothered yet; and there was other stuff, day lighting and so on, that saved about 90 percent of the household electricity along with 99 percent of the water heating energy.

All those savings paid for themselves in the first 10 months. I know that is a long time to wait, but this is 1983 technology. Now you could do a lot better. The household electric bill for 4,000 square feet is $5 per month. We coined the term "negawatt" to describe such energy savings. It originally appeared as a typo in a document, but we liked it and found it appropriately descriptive, and now it is widely used.

Again, it is not a very complicated idea, but it is asking the right question about what you are optimizing. You are not just optimizing insulation, but rather the whole house. They have done the same trick in some houses for Pacific Gas and Electric, in climates up to 113 or 115°F with no cooling system or heating system. The houses cost a little less to build if this is a general practice, not a one-time experiment, and they are equally more comfortable. In another example of the same idea, we figured out in renovating a 20-year-old all-glass-and-no-windows office tower near Chicago how you could do the normal 20-year renovation a little differently: spend the money on the right things, and save three fourths of the energy while you were at it, and make the building much more pleasant and comfortable.

By the way, we find, typically, if you give people better thermal, visual, and acoustic comfort—that is, if they can see what they are doing, hear themselves think and feel more comfortable—they typically do about 6–16 percent more and better work, and that's worth about 6–16 times as much on your bottom line in a typical office as eliminating the entire energy bill, because in an office you pay 100 times as much for people as for energy.

Now let me give an industrial example that makes the same point as the example of the buildings where we save a lot of money by getting rid of superfluous equipment. We saved energy and we saved equipment, so we had two benefits for the price of one. Just as in our house, the arch that holds up the middle of the building has twelve different functions, but we only paid for it once. That is a pretty good way to get cheap benefits: get a lot of them, and only pay for them once.

In this industrial case, Jan Schilham, a brilliant Dutch engineer with the firm Interface, Inc., was designing an industrial pumping loop for an American carpet factory in Shanghai. I told him a few tricks we learned from an engineering wizard in Singapore, Eng Lock Lee of the firm Supersymmetry. Schilham applied them to the pumping loop, which had been designed by the top specialist firm and was supposedly optimal to start with. When he got through, instead of 95 horsepower of pumps, he was using 7 horsepower, a reduction of 92 percent, or a factor of 12. It cost less to build and worked better in all respects. How do you do that? Well, you just change the design mentality. You stop having an old idea. The old idea in this case is that you should decide how big the pipes should be by comparing the extra cost of fatter pipes with the value of the pumping energy saved over time, because as you make the pipe bigger, the friction goes down as almost the fifth power of diameter. That sounds pretty logical. But it is the same fallacy as in the house insulation, because we forgot about the capital cost of the pumping and drive equipment that have to be big enough to fight the friction.

In other words, we were optimizing the pipes in isolation, and thereby "pessimizing" the system. If you want to optimize the system, you have to look at all of it, including the pump, motor, inverter, and electricals. If you do that, instead of using little skinny pipes with lots of friction, you use big fat pipes with very little friction. You make the pipe twice as fat, the pumping energy goes down in order of magnitude, and the equipment gets that much smaller. You save more money making the equipment smaller than you pay extra for the fatter pipes, so the whole thing gets cheaper. In other words, you use big pipes and small pumps instead of small pipes and big pumps.

The other thing we did was to lay out the pipes first, then the equipment, not the other way around. Normally, people plunk down the equipment to be connected in some arbitrary traditional place, and tell the pipe fitter to come connect point A to point B. But usually the equipment is in the wrong place, facing the wrong way, at the wrong height, and there is a lot of stuff in between, so the pipe has to go through lots of curlicues to get there, increasing friction by a factor from about 3 to about 6. It makes a lot more sense to have short straight pipes than long crooked pipes; and, again, it results in less friction, smaller equipment, less capital cost, and less energy fighting the friction forever. Also, it is easier to insulate short straight pipes. So we also saved in this case 70 kilowatts of heat loss with the two-month payback on thicker insulation.

None of this is rocket science, right? This is just good Victorian engineering, lately forgotten. But it gives you multiple benefits for single expenditures—in that case, less equipment, less capital cost, and less energy cost, and there are a lot of other benefits, too. It is more reliable, easier to control, takes up less space, and makes less noise. It turns out, actually, superwindows like those in our house have ten different benefits, premium efficiency motors have sixteen, and dimmable electronic ballasts for fluorescent lights have eighteen. Why are we counting just one? When you have that many benefits to play with, it is not hard to make big savings become cheaper than small savings, and the savings really add up. You can save one fourth of the electricity in the country retrofitting lighting systems, another fourth retrofitting motor systems, and another fourth on everything else. There's three fourths gone. Not bad, since we spend $200 billion a year on electricity.

Now let me give a more interesting example of how to apply whole-system thinking to tunnel through the cost barrier—that is, what we can do with cars. Ed Colt once accidentally invented a wonderful kind of graph (figure 4). Apparently he hit the wrong button on the spreadsheet routine or something and came up with this. I use it here to show the history of movements of world oil prices over 113 years. It is a very nice way to show this Brownian random walk, just like any commodity price. All that happened after 1973 with the first oil shock is that the volatility tripled.

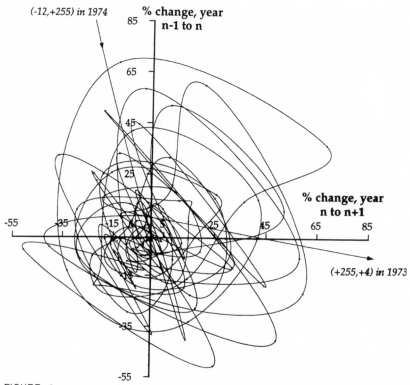

(-12,+255) in 1974

% change, year n-1 to n

% change, year n to n+1

(+255,+4) in 1973

FIGURE 4

Brownian-motion random walk of world real oil prices. (Rocky Mountain Institute)

There could have been two times you experienced long gas lines: once in 1973, when the oil price shot up, and therefore demand grew more slowly than it had before; or the second oil shock in 1979, when there was the worst price jump after the Shah of Iran fell. In fact, at one time the price went up so much that demand actually went down, to the point where— guess what?—the price came down again, and then demand went up again. It kind of does what the textbooks say. But if price is perfectly random, which it seems to be historically, you do not really want to depend on it for public policy goals. If we had kept on saving oil as rapidly after 1986 as we did for the previous 9 years, we would not have needed a drop of oil from the Persian Gulf since then.

The biggest part of our oil thirst has to do with cars, and we did in fact have a very successful policy called Corporate Average Fuel Economy (CAFE) standards, which was largely or wholly responsible for cutting in half the fuel intensity, the gallons per mile, of our cars, except then we stag-

FIGURE 5
A hypercar. (Rocky Mountain Institute)

nated. But there are a lot of good technologies that have accumulated meanwhile. Instead of just trickling them into the car stock, we could put more of them in at once. The Partnership for a New Generation of Vehicles, the partnership between the Big Three automakers and the federal government, aims to come up with a triple-efficiency prototype over a ten-year period. They will do that, and better. But I want to tell you about a different prototype, because having not a threefold, but a car 4–8 times as efficient turns out to be cheaper, easier, and faster than one 3 times as efficient, and it will ultimately save as much oil as OPEC now sells.

We call it a Hypercar, and it integrates a number of different technologies, dozens of them, including ultra-light auto bodies made of advanced composites like carbon fiber, using new manufacturing techniques; very low-drag design, using new simulation and measurement techniques for aerodynamics; very efficient tires with better polymers and design; and hybrid electric propulsion. When you put these things together, something quite remarkable happens. I am not, of course, talking about the standard battery electric car. We are building on the same technical foundation, and battery cars have gotten a lot better, but battery cars are cars for carrying mainly batteries, but not very far, and not very fast, or else they would have to carry even more batteries. The reason for that, of course, is that batteries have only about 1 percent as much useful energy per pound as gasoline. You cannot get the considerable advantages of electric propulsion without the disadvantages of batteries. A better way to design for fuel efficiency is first to make the car need only half as much power to make it go by making it 2–3 times lighter with the advanced polymer composite body. That results in 2.5–3.5 times less air drag, which does not remove your stylistic flexibility; it just means that things get smooth.

The first thing you do is make the underside of the car as smooth as the top side, because the air doesn't know which side it is on. We have smooth seams and so on, so the laminar flow adheres as far back as possible before detaching into turbulent eddies. You have better tires that are small, hard, and narrow, and you are not pressing down on them as hard, because the car is lighter, but it will have just as smooth a ride. You also have very efficient accessories. When you have in these ways made the car twice as efficient, then you give it a hybrid electric drive, which means you run the wheels with special electric motors, but you use a small engine turbine or fuel cell on board to make the electricity from fuel as needed, buffering the electricity through a small, high-powered storage device just to recover braking energy. The whole power system is controlled by sophisticated software. You get much more than a sum of the parts through this kind of whole system design.

If you just make the car ultra-light and slippery, you get an improvement of a factor of 2 or 2.5 in fuel economy. Making it hybrid electric gives you typically a 50 percent improvement, but both together give you improvements of a factor from 3.5 to 5, then to 8, and then even more. So a six-seat Taurus-class car, with no compromises, depending on your choice of technology, will get from about 80 to more than 180 miles per gallon. A four-to-five-seater will get from about 100 to 200+ miles a gallon. Emissions go down by a factor ranging from 10 to 1,000 or more. You can burn any kind of fuel, and it becomes very convenient to burn compressed gases, like natural gas or hydrogen, because a small, light, cheap tank will take you a long distance if your car is this efficient. You can make the car as safe as present models or safer, even in hitting heavy steel cars or sport-utility vehicles. Because these ultra-light materials also can absorb 5–10 times the crash energy per pound of steel, and do so more smoothly, the design and materials can overcome their mass disadvantage.

Because the car is slippery, it loves to go fast. Because it is ultra-light, it's very agile. Because of the inherent engineering properties of the materials, which do not dent, rust, fatigue, and so on, it is very durable and has better life cycle properties. It is also quieter, more comfortable, more beautiful, and nicer in every way to the driver. It should certainly be competitive in cost, and there is increasing evidence that it may cost less to make. These are expensive materials, but very cheap to fabricate, just the opposite of steel cars. The car itself gets radically simplified when it is that light. So it is the same car, only better for the user, and we think people will buy it because it is a better car, not because it saves fuel, just as they now buy compact discs instead of vinyl records.

We are doing a complete end-run around the scholastic debates about how many price elasticities can dance on the head of a pin, and how much do you need to put up the gasoline price to induce people to climb into a small, sluggish, unsafe, squinchy box. That is not what this is about. This is a better car. It is not even small. But it is completely different in what it is and how it is made and sold. It may amount to the biggest change in industrial structure since microchips. In fact, we think it will be the end of the auto, oil, steel, aluminum, electricity, and coal industries as we know them (which account for about one-third of the gross domestic product) and the beginning of new industries that are more benign and more profitable.

The main barriers to doing this are not any longer technical or economic, they are mainly cultural, such as the "metal mindset" design of single components rather than integrated vehicles, and basic decisions on accounting rather than economic criteria—unamortized assets rather than sunk costs. By the way, Hypercars, although they buy time, cannot solve the problem of too many people driving too many miles in too many cars, and they will probably make that a bit worse by making driving even cheaper and more attractive. So there is an important parallel agenda of getting real competition going in honest prices between all ways of getting around, or even, for example, being already where you want to be, so you don't need to go somewhere else—or just moving the electrons, and leaving the heavy nuclei behind.

I would like to give you a little summary of how far this idea has developed. As a recovering physicist, I have been thinking in the background for 20 years about why it is that today's cars, after a century of devoted engineering effort, use only 1 percent of their fuel energy to move the driver. This is really not very gratifying. I think we ought to be able to do better than that. The synthesis was first published in 1991, pursuant to a meeting in the same year of the National Academy of Sciences on fuel-efficient cars.

We spent a couple of years with General Motors and others, confirming that it would work and was a good idea, but then we faced the inventor's dilemma: How do you get a thing like this on the market quickly, and well, in high volume and quality? The normal method is to patent and auction the intellectual property, and hope that the single buyer succeeds with it and doesn't sit on it. But you only have one shot at success.

We thought it would be more fun to choose the open software model and put the work prominently in the public domain, where nobody can patent it, and get everyone fighting over it. We thought that could work without government intervention because of two powerful market forces. Customers want superior cars, and manufacturers want competitive advantage. The competitive advantage that they get turns out to be decisive. They get up to a tenfold decrease in the concept-to-street product cycle time, the number of parts in the body, the assembly space and effort, and, above all, the investment required for tooling and equipment, which is the main source of financial risk and the main barrier to market entry.

As a result of people's going after those enticing inducements to enter the market, by the end of 1997 about $3 billion had been committed to this line of development by proprietary programs and several dozen firms worldwide. About half of those firms were not automakers, but intended to become so—for example, car parts makers, who have the automakers' capability, and the large electronics companies, who have all the capabilities needed, except they have never made something with wheels. The skills needed to do that can all be bought, allied with, or hired in the market place. Any first-tier supplier will happily do it for a fee, you can badge it however you want, and soon you have a formidable, real or virtual, Hyper-car company.

The job of our little non-profit in this, one of ten areas we work in, is to support these developments technically and strategically in a compartmentalized fashion, keeping the secrets of each company from the others, but also in a clearly non-exclusive fashion, telling them that we are working with all their competitors. In this way, we maximize competition among them to encourage the others. We help in other ways as well. We were saying for quite awhile, originally to much skepticism, that early models could well start to enter the market around 1999, plus or minus a year. Skepticism about that has evaporated, so let me give you a little report on what has happened.

The story starts back in November of 1996, when GM made a little-noticed announcement. It was developing cars with half the weight and half the drag of standard cars, and with hybrid drive. Those are Hypercars in all but name; they even used the word in some TV ads in Thanksgiving television football games. There was a leak in Japan that Toyota had developed, and would sell in Japan in late 1997, in volume production, a doubled efficiency Corolla-class hybrid sedan, which was then officially announced in the *Wall Street Journal* in March. The *Wall Street Journal* also announced that Ford, by October of 1997, would be testing about a dozen so-called P2000 mid-size sedans of the Contour class, based on aluminum, with two-fifths less mass than usual, doubled or tripled efficiency, ultra-low emissions, and hybrid drive.

The following month, Ballard and Daimler-Benz put a third of a billion dollars into accelerating fuel cell commercialization for cars. The Toyota hybrid started to get tested, and it was well received. Meanwhile, Chrysler showed an emerging markets car, injection molded out of plastic like a toy car, that snaps together. It was half the weight of a Neon subcompact, but roomier, 15 percent cheaper, with one-fifth the investment and one-seventh the factory space. It had doubled efficiency without hybrid drive and tripled with it. Honda started to do some interesting things. Toyota finally announced its December launch of its hybrid in October of 1997, and their president made some waves when he projected that hybrid electric propulsion would gain a one-third share of the world car market by 2005. Meanwhile, other companies started bringing out or announcing tripled efficiency cars.

That Toyota hybrid was released at the end of 1997. (The Prius has now been launched in the United States.) A couple of thousand were pre-sold the first day, before there were any to sell or to show. GM said, we'll be second to none, watch the Detroit show next month. Ford added an even bigger amount of money to the fuel cell project. Other manufacturers also showed interesting things. The empire struck back in January 1998, when GM unveiled at the Detroit show three stretch versions of its EV1 battery electric car, but with hybrid drive. They were four-seaters, run by a gas turbine, a diesel engine, and a fuel cell, and even though they weighed a lot, they still had quite impressive performance. Production-ready hybrids with fuel cells were promised by 2004, if not earlier. This announcement got a lot of attention. Meanwhile, a two-fifths lighter Ford car was said to be coming on the market in 2000 but is not yet available. Chrysler showed a US version of their molded plastic car and said it could be crash-worthy here, predicting it would cut factory investment threefold and total car cost twofold, and enter production as soon as 2003.

At any rate, there was obviously an explosion of activity, and if these were the kinds of things that were being announced out of the secret programs, you can imagine there was a lot that is still behind the curtain. If you wonder what they are up to back there and give your imagination pretty free reign, you will be about right. The most surprising development, for me, in that line of work is the unexpected technology fusion that is going to lead to profound changes in the electricity system. It works, in brief, like this: The way to run a Hypercar is with a fuel cell. That is a gadget that is a bit like a battery, but it is an electro-chemical device into which you feed hydrogen and air. What comes out is electricity, pure hot water you can drink, if you don't scald yourself, and nothing else. It is extremely reliable, which is why it is used in space stations and space shuttles. It is silent and completely clean.

To make fuel cells cheap enough to go in a car, you need to make a lot of them. Manufacturing cost predictions at the end of 1998 were around $800 a kilowatt. The cost needs to get down to about $100 a kilowatt to go in a Hypercar. They would need to get twofold cheaper than that to go in a regular car, but Hypercars only need half as many kilowatts, so you can pay twice as much per kilowatt. How to get them cheap enough to go in cars? You make a large quantity for buildings, because 170°F hot water also comes out of the fuel cell, and you can use that for heating, cooling, and so on; those building services are worth about enough to pay for natural gas and a gadget called a reformer to change it into hydrogen in the building to feed to the fuel cell. If your fuel is already paid for in that way, then the electricity is extremely competitive, even at early prices.

Once you've done that, and you have the price down to where you can put the fuel cells in cars, which happens very quickly, then why don't you lease some of those fuel cell Hypercars to the people who work in the buildings that already have fuel cells in them? Then they can drive to work, and in the parking lot plug into the power system and the hydrogen line that brings out extra hydrogen from the reformer in the building, because it is not kept fully occupied all the time, and it can store a little on the side for this purpose. Then while you are sitting at your desk, your previously idle, second biggest household asset has just turned into a profit setter and is sending about 20 kilowatts of premium power back to the grid at the daytime price, which is pretty good. That will earn you enough money to pay for half the cost of leasing the car.

It does not take many people doing this to put any remaining coal and nuclear plants out of business, because the Hypercar fleet will ultimately have about 5 times the generating capacity of the national grid. You start to see why I think oil is likely to become non-competitive even at low prices before it becomes unavailable even at high prices. There are, of course, a lot of specific obstacles that get in the way of doing this sort of thing. You will find a lot of them classified in showing how to turn each one into a business opportunity in a paper Hunter and I wrote in 1998 titled "Climate: Making Sense *and* Making Money."

I think it is worth concluding on a slightly different note of where resource efficiency fits. Hunter and I have written a book with business author Paul Hawkin, titled *Natural Capitalism,* which, I think, provides the context we have been seeking for what resource efficiency is about and why are we doing it. It isn't just to prevent pollution somewhere or save a resource somewhere. It is part of a much deeper historical process. You could say that what we have now is unnatural capitalism. It is a temporary aberration, not because it's capitalist, but because it defies its own logic by liquidating without valuing the largest source of capital, namely, the natural capital that provides our ecosystem services, for which there is no known substitute at any price, and the base of natural resources that the economy consumes. Of course, capitalists have long valued physical and financial capital, but they have not, in general, valued human and natural capital, which are much bigger, more valuable, more important, less substitutable.

What would happen if we had capitalism that actually valued all forms of capital? It would look very different because it would be dealing with a different pattern of scarcity. Economics tells you to economize on your scarcest resource. If somebody had gone to Parliament in Britain in 1750 and said, we are going to make people more than 100 times as productive, he would have been laughed out of the room. But that is exactly what we did, and it made sense at the time, because at that time, people, and indeed technology and capital, were relatively scarce. What was apparently abundant were resources and places to put waste. So we made people more than 100 times as productive, and they were no longer the scarce resource.

Now we have abundant people and scarce nature. If you apply the same impeccable economic logic to this new pattern of scarcity, then it makes sense to make natural resources 10 or 100 times as productive. By tunneling through many, many cost barriers, indeed we can do that, not just at reasonable cost, but profitably. It is cheaper to do it than not to do it. It is cheaper to save the resources than to buy them, just as protecting the climate is profitable because it is cheaper to save fuel than to buy it, let alone burn it.

Advanced resource productivity is only one of four main elements of natural capitalism. Another one is to redesign industry on biological lines with closed loops and no waste. As Bill McDonough says, waste equals food. Everything's waste is somebody else's food. Therefore nothing is left over, everything is used. Third, we need to change the economy from the episodic acquisition of goods to a continuous flow of service and value at the pull of the customer. Fourth, as any prudent capitalist would do, we need to reinvest in restoring, sustaining, and expanding the stock of natural capital, our scarcest resource. These are, in a way, not such novel ideas, but they are very powerful ideas, and, I think, especially so because they are not just a better way to make money. They also get at some of our most profound social problems. Lack of work, lack of hope, and shortages of satisfaction and security are often not just isolated pathologies but result from the intimate connections among the waste of resources, of money, and of people. The solutions are equally intertwined. For example, if you fire the unproductive tons, gallons, and kilowatt hours, you then have a chance to keep the people, who will have more and better work to do.

I think historians may look back and say that the end of the cold war, via the fall of communism, was indeed a landmark in the intellectual transition at the close of the twentieth century. But equally significant, I think, will be the end of the war against the Earth and the rise of natural capitalism. Many businesses, in practice, are already gaining huge competitive advantages from applying these principles. Companies that don't try hard to do so will not be a problem, because they won't be around. Those firms that survive the transition to natural capitalism will, I think, thereby be creating the salvation of nature, the catalyst for a just society, and the new cornerstone of commerce. That would be, indeed, a useful invention.

PORTRAIT OF INNOVATION: SUBHENDU GUHA

MARTHA DAVIDSON

It was both a fascination with science and a concern for societal problems
that led Subhendu Guha to new discoveries about the properties of amorphous silicon and the consequent invention of flexible solar shingles, a
state-of-the-art technology for converting sunlight to electricity. Yet if
asked about the secret of his creativity, Guha would probably say the key
was reading. "Reading," he says, "is extremely important, because that
helps you in making decisions. As you read more and more, you will ask
questions. And you will try to solve those questions. And you get the
biggest thrill when you realize you have thought of something new, and
it works."

Guha, a physicist who is executive vice president of United Solar Systems Corporation in Troy, Michigan, began his life and his career in Calcutta, a major city in eastern India known for its rich literary tradition. As
a child, he was as captivated by literature as he was by science or technology. He read everything he could get his hands on, from poetry to textbooks, but he was particularly drawn to the work of Rabindranath Tagore,
the great Bengali writer, composer, and artist, winner of the 1913 Nobel
Prize for literature.

"The writing, music, and painting of Tagore had and still have the most
profound effect on me," Guha explains. "While I cannot readily see any
direct connection between my literary interests and my scientific work, . . .
I feel that creativity is universal. Reading, listening, or seeing creative
expressions in any form must have a subtle way of triggering your own
creativity in an area which you enjoy the most." Looking back, he realizes
that it was through reading that he broadened his own horizons sufficiently
to make informed choices in his path of study.

For those who seek to express their creativity in the realm of invention, Guha offers this advice:

Don't be afraid to try new ideas, and don't be disheartened if it doesn't work the first time. . . . To put an idea into practice can sometimes take a very long time. And there is a difference between discovery in a fundamental area and invention of a product. If you are a scientist, you may discover something which is not used in a practical way, but that is also immensely important.

There is no set rule. Sometimes, invention comes from serendipity. You may stumble on to it, but you must have the knowledge and wisdom to recognize its importance, and pursue it to create a worthwhile concept or a product. . . . Quite often, on the other hand, you think of something and argue out that it will work. Then you test the hypothesis and demonstrate that the concept really does work.

Guha grew up in a large family, with four brothers and five sisters. His father was a respected attorney. His mother, a full-time homemaker, was a brilliant woman whose innate inventiveness made a strong impression on her son. It was not until he began to study physics at Calcutta's prestigious Presidency College that Guha became serious about science. There and at the University of Calcutta, where he later did graduate work, Guha was exposed to a rigorous approach to research, which he found very exciting. "In physics," he says, "you ask really fundamental questions, and try to get answers to them. I got interested in physics when I was in Presidency College, but I got interested in semiconductors only when I was pursuing my Master's degree at Calcutta University. Semiconductors are the building blocks of electronic devices like transistors and integrated circuits. In the early sixties, new rules in semiconductor physics were being formulated, and a new generation of devices was being made. That intrigued me."

After earning a Ph.D. from the University of Calcutta in 1968, he joined the Tata Institute of Fundamental Research, an internationally renowned academic research institution in Bombay. Other significant developments soon followed that affected the direction of his life and work. In 1971, he married Jayashree Bose, who has been extremely supportive of his research and accepting of the long, irregular hours that his scientific work has demanded. A son, Aveek, was born 2 years later. It was at that time, too, that Guha began to investigate the use of semiconductors to convert sunlight into electricity. This field of research appealed to him for two reasons: he found the problems intellectually challenging, and it was work that could address societal problems.

Semiconductors are a class of materials whose capacity for conducting electrical charges falls between those of true conductors (such as copper and other metals) and those of insulators (such as rubber or glass). There are a number of semiconductors that can be used for photovoltaics (the conversion of light to electricity), but Guha's concern for societal and environmental problems led him to focus on amorphous silicon, an element found in sand that can be applied as a thin film to produce a photovoltaic material. Other semiconductors that can be used for thin-film photovoltaics require the use of certain toxins in their production, but amorphous silicon presents no hazards to humans or to the environment in its manufacture or use.

In an amorphous material, the atoms are not placed in a regular orderly manner, as they are, for example, in a crystal. In the early 1970s, the physics of amorphous materials was not well understood. In the United States, Stanford Ovshinsky and his collaborators had begun pioneering work in this area of physics, generating a great deal of interest. New theories were being formulated for amorphous solids, and new devices were being built to take advantage of their inherent properties. The possibility of doing cutting-edge research in an area that had practical applications appealed strongly to Guha.

In 1974–75, Guha did a year of postdoctoral training at the University of Sheffield in England, pursuing his interest in amorphous silicon. Returning to the Tata Institute, he began to explore a way of producing amorphous silicon that had not been tried before. His research led him to believe that by adding hydrogen in the production process, a more useful form of amorphous silicon could be created. He explains the reasons for his experiments as follows:

In the late seventies, it was recognized that amorphous silicon has potential as an inexpensive solar cell material. Till that time, solar cells were made only with single crystals. They were reliable but expensive, since you have to take a great deal of care in preparing these materials in which the atoms must be arranged in an ordered structure. The first amorphous semiconductor solar cells were made using amorphous silicon prepared by breaking up a gas called silane (silicon hydride) in an electric field. The quality of the material, however, became poorer when exposed to light. Solar cells, of course, work only in light, and this caused serious doubts among the physicists whether these materials could ever be useful for making efficient solar cells.

Rather than using only silane as the starting material, I diluted the gas with hydrogen. The reason at that time was twofold: the dilute mixture is less costly and also less hazardous than pure silane. To our surprise, however, we found that the new material was much more stable to light exposure. We did a series of very systematic experiments to prove that was indeed the case, gave some possible explanations, and submitted the paper to the *Journal of Applied Physics* in the United States.

That paper, published in 1981, was a major breakthrough in the field, and today most amorphous silicon produced anywhere in the world is made with hydrogen, just as Guha had described.

Although he took pride in this important achievement, Guha did not cease his exploration of the possibilities of this material. Up to this point, in graduate school, at the Tata Institute, and in Sheffield, he had been involved in academic research, expanding the frontiers of scientific knowledge. He enjoyed the research immensely, yet he was increasingly curious about the practical applications of his work. He knew that Stanford Ovshinsky and his wife, Iris, had founded a company called Energy Conversion Devices (ECD) to explore and promote the use of amorphous materials in energy conversion and information storage. The company was located in Michigan. Guha wrote to Ovshinsky and asked if he could spend a sabbatical year at ECD. Ovshinsky welcomed Guha but advised him to make it a two-year sabbatical. One year would not be adequate for the kind of work he intended to do there.

Guha arrived in the United States with his wife and son in 1982. His sabbatical at ECD centered on experiments to understand some fundamental properties of amorphous silicon. "I was also interacting with other groups who were more involved in the technology development," he recalls. "I started working with them and used the fundamental tools that I had to solve technology-related issues. In the process I invented new materials and new device designs which led to obtaining higher efficiencies in solar cells."

By the time his two-year sabbatical was drawing to a close, Guha and his associates in Michigan were making real progress on several fronts. He decided to resign from the Tata Institute and continue his work with ECD. He had become hooked by the challenges and broad scope of high-tech industrial research, which demanded more focused, practical results than did research in the academic realm, yet also included theoretical and experimental science. One problem in particular was absorbing him. He explained:

The degradation of properties of amorphous silicon on light exposure was becoming a serious hindrance to improving the light-to-electricity conversion efficiency of the solar cells and was considered a major obstacle to the successful commercialization of this technology. By the mid 1980s, several Japanese groups were exploring the use of hydrogen dilution to improve the stability of the material against light exposure. Since the cause of this light-induced degradation was assumed to be somehow related to the presence of hydrogen, how the use of excess hydrogen in the starting gas mixture can reduce degradation was very puzzling.

Several industries took a two-pronged attack. The first aim was to reduce the degradation by optimizing the amount of hydrogen dilution and other deposition parameters empirically; simultaneously, a dedicated research effort was directed to have an understanding of this phenomenon.

Guha and his colleagues were addressing the problem both empirically and analytically, with considerable success. In 1987, the US Department of Energy, which was interested in the potential of solar energy, requested proposals in the field of solar cells using new thin-film materials. Guha's team responded, and their project was awarded a three-year, $6.26 million cost-sharing grant through the National Renewable Energy Laboratory in Golden, Colorado. It was the beginning of a relationship with DOE/NREL that continues to this day.

Gradually, Guha's responsibilities expanded beyond research and development to include administration, manufacturing, and sales as an executive of United Solar Systems, a joint venture company started in 1990 by ECD and Canon of Japan to manufacture solar cells.

At that time, conventional solar energy technology consisted of photovoltaic cells made with crystalline silicon. Crystalline silicon, like amorphous silicon, is a semiconductor. When sunlight falls on a semiconductor, it is absorbed to create charges (positive or negative) that are separated and then channeled to produce an electric current. The regular atomic structure of the crystalline silicon actually makes absorption of light more difficult than in the looser arrangement of amorphous silicon. The conventional solar cells were therefore thick, required a rigid and heavy support, and were expensive to produce. They were not cost effective, since it took 7 or 8 years of operation to recoup the energy spent in their production. ECD and United Solar Systems were interested in developing lighter, less expensive solar cells using amorphous silicon, which could be 100 times thinner than crystalline silicon with equal capacity for light absorption. Guha notes that hydrogen dilution was the key:

Using hydrogen dilution, we have made the highest efficiency solar cells and panels. Much of the work is proprietary, and that is why we have a comfortable lead over anybody else in the world in terms of cell efficiency. . . .

We have also published extensively, explaining the role of hydrogen dilution. We have shown that the use of hydrogen dilution results in a better structure of the material; it is still amorphous but has a better order than the conventional amorphous material.

From a practical point of view, hydrogen dilution is a necessary requirement for obtaining high efficiency cells. Every industry is using it, and new understanding is still emerging.

In 1994, while giving a presentation about the company's products, Guha showed a slide of rooftop solar cells. An architect in the audience exclaimed "But it's so ugly! Who would want that on their house?" When Guha returned to his office in Troy, that comment was on his mind. He called a staff meeting and proposed that they find a way to make solar cells look more like standard roofing material. Perhaps, he suggested, the cells could even be incorporated into roofing shingles rather than mounted separately on a metal frame.

His group of 25, including researchers and other staff members, immediately began working on the problem. They studied roofing shingles and experimented with production processes. Collaboration was essential, according to Guha:

Science has become very sophisticated these days, and quite often you need people belonging to different disciplines to develop a concept to its fruition. This is why teamwork is so important. The idea was mine in terms of developing a solar panel which will look like a conventional shingle and can be installed the same way. Kais Younnan is a mechanical engineer who looked at the logistics of how the shingle can be designed such that it resembles the ordinary shingles, and still will work as a solar cell. Troy Glatfelter worked on different materials that will complete the solar panel so that it can withstand the rigors of the outside weather for 20 years or more. Jeff Yang was the leader of the team who coordinated the efforts of Troy, Kais, and others, and also made many original contributions.

The product they developed is a photovoltaic panel, 7 feet long and a foot wide, that is lightweight and flexible, yet rugged and durable; it has no moving parts and is easy to install. Produced in a brownish color with subtle variations, it blends in well with conventional roofing shingles. This innovation was not in the basic technology of photovoltaics, but in the design, materials, and production process.

Manufacturing begins in Michigan, with a half-mile-long, one-foot-wide roll of stainless steel that is fed through four machines in an automated system. First, the steel is washed to remove surface dirt. Then two layers of a reflective coating are applied. Next, layers of amorphous silicon and amorphous silicon-germanium alloys are applied to create a kind of triple-decker sandwich: each layer absorbs a different photon-energy or wavelength of light (blue, green, or red). The germanium concentration is applied in a gradient to obtain higher efficiency, based on a concept developed by Guha and confirmed through experimentation.

The long strip is then cut into one-foot lengths, which are shipped to a plant in Tijuana for final assembly. There, the pieces are joined to form 7-foot-long shingles, which are wired and encapsulated with a plastic protective coating.

The flexible panels can be mounted on a roof with nails in the same way that ordinary roofing shingles are attached. The only difference is that a small hole must be drilled in the roof every 7 feet so that wires can be dropped from the panels into the building. Then an electrician can hook the wires up to the building's electrical boxes to channel energy to house circuits, batteries, or appliances.

Flexible solar shingles went on the market in 1997. United Solar Systems now holds approximately 160 United States patents on the design, technology, and production process and has become the world's largest manufacturer of roof-mounted solar cells.

While amorphous silicon solar cells are not yet as powerful as conventional crystalline silicon cells (the 13 percent stable energy conversion rate of United Solar's shingle cells is still less than the 17 percent rate for crystalline cells), they are actually more cost-effective, since both their materials and manufacturing costs are far lower than the production costs of standard solar cells. Moreover, their energy conversion rate is rising, as United Solar Systems continues to refine and improve the product as well as to lower its production cost. United Solar Systems holds all the world records for thin-film solar cells, with an initial cell efficiency of 14.6 percent (the lower stable rate of 13 percent is the effect of light-induced degradation; without hydrogen dilution, the drop in efficiency would be much greater). The next best rate is from a Japanese company, whose cells have a stable rate of 10.6 percent.

Flexible solar cells are still prohibitively expensive in communities where there is an established infrastructure for producing and distributing electrical power. The world price at present is about four dollars per watt. Most of the cost is from overhead, not from materials, and will decrease rapidly as production capacity increases. United Solar Systems plans to build a 25-megawatt production machine to meet increasing market demand. When a machine with an annual capacity of 100 megawatts is available, the consumer cost will be less than a dollar per watt, making the solar panels competitive with conventional energy sources.

Nevertheless, in areas of the world where power plants have not yet been built, these cells even today offer a lower-cost alternative to diesel generators. In fact, they have already been used in many communities in Egypt, Mexico, and other countries where power grids do not serve remote regions. They also have great potential, as their conversion efficiency improves and their production cost decreases, to provide safe, clean, renewable energy for home use in sunny states like Arizona and California. Eventually, this source of energy could provide an important alternative to fossil fuels, reducing both pollution and dependency on foreign suppliers.

Guha and United Solar Systems are now exploring other applications for their technology, using even lighter and more flexible materials. They have produced nylon-backed cells that can be attached to a boat or tent, or folded and carried in a backpack for military use or camping. By using a thinner stainless steel or by replacing the stainless steel with a tough, very thin, extremely light plastic film, they are developing photovoltaic cells that can be used on satellites, space stations, or exploratory space vehicles. United Solar was awarded a contract by NASA and the US Air Force for further research.

The importance of Guha's innovations has been recognized not only by the US Department of Energy, but by others who are concerned about environmental technologies. In 1996, the flexible solar shingles received *Popular Science*'s Grand Award for "The Best of What's New." In 1997, they were selected by a panel of environmental experts for *Discover*'s Technology Innovation Award, which recognizes contributions that "improve the quality of our everyday life and alert us to what is next from the frontiers of human achievement and ingenuity." And in 1998, *R&D Magazine* identified United Solar System's work on high efficiency amorphous silicon solar cells as one of the hundred most significant advances of the year.

Guha's enthusiasm for science has not diminished with his success, although his time for it has become more constrained. His administrative, manufacturing, and other business responsibilities at United Solar Systems demand much of his time and energy; only about 25 percent of his day is devoted to research. However, he is philosophical about the dilemma, realizing that even in the academic realm, non-academic responsibilities impinge increasingly on science. He reserves Saturdays for his own projects, to review his research and determine their direction. Much to the chagrin of his family, he does not feel a need for vacations. He thrives most when absorbed in his work, finding great enjoyment in the research process itself. His creative innovations, while benefiting society and the environment, have also given him a deep personal satisfaction that seems only to fuel his passion for the pursuit of science.

The old adage that "one man's trash is another man's treasure" offers an oversimplified definition of the goals of industrial ecology. Industrial ecology, the historian Christine Rosen explains, "is founded on the recognition that business and nature are inseparable parts of a single interactive 'industrial ecosystem.'" Rosen traces the history of early industrial waste management—typically, recycling and "end-of-pipe" methods meant to treat the effluents of factories—and shows how the principles of industrial ecology have grown from the inadequacies of earlier systems. Rosen's brand of environmental business history, she argues, "will help managers and policy makers to better understand the advantages of making the transition to new, more effective, more systematic ways of mitigating industry's harmful impacts."

Braden Allenby is one of the founders of the field of industrial ecology. In his essay, he offers examples of how waste from one industry becomes raw material for another. The systems view of industrial ecology links technology, human culture, and economics, emphasizing that all three are interconnected in their effects on the environment. There is another component to effective stewardship of the planet, though, Allenby argues. "While industrial ecology may be able to help understand what is possible within technological, population, and economic constraints, it is institutions and values which must determine the answer to that question."

Robert Socolow, a physicist, has devoted much of his life to uniting his love of science and his sense of social responsibility. He rejected a career in arms control ("I so disliked weapons that I couldn't force myself to learn about them"), forged close ties with Russian and North Vietnamese scientists in the pre-Glasnost era, and edited *Patient Earth* (1971), an early text designed to teach students in science about environmental issues. In the early 1990s, Socolow became an advocate of industrial ecology; one example of his work in this area is a recycling plan for lead batteries.

Rosen, Allenby, and Socolow see promise in applying the principles of industrial ecology. "It is important," Allenby notes, "to remember that industrial ecology is an integrative, not a reductionist, field. It focuses on a comprehensive, holistic understanding of systems rather than the reductionist approach of developing more and more knowledge about increasingly specific subsystems, and, . . . cuts across a number of disciplines in doing so." Recognizing the importance of connections, he argues that "the distinction between a 'human' system and a 'natural' one is itself somewhat artificial in many cases."

holism vs reductionism

INDUSTRIAL ECOLOGY AND THE TRANSFORMATION OF CORPORATE ENVIRONMENTAL MANAGEMENT: A BUSINESS HISTORIAN'S PERSPECTIVE

CHRISTINE MEISNER ROSEN

Industrial ecology is an important new approach to thinking about how to manage business's interface with the natural environment. I want to speak about it from my perspective as someone who both teaches in a business school and is a historian studying the history of pollution control in the United States. I argue that industrial ecology is a potentially transforming concept not only for business managers but also for historians who study business and the environment.

Business historians have been remarkably uninterested in business's impact on the environment. Most early work in business history focused on people like John D. Rockefeller, the architect of the American petroleum industry. Historians studied the history of business from the vantage point of the men who created big business, viewing the development of business as the history of "robber barons" or "industrial statesmen." They gave very little attention, if any, to the environmental impact of industrial development, even insofar as environmentally destructive industries like the petroleum industry were concerned. In the 1960s, Alfred Chandler transformed the field of business history into something much more sophisticated and much less concerned with individual businesspeople. Chandler and the generation of business historians he inspired have focused on trying to explain the internal evolution of large, modern, vertically integrated firms. They have examined in great depth the strategies and internal organizational structures that managers developed in order to take advantage of economies of scale and scope. These historians also paid little attention to business's environmental impacts or to the management of those impacts.[1]

Until recently, environmental historians have also been remarkably disengaged with the issue of how people in business managed their organizations' interface with the environment. As Richard White points out, many environmental historians still view nature as something very separate from culture and society. If you are a historian and you study the environment, you don't study people. That is changing now. For many environmental historians, however, business is still a sort of black box, a dark and evil capitalist entity, not a part of the natural world, not something environmental historians need to study, and better left to urban and business historians and historians of technology.[2]

The concept of industrial ecology is based on a completely different vision of business's relationship to the natural world. It is founded on the recognition that business and nature are inseparable parts of a single interactive "industrial ecosystem." Biologists examining natural ecosystems observe that in nature living organisms are knit together with one another and with the natural world, drawing nourishment from the bodies and wastes of other organisms as well as from the water and minerals in the soil and the energy produced by the sun. So it is for industrial ecologists. They see that business is also woven into the natural world. Business enterprises feed on the natural resources found in the earth, on energy ultimately derived from the sun, wind, or geological forces deep within the earth, and on the manufactured inputs of their industrial supply chains. They return their wastes to the earth, the seas, and the atmosphere.

Industrial ecologists like to use flow diagrams to explain this because it helps convey this systems understanding of the way industry and nature relate to each other. Figure 1 is a flow diagram of a simplified industrial ecosystem. It focuses on the materials and energy flows that course through an industrial ecosystem as goods are manufactured. The flows start with the extraction of natural resources from the earth. From here materials and energy flow into the industrial sphere, where they are converted into the inputs used to manufacture products and then into the products themselves. In this form they are then distributed to and used by consumers, who ultimately dispose of them, returning the materials and energy that composed the goods back into the earth.[3]

Figure 2 illustrates this idea in more detail. It is so complex that it is difficult to take in quickly, but it is worth giving a good look because it graphically conveys the fact that the industrial ecosystems encompass all aspects of the interface between industrial society and the natural world. The parts of the diagram labeled "Households & 'Personal Consumption'" and

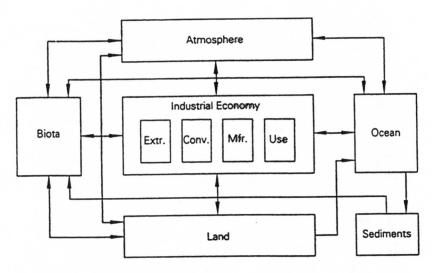

FIGURE 1

Flow of materials and energy through an industrial ecosystem from the perspective of the product life cycle. (Hardin Tibbs)

"Scenery, 'Environmental Conditioning' " refer to how ordinary consumers use the materials and energy of the natural world. "Agriculture & Forestry" and "Mining & Drilling" are also broken out, as are the forms of natural resources that people use and the ways they utilize them. All these functions and resources are bound up in the materials and energy flows of a generic industrial ecosystem. This diagram provides an in-depth sense of industrial ecology, for it depicts more of the social and environmental context in which business operates and goods are manufactured and used.[4]

Wastes are generated at every stage of this systems flow. The process by which resources are extracted from the natural world generates wastes, as does the manufacturing process. The same is true of distribution and consumption, as the diagrams make clear. These wastes may take the form of air or water pollution or solid waste. These emissions can cause significant disruptions to natural ecosystems. Mitigation of these impacts requires management at every stage of the production and consumption cycle, not just the manufacturing stage.

In my view, it is absolutely essential that historians start putting business into the larger eco-industrial context embodied in these concepts and diagrams. Business historians need to broaden their analysis of the evolution of business institutions and management strategies to incorporate analysis of

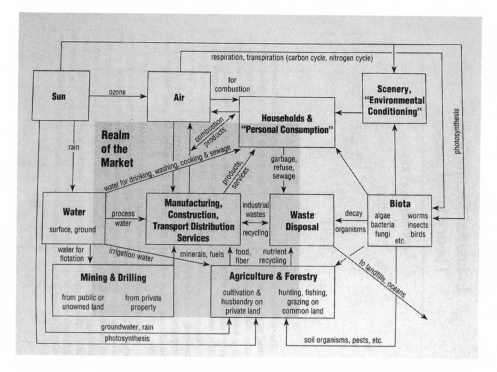

respiration, transpiration (carbon cycle, nitrogen cycle)

Sun — ozone → Air

for combustion

Scenery, "Environmental Conditioning"

Realm of the Market

rain

Households & "Personal Consumption"

combustion products

water for drinking, washing, cooking & sewage

products, services

garbage, refuse, sewage

photosynthesis

Water

surface, ground

process water

Manufacturing, Construction, Transport Distribution Services

industrial wastes

recycling

Waste Disposal

decay organisms

Biota

algae   worms
bacteria insects
fungi   birds
etc.

water for flotation

irrigation water

minerals, fuels

food, fiber

nutrient recycling

to landfills, oceans

Mining & Drilling

from public or unowned land   from private property

Agriculture & Forestry

cultivation & husbandry on private land   hunting, fishing, grazing on common land

groundwater, rain photosynthesis

soil organisms, pests, etc.

FIGURE 2

The interface between industry, consumers, and the natural environment in an industrial ecosystem. (reprinted with permission from *The Greening of Industrial Ecosystems*, courtesy of National Academy Press)

business's interactions with the natural environment. We need to examine the history of four crucial factors:

• the roles played by the distribution of natural resources, topography, climate and other natural environmental forces in the evolution of business firms and economic activity
• the effects of industrial activity on biological ecosystems, climate, natural materials cycles, and the physical landscape
• how business managers, ordinary citizens, and more broadly society through government have managed (or not managed) business's impacts on and interactions with the natural world
• the meanings society has imputed to such impacts and interactions and its efforts to manage them over time.[5]

My own interest, both as a historian and as a teacher in a graduate school of business, is primarily in the third factor, the management of business's environmental impacts. This is the point where the traditional concerns of business historians for institutional structures, organizational reward systems, and management strategies must come into and be integrated with environmental history.

I argue that the concept of industrial ecology represents a potentially transforming new way of thinking about how to manage business, as well as an innovative new way to approach the subject of doing business history. Let me start with a little of my own history.

When I began my book on the history of pollution control, I thought it would be easy to write because I figured there was no history. I knew factories polluted back in the nineteenth century and the early twentieth, but, like most people, I assumed that it wasn't really that bad and that nothing much had been done about the problem until the passage of strict federal pollution regulations in the 1970s. What I found, however, was that the history of pollution control was much more interesting and far more complex and ambiguous than I had realized. I discovered that business had polluted a great deal and that people had been struggling to deal with this pollution since the very beginnings of the Industrial Revolution.

One way to get a sense of what this early pollution was like is to look at pictures from the time. In an image of the McCormick Reaper Works in the 1860s (figure 3), you can see smoke, the nineteenth century's primary form of air pollution, pouring out of the smokestacks. This factory was close to Lake Michigan; you can see the ships in the background. Undoubtedly it was producing water pollution as well.[6]

Unfortunately, I've found that a lot of pictures from this time period provide highly sentimentalized views of industry that mask the reality of industrial pollution. Figure 4 is fairly typical. It shows a nineteenth-century chemical factory, the Analine and Chemical Works of Rumpff & Lutz. Like many other images of factories, this picture was designed to function as a form of company advertising. It comes from the cover of a piece of sheet music commissioned by Rumpff and Lutz to commemorate their business. Other companies put images of their factories on company calendars and product labels. Needless to say, the purpose was to show the factory in a positive light. Look at the bucolic way in which this chemical plant is depicted. Although it was undoubtedly stinking up the entire area with its

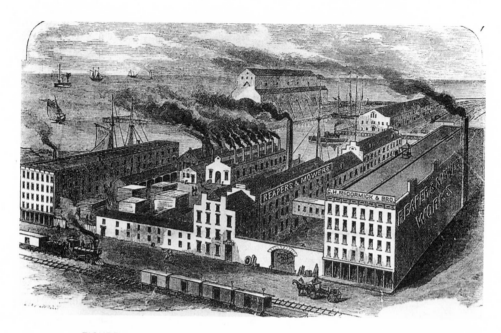

FIGURE 3

Smoke pouring from Chicago's McCormick Reaper Works in the 1860s. (Wisconsin Historical Society)

poisonous fumes and polluting the water, it is shown in a lovely country setting. Sailboats are floating by, their sails billowing in the breeze, on what appears to be a lovely little stream. There is a lot of greenery around. The picture shows smoke coming out of the factory's smoke stacks, but, of course, at the time such smoke was viewed as a sign that a business was operating at full capacity and as such was a symbol of its success. You cannot get a realistic sense of the foul smells and the other actual environmental impacts of such businesses from pictures like this.[7]

Case law reports are a much better place to look for information about industrial pollution. Private and public nuisance law allowed individuals and communities to sue businesses that generated industrial pollution. The reports generated by the litigation provide a good sense of what industrial pollution was like back in the nineteenth and early twentieth centuries, what people did not like about it, and what they wanted to do about it. The case reports show that people were particularly distressed by pollution that they could see or hear or smell or taste, or that was damaging their property. People sued businesses burning soft bituminous coal that emitted thick

FIGURE 4

Analine and Chemical Works of Messrs. Rumpff & Lutz. (J. Clarence Davies Collection, Museum of the City of New York)

sooty smoke and cinders, blackening everything in their homes and places of business. They sued slaughterhouses, rendering businesses, chemical plants, gas works, and other factories to stop them from emitting nauseating stenches and poisonous fumes. They sued textile mills, coal mines, and other plants that discharged wastes into streams, contaminating and discoloring water used for drinking, household use, watering of farm animals, and even manufacturing. Many plaintiffs went to court to obtain injunctions to force polluters to shut down or abate their nuisances. Others sued for compensation for the damages and inconveniences they suffered as a result of pollution.

Case law reports provide a great sense of how people approached the problem of controlling such pollution. Pollution nuisance litigation increased steadily during the nineteenth and early twentieth centuries. I have collected hundreds of cases. What I found was that, in the nineteenth century, the judges who decided such cases decided in favor of the plaintiffs about half the time. They issued damage awards that forced the owners of polluting businesses to pay compensation to plaintiffs. More important, they also issued injunctions that required them to curtail their pollution, or even shut down altogether.[8]

From the middle of the nineteenth century on, municipalities also attempted to regulate various kinds of industrial pollution. Public health reformers used early forms of zoning to try to move businesses that emitted noxious stenches out of densely settled areas in a number of cities. Women's organizations and business reform groups agitated for smoke regulation, noise regulations, and public works to deal with water pollution in many places. Groups of neighbors mounted protests in efforts to force some especially noxious businesses, like bone boiling establishments, to improve their operations.[9]

What is important from the perspective of understanding the history of industrial pollution management is that these legal and political developments led to the development of pollution control technologies. Recognition on the part of some enlightened managers that pollution was a sign of inefficient production also spurred this kind of innovation. Technologies for abating smoke pollution proliferated during the late nineteenth and early twentieth centuries. Figure 5 is a schematic diagram of a piece of smoke-abatement equipment from the turn of the century, in this case a cinder catching apparatus. All sorts of smoke consuming devices were invented in

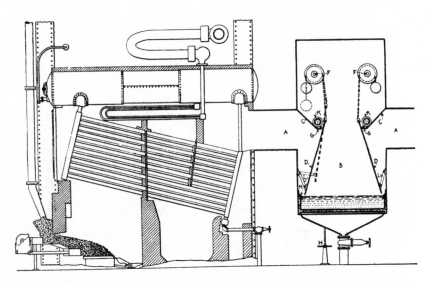

FIGURE 5
Smoke-consuming device. (*The American City,* February 1914)

the late nineteenth and early twentieth centuries. The apparatus shown in figure 5 was a typical end-of-pipe solution to pollution that involved a flue system that blew the cinders in a factory's smoke into the water at the bottom of the device before the smoke left the stack.[10] Figure 6, which is from an article on smoke abatement, depicts a factory before and after the installation of a smoke-consuming device. People have always loved before-and-after pictures. This one was published to give visible proof to doubters that it was technically feasible to install smoke-abatement equipment in a factory that clearly eliminated its visible black smoke.[11]

Engineers also devised technologies for controlling some forms of industrial water pollution. The earliest water pollution control technologies were simple screens and weirs for trapping large sized particles of waste at the point at which factory effluents were discharged into streams and other waterways. Over time, industrial and sanitary engineers invented increasingly elaborate methods for treating industrial water pollution. These technologies were especially important at some of the big chemical companies, like Dow Chemical, that discharged huge quantities of waste into watercourses. In the 1940s, for example, Dow's huge chemical plant in Midland, Michigan manufactured more than 400 different chemicals, requiring the daily disposal of some 200 million gallons of waste water, 70 tons of refuse,

FIGURE 6
Before and after smoke abatement. (*The American City,* January 1912)

FIGURE 7

Trickling filters at Dow Chemical Plant in Midland, Michigan. (reprinted with permission from *Industrial & Engineering Chemistry* 39, © 1947 American Chemical Society)

and millions of cubic feet of vented air. During the 1920s, the 1930s, and the 1940s, Dow developed a state-of-the-art system for treating phenols in this wastewater stream, largely in an effort to stave off regulation. (Phenolic waste interested regulators because it gave water a foul taste and color.) The treatment facility included a 30-acre, 45-million-gallon settling pond, a biological oxidation clarifier, four trickling filters, an activated sludge plant where the phenolic wastes were further oxidized, and 50 acres of effluent ponds.[12]

Clearly, then, industrial and sanitary engineers began developing technologies for managing industrial air and water pollution—and managers invested in installing them—long before the 1970s. There is a history to the management of industrial pollution control of which most people are unaware.

Significantly, however, early-twentieth-century engineers had a very different way of thinking about what they were doing than today's industrial ecologists. They focused on dealing with industrial air and water pollution at the level of the individual manufacturing plant, primarily through the development and installation of end-of-pipe pollution-abatement and treatment technologies. Their focus on the factory is evident in industrial waste flow diagrams drawn in the 1930s and the 1940s to describe the pollution control technologies they developed. Figure 8 depicts the waste flow

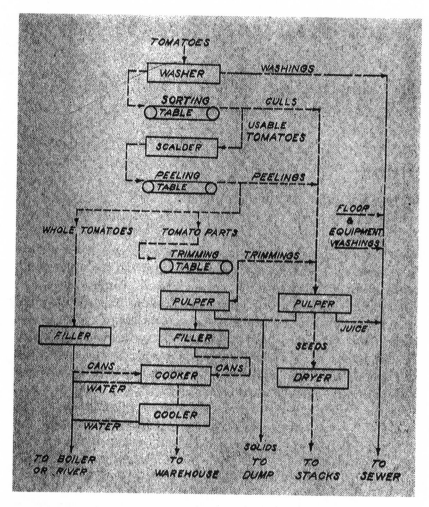

FIGURE 8

Flow diagram of tomato cannery. (reprinted with permission from *Industrial & Engineering Chemistry* 39, © 1947 American Chemical Society)

from a tomato factory where there was no pollution treatment going on. You can see the factory wastes going directly from the production process to the sewer, to the stacks, to the dump, and to the river.

In contrast, as I pointed out earlier, today's industrial ecologists conceptualize industrial waste streams from the vantage point of the material and energy flows moving through entire industrial ecosystems. As shown in figures 1 and 2, industrial material and energy flows start with the extraction of materials and energy from the earth and proceed through various stages of the manufacturing process and continue through to distribution and use by end consumers. Equally important, wastes are generated at each of these stages. These wastes move into the natural environment throughout the industrial ecosystem from every point along the continuum from production to consumption—not just from the end of a factory's water effluent pipe or smokestack.

This systems perspective makes it possible for today's environmental managers to think about pollution control in highly innovative ways. Instead of thinking of pollution control as an end-of-pipe problem at a particular factory, they seek creative ways to minimize wastes across industrial supply chains. Taking their cue from nature, they also seek to apply the closed-loop model of biological ecosystems to the problem of managing industrial waste.

What does this mean? It means a great deal. For example, I teach students in my MBA course that we have to get away from thinking about waste management at the plant level and start thinking about it at a systems level. The operant industrial ecology metaphor is that of the closed-loop biological ecosystem in which bacteria decompose the wastes of living organisms, drawing nutrition from these wastes and returning nutrients bound up in the organisms' excreta and dead bodies back to the soil, where they can be absorbed again by other living organisms. Industrial ecologists urge managers to find ways to turn their wastes into materials that can be utilized in production elsewhere in an industrial system.

The classic example of this idea is "industrial symbiosis," or waste exchange, developed in Denmark starting in the 1970s. Figure 9 is a schematic diagram of the Kalundborg industrial symbiosis. The most important players are Denmark's largest electric power plant, the Asnaes Power station; Denmark's largest oil refinery, the Statoil Refinery; a multinational biotech plant, Novo Nordisk; a large plasterboard factory, Gyproc;

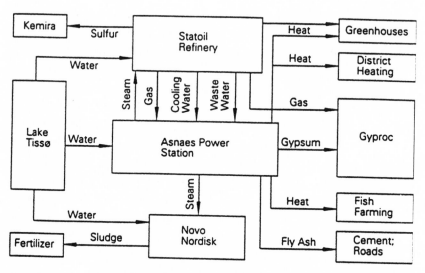

FIGURE 9

Flow diagram for Kalundborg industrial ecosystem. (Hardin Tibbs)

and the city of Kalundborg. Rather than burning off surplus gas, the refin-ery now sends it to the power plant and the plasterboard factory, which use it as a fuel in their production processes. The refinery also sends excess cool-ing water and wastewater to the power plant to help it meet its cooling needs and turns waste sulfur into sulfuric acid, which it sells. The power sta-tion, in turn, sells waste steam to the city, the biotech company, and Statoil for heating. This has allowed the city and the Danish government to replace about 3,500 air-pollution-emitting oil furnaces. By developing a way to use salt water from the local fjord for some of its cooling needs, the power sta-tion reduced its withdrawals of fresh water from nearby Lake Tisso. It sends the resulting water waste by-product, hot salt water, to a local fish farm. The power plant also processes waste SO_2 into calcium sulfate (gypsum), which it sells to the plasterboard company. It sells its desulfurized fly ash to a cement company. The biotech facility, in turn, sells sludge from its manu-facturing process to nearby farms for use as fertilizer. This web of waste exchanges has benefited the firms involved by lowering their costs and cre-ating new revenues. It has also reduced air, water, and land pollution in the region.[13]

The industrial ecology concept also encourages waste minimization through design for environment (DFE). Instead of focusing on designing end-of-pipe systems for treating air and water pollution, industrial ecology encourages industrial designers to focus on redesigning products and manufacturing processes so as to reduce the burden our consumer society puts on the natural environment. This can be done by substituting less toxic chemicals for more toxic ones in product design and manufacturing processes. Products can also be designed to use materials and energy in ways that reduce the emission of wastes throughout the product life cycle, from resource extraction through end product disposal.[14]

Products can also be designed to be easily repaired, remanufactured, or taken apart and recycled, in order to extend the usefulness of the materials and energy bound up in them. Although it is important to take into consideration the environmental impacts of all the resources used to repair or recycle a product, designing products so that they can be reused or recycled can protect the environment by reducing the need to extract raw materials from the natural world in the first place.[15]

To be effective, DFE and the Kalundborg waste exchange approach to waste management compel managers to move away from conventional short-term, arm's-length relationships with their suppliers to closer, longer-term relationships. This is necessary not only for companies developing innovative relationships to buy or sell wastes as inputs into production, but also for those interested in doing environmental design. Because many industries are increasingly vertically disintegrated, many companies, like most of those in the computer industry, do not manufacture or even do all the design work on the products they sell. They simply do the final assembly and slap their labels on the goods. To reduce pollution emissions associated with resource extraction and manufacturing, they must therefore work with the companies in their supply chains to help them improve their environmental performance and develop the capability to do DFE. This is especially important where suppliers are located in developing countries with weak pollution regulation and little practice in green design.[16]

Similarly, companies that seek to use recycled materials in production must develop systems for acquiring the appropriate recycled materials. If they wish to remanufacture or recycle their own products they must develop mechanisms for taking back their old equipment. This necessitates that they develop closer relationships with their customers as well as with the suppliers who will be utilizing recycled materials in production.

Although these innovations may sound incredibly idealistic and economically unrealistic, it is important to realize that regulatory developments are taking place that support this sort of systems change. Germany has pioneered the use of product and packaging take back regulations that have stimulated important advances in green packaging design as well as in DFE in the electronics industry. Various northern European countries have also established voluntary eco-labels, like Germany's well known Blue Angel label, that help green consumers identify and purchase more environmentally benign products. In addition, various industry trade organizations have established rules and processes by which firms can certify that they have implemented environmental management systems that meet the organization's standards. The best known of these voluntary industry standards is the International Organization for Standardization's 14000 standard, but the British Standards Organization has also promulgated its own set of standards and the European Community has implemented a standard known as EMAS. Admittedly, these advances are mostly taking root in Europe, where the environmental movement is far more advanced than in the United States. But they support the development of a marketplace that rewards corporations for selling environmentally sound products and establishing stronger corporate environmental management systems. This creates incentives for managers and engineers to adopt the principles of industrial ecology for competitive advantage.[17,18]

What does all this mean for historians of business and the environment? First, it means that business is possibly at the cusp of a major historical change, a paradigm shift. To help corporate managers, policy makers, and the public as a whole better understand what is going on, we historians must start trying to put this change into historical perspective. This is why I am working on a history of industrial pollution control in the United States that will go back to the Industrial Revolution. We have our jobs cut out for us, trying to explain the history of how business got to the point it is today in its management of its environmental impacts. Second, industrial ecology offers us historians an important and very useful conceptual framework that we can use to think about and explore the nature of business's interaction with the natural world over time in all its many dimensions. The concept reminds us that the history of business and the environment includes the history of the extraction of natural resources for economic production, the design, distribution, use, and disposal of consumer goods, and the history of factory pollution control. It encourages us to recognize that business firms are far more than organizational structures managed by people pursuing efficiency and profit maximization in an economic marketplace. Equally important, they are parts of a broader socio-environmental system—intimately bound up in networks of relationships and mutual interactions with all parts of the natural world and society. To fully understand the history of industry's impact on the natural world, we will thus have to do a lot more than study the history of the management of industrial pollution and waste. We will have to examine all aspects of the interface between our business system and the natural environment, including the history of advertising and the rise of modernity's culture of consumption.

My hope is that this sort of broad historical inquiry will give managers and policy makers the perspective that will help them understand the limits of the old, conventional end-of-pipe approaches to pollution control. I also hope that this kind of environmental business history will help managers and policy makers to better understand the advantages of making the transition to new, more effective, more systematic ways of mitigating industry's harmful environmental impacts based on the concept of industrial ecology.

NOTES

1. Matthew Josephson, *The Robber Barons: The Great American Capitalists* (Harcourt Brace, 1934) is classic of the Robber Baron tradition. For the debate between the Robber Baron and Industrial Statesmen schools, see Peter D. A. Jones, ed., *The Robber Barons Revisited* (Heath, 1968). For more on J. D. Rockefeller, see Ida M. Tarbell, *History of the Standard Oil Company* (McClure Phillips, 1925); Allan Nevins, *Study in Power: John D. Rockefeller, Industrialist and Philanthropist* (Scribner, 1953). An excellent review article that discusses business history before and after Chandler is Richard R. John, "Elaborations, Revisions, Dissents: Alfred D. Chandler Jr's *The Visible Hand* after Twenty Years," *Business History Review* 71 (1997), summer, 151–200. See also Jeffrey K. Stine and Joel A. Tarr, "At the Intersection of Histories: Technology and the Environment," *Technology and Culture* 38 (1997), July, 618–625.

2. For examples of this kind of thinking, see the articles by Donald Worster, Alfred Crosby, Richard White, Carolyn Merchant, William Cronon, and Stephen J. Pyne in "A Roundtable: Environmental History," *Journal of American History* 76 (1990), 1087–1147. See also Marcy Darnovsk, "Stories Less Told: Histories of US Environmentalism," *Socialist Review* 22 (1992), October-November, 26–28; Richard White, "American Environmental History: The Development of a New Field," *Pacific Historical Review* 54 (1985), August, 330. For evidence that things are changing, see the essays in *Out of the Woods: Essays in Environmental History,* ed. C. Miller and H. Rothman (University of Pittsburgh Press, 1997). See also *Uncommon Ground: Rethinking the Human Place in Nature,* ed. W. Cronon (Norton, 1996).

3. Hardin Tibbs, *Industrial Ecology: An Environmental Agenda for Industry* (Arthur D. Little, 1991), 13.

4. Robert U. Ayres, "Industrial Metabolism: Theory and Policy," in *The Greening of Industrial Ecosystems,* ed. B. Allenby and D. Richards (National Academy Press, 1994), 24.

5. For more detailed discussions, see Christine Meisner Rosen, "Industrial Ecology and the Greening of Business History," *Business and Economic History* 26 (1997), 123–137; Christine Meisner Rosen and Christopher C. Sellers, "The Nature of the Firm: Towards an Eco-Cultural History of Business," *Business History Review* 73 (1999), spring, 577–600.

6. Harold M. Mayer and Richard C. Wade, *Chicago: Growth of a Metropolis* (University of Chicago Press, 1969), 52.

7. Jerry E. Patterson, *The City of New York: A History Illustrated from the Collections of the Museum of the City of New York* (Abrams, 1978), 169.

8. Some of my findings are analyzed in "Balancing Doctrine in Pollution Nuisance Law: Differing Perceptions of the Value of Pollution Abatement Across Time and Place, 1840–1906," *Law and History Review* (fall 1993); "The Costs and Benefits of Pollution Control in Pennsylvania, New York, and New Jersey, 1840–1906," *Geographical Review* 88 (1998), April, 219–240; and "Noisome, Noxious, and Offensive Vapors, Fumes, and Stenches in American Towns and Cities, 1840–1865," *Historical Geography* 25 (1997), 49–82. See also Peter Karsten, *Heart Versus Head: Judge Made Law in Nineteenth Century America* (University of North Carolina Press, 1997), 134–143, 465–515; Robert Bone, "Normative Theory and Legal Doctrine in American Nuisance Law: 1850 to 1920," *Southern California Law Review* 59 (1986), September, 1104–1226; Craig E. Colten, "Environmental Justice in the American Bottom: The Legal Response to Pollution, 1900–1950," in Andrew Hurley, ed., *Common Fields: An Environmental History of St. Louis* (St. Louis: Missouri Historical Society Press, 1997), 163–175; Paul M. Kurtz, "Nineteenth Century Anti-Entrepreneurial Nuisance Injunctions—Avoiding the Chancellor," *William and Mary Law Review* 17 (summer 1976), 621–669; Jan G. Laitos, "The Social and Economic Roots of Judge-Made Air Pollution Policy in Wisconsin," *Marquette Law Review* 58 (1975).

9. Christine Meisner Rosen, "Businessmen Against Pollution in Late Nineteenth Century Chicago," *Business History Review* 69 (1995), autumn, 351–397; Rosen, "Noisome, Noxious, and Offensive Vapors, Fumes, and Stenches," 65–77; Maureen Flannagan, "Gender and Urban Political Reform: The City Club and the Woman's City Club of Chicago in the Progressive Era," *American Historical Review* 95 (1990), October, 1032–1050; Andrew Hurley, "Busby's Stink Boat and the Regulation of

Nuisance Trades, 1865–1918," in Hurley, *Common Fields,* 145–162; Martin V. Melosi, ed., *Pollution and Reform in American Cities, 1870–1930* (University of Texas Press, 1980); William J. Novak, *The People's Welfare: Law and Regulation in Nineteenth Century America* (University of North Carolina Press, 1996), 217–233; Harold L. Platt, "Invisible gases: smoke, gender, and the redefinition of environmental policy in Chicago, 1900–1920," *Planning Perspectives* 10 (1995): 67–97; David Stradling, Civilized Air: Coal, Smoke, and Environmentalism in America, 1880–1920 (Ph.D. dissertation, University of Wisconsin, 1996). Joel A. Tarr, *The Search for the Ultimate Sink: Urban Pollution in Historical Perspective* (University of Akron Press, 1996). Joel A. Tarr and Carl Zimring, "The Struggle for Smoke Control in St. Louis: Achievement and Emulation," in Hurley, *Common Fields,* 199–220.

10. "Abating New York's Smoke Nuisance," *The American City* 10 (1914), February, 187.

11. For other before-and-after pictures, see John A. Switzer, "Economic Aspects of the Smoke Nuisance: Chimneys Need Not Be 'Smokestacks,' " *Scientific American Supplement,* No. 1888 (March 9, 1912), 153–155; "Automatic Stokers as Smoke Preventers," *The American City* 9 (1913), 254, 255, 281.

12. I. F. Harlow and T. J. Powers, "Pollution Control at a Large Chemical Works," *Industrial and Engineering Chemistry* 39 (1947), May, 572–577. See also I. F. Harlow, "Waste Problems of a Chemical Company," *Industrial and Engineering Chemistry* 31 (1939), November, 1346–1349; Thomas Powers, "The Treatment of Some Chemical Industry Wastes," *Sewage Works Journal,* 17 (1945), March, 330–337.

13. John Ehrenfeld and Nicholas Gertler, "Industrial Ecology in Practice: The Evolution of Interdependence at Kalundborg," *Journal of Industrial Ecology* 1 (1997), winter, 67–79; Henning Grann, "The Industrial Symbiosis at Kalundborg, Denmark," in *The Industrial Green Game,* ed. D. Richards (National Academy Press, 1997), 117–123; E. A. Lowe et al., *Discovering Industrial Ecology: An Executive Briefing and Sourcebook* (Battelle Press, 1997), 129–158. See also Marian R. Chertow, "The Eco-Industrial Park Model Reconsidered," *Journal of Industrial Ecology* 2 (1998), summer, 8–10. Hardin Tibbs, *Industrial Ecology: An Environmental Agenda for Industry* (Arthur D. Little, 1991), 8.

14. Braden R. Allenby, "Integrating Environment and Technology: Design for Environment, in *The Greening of Industrial Ecosystems,* ed. Allenby and Richards; T. E. Graedel and B. R. Allenby, *Industrial Ecology* (Prentice-Hall, 1995); Graedel and Allenby, *Design for Environment* (Prentice-Hall, 1996); Janine C. Sekutowski, "Greening the Telephone: A Case Study," in *The Greening of Industrial Ecosystems,* ed. Allenby and Richards; Joseph Fiksel, *Design for Environment: Creating Eco-Efficient Products and Processes* (McGraw-Hill, 1996); Livio D. DeSimone and Frank Popoff, *Eco-Efficiency: The Business Link to Sustainable Development* (MIT Press, 1997), 118–125.

15. Walter R. Stahel, "The Utilization-Focused Service Economy: Resource Efficiency and Product Life Extension," in *The Greening of Industrial Ecosystems,* ed. Allenby and Richards, 178–190.

16. DeSimone and Popoff, *Eco-Efficiency,* 127–133; "Approaches to Integrating Suppliers in EH&S," *Business and the Environment* 10 (1999), June, 2–4; Christine Rosen, Janet Berkovitz, and Sara Beckman, The Dawning Challenge of Environmentally Sound Supply Chain Management: A Study of Progress in the Computer Industry (working paper, Haas School of Business, University of California. Berkeley, 1999).

17. Graedel and Allenby, *Industrial Ecology,* 316–324; Braden R. Allenby, *Industrial Ecology: Policy Framework and Implementation* (Prentice-Hall, 1999).

18. An excellent source of information on current developments in European environmental regulation is *Environment Watch: Western Europe.* On ISO 14000, see *Business and the Environment's ISO14000 Update.* Both are monthly newsletters published by Cutter Information.

INDUSTRIAL ECOLOGY

BRADEN ALLENBY

Industrial ecology is a new way of looking at the interactions between human and natural systems. It represents a significant augmentation, rather than replacement, of existing approaches to environmental issues. In this light, two fundamental premises of industrial ecology should always be emphasized. First, reflecting the reality of the world within which human and environmental systems operate, industrial ecology always takes a systems view. Second, industrial ecology is not about the environment: it is about technology and the evolution of human culture and economic systems. These principles sound simple, almost trivial, but when applied they make industrial ecology a significant, almost radical, improvement over current practices.

One must begin, however, by asking whether industrial ecology is justified in the first place. Why is there a need for fundamental change in how environmental issues are approached? In short, this need arises because of a premise that underlies current environmental regulation and practices: that environmental perturbations can be mitigated by treating them only as they are manifested, without serious regard for the economic activity that caused them. This approach treats environmental considerations as if they are "overhead," not "strategic," for consumers, producers, and society. ("Overhead" activities are ancillary to the primary activity of an individual, a firm, or society, while strategic activities are integral to the primary activity.) Thus, environmental problems have been widely perceived as local in space and time, and frequently associated with a single substance, such as the pesticide DDT, polychlorinated biphenyls (PCBs), or heavy metals. Social responses are still ad hoc, reflecting a strong bias toward remediation of existing localized problems—individual airsheds or watersheds, or specific waste sites—even though these in many cases pose little, or easily managed, risk. Compliance activities, especially in the United States, are usually focused on limiting emissions of undesirable materials through the use of emission control technologies—the so-called end-of-pipe approach. Even "pollution prevention" programs generally focus on relatively simple adjustments of existing production technologies. Virtually all environmental regulation and management also focus on manufacturing, which is only one life-cycle stage in a product's passage through the economy, and frequently not the one generating the most significant environmental impact. More fundamentally, service sectors, which account for about 60–70 percent of most developed countries' economies and in many cases offer the potential for discontinuous improvements in environmental performance across the economy as a whole, are virtually invisible.[1]

service sectors

It is not so much a question of whether the current paradigm of environmental regulation is working. Indeed, as implemented in most developed countries, it has demonstrably resulted in cleaner air and water, and less toxic loading of the environment, and such emissions reduction initiatives are of unquestionable value in reducing environmental insults in many developing countries. The question, rather, is whether the paradigm is adequate to respond to the new information and data that have accumulated since it was first developed, or whether it must, in turn, be subsumed into a broader approach. The answer to this latter question is clearly yes. The ad hoc focus on symptoms of economic activity—specific media impacts, or waste sites—is augmented and made more efficient, not replaced, by a far more comprehensive approach which focuses on production and consumption patterns throughout the economy.

Table 1 compares the prevailing "overhead paradigm" and the evolving industrial ecology paradigm. To begin with, the risks that each approach addresses are profoundly different in both scale and complexity. Remediation and compliance aim at the reduction of localized risks, often defined only in terms of human risk, while industrial ecology addresses not only those, but also the environmental perturbations threatening sustainability, such as loss of biodiversity, global climate change, stratospheric ozone depletion, and global degradation of water, soil, and air resources. Mitigating the latter requires fundamental changes in technology and in economic and cultural behavior, not just the establishment of a fund to support cleanups. Reflecting the greater complexity, the disciplines involved in the traditional environmental approach are relatively limited, and as reductionist as environmental science allows. With industrial ecology, on the other hand, a far broader and more integrative knowledge base is required: in par-

TABLE 1
Technology and the environment: characteristics of principal approaches

Primary activity	Time frame	Primary risks addressed	Focus of activity	Relation of environment to economic activity	Disciplinary approach
Remediation and compliance	Past/present	Local and visible environmental and human risk	Individual site, media, or substance	Overhead	Toxicology and environmental science; reductionist
Industrial ecology	Present/future focus	Global climate change; loss of biodiversity and habitat; degradation of air, water, and soil resources	Materials, products, services, and operations over life cycle	Strategic and integral	Engineering; physical sciences; biological sciences; social sciences; law and economics; highly integrative

ticular, sophistication in engineering, business and technology fields is required. This, of course, reflects the basic premise that technology—which is, after all, the means by which humans interact with the environment—is a crucial theme of industrial ecology. More subtly, remediation and compliance programs generally assume a complete understanding of the systems involved, so that a specific regulation can be targeted to have the desirable effects without causing any unanticipated side effects elsewhere. If properly implemented, this is an appropriate approach to specific, well-defined, localized hazards. For the kinds of complex natural and human systems with which industrial ecology deals, however, it is inappropriate, the more so because it is often unconscious.

This highlights one of the principal lessons that modern environmentalism and industrial ecology teach, but one which is often ignored: our lack of knowledge about these complex systems and their interactions is profound. In fact, in many cases we lack the data and understanding to even ask the right questions. In industrial ecology, this is frequently illustrated in assessments of environmental impacts of technologies or materials over their life cycle, which tend to reveal significant surprises. One example is the issue of paper versus plastic drinking cups, which was a fairly contentious debate until a life-cycle study by the government of the Netherlands surprised everybody by demonstrating that, under most circumstances, both paper and plastic disposable cups were preferable to ceramic ones, a quite counterintuitive result.[2] Another life-cycle study indicated that, from an environmental perspective, it was probably better to discard a polyester blouse after one wearing and buy a new one than to wash it (because of the energy and material inputs required by washing and drying).[3]

[handwritten margin note: surprising life-cycle study results]

A CONCEPTUAL FRAMEWORK FOR INDUSTRIAL ECOLOGY

True to the systems approach, industrial ecology is best illustrated by locating it within a conceptual framework that, at least in broad outline, links goals and methodologies. One such framework is presented in figure 1; it should be taken, of course, as illustrative and not definitive at this point.[4]

The highest level is the vision of sustainable development. This was defined by the Brundtland Commission (formally the World Commission on Environment and Development), which originated the term, as "devel-

[handwritten margin note: Level 1: sustainability]

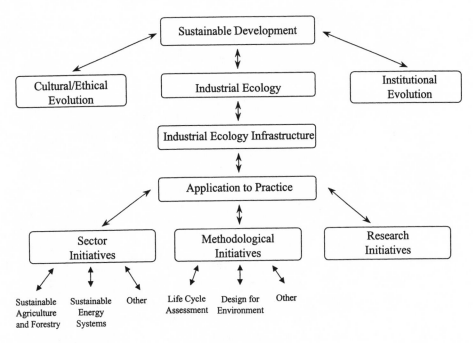

FIGURE I

A conceptual framework for industrial ecology.

opment that meets the needs of the present without compromising the ability of future generations to meet their own needs." It embeds within it a number of values, including redistribution of wealth within and among human generations, a need to aggressively restrict population growth, and equal rights for women. It is also anthropocentric: that is, it defines sustainability in terms of the primacy of humans over other forms of life, a position accepted widely but not universally.

It is also a very ambiguous concept. Most important, there are many possible global systems which would be sustainable over some finite time period. For example, an alternative sustainable world might be one where elites barricade themselves and continue to enjoy a materials-and-energy-intensive high quality of life, while sustainability over time is maintained by low levels of biodiversity and mortality rates among the poor. Sustainable development is not only one of many futures; it may not be the most probable (indeed, cynics would argue that modern trends, such as increasing income disparity within and among nation states, are away from, not toward, it). Less fundamentally, existing data cannot prove or disprove the implicit assumption that present economic development can continue without compromising the ability of future generations to meet their own needs—that is, whether sustainability as defined by the Brundtland Commission is even possible. Equally important, it is not clear what sustainable development might mean operationally. From a systems perspective, for example, it is meaningless to analyze small subsystems in terms of sustainability when "sustainable" can be defined only in terms of the global system. Thus, for example, a firm that chooses to call itself sustainable may mean that it has adopted sustainability as its vision, but it cannot be sustainable in actuality so long as it is integrated into an unsustainable global economy. It would be equally implausible to call a product, a process, or a community sustainable at this point in the evolution of the Industrial Revolution. One cannot make such assertions of sustainability without understanding the global human and natural systems as an integrated whole and without being able to project their stability through time, which is clearly not the case now and will not be for the foreseeable future.

This does not mean, however, that sustainable development cannot provide the general goal toward which environmentally preferable activities can be directed. Thus, for example, the Netherlands has adopted the goal of becoming sustainable within one generation. This is not because the people of the Netherlands (a small, heavily exporting nation that is only a part of regional European watersheds and airsheds) believe that their country can be sustainable in a global sense. Rather, defining their goal in such a manner has allowed them to do path-breaking research into what sustainability might mean for a modern, developed country, and what metrics might be appropriate for determining progress toward those goals.[5] Similarly, the Institute of Electrical and Electronics Engineers, the world's largest technical professional society, considers sustainable development a worthy vision, even while noting in its 1995 White Paper on Sustainable Development and Industrial Ecology that "standing alone, [sustainable development] cannot guide either technology development or policy formulation."

The second level, industrial ecology, is the multi-disciplinary study of industrial systems and economic activities and of their linkages with fundamental natural systems. It provides a theoretical basis and an objective understanding upon which reasoned improvement of current practices can be based. Important disciplines contributing to industrial ecology include the physical and biological sciences, engineering, economics, law, anthropology, policy studies, and business studies.[6] Even with this broad scope, however, it is important to note that the study of industrial ecology alone is not sufficient to support the achievement of the vision of sustainable development, which is heavily normative. Rather, progress toward any desirable sustainable world requires concurrent evolution in ethics and institutions. In part, this is because any global sustainable state relies on political, cultural, and religious systems for its definition and achievement.[7] Industrial ecology, on the other hand, is an objective field of study, relying on traditional scientific, engineering, and other disciplinary research for its development; it provides scientific and technological understanding, but, standing alone, it cannot define what is at bottom a values decision about what kind of world we as humans desire.

The third level, the industrial ecology infrastructure, is society's response to this question: "Assuming that private and public firms and consumers can be encouraged to behave in environmentally appropriate ways, what must the state and society in general provide so that they may do so?" It thus includes developing and implementing the legal, economic, and other incentive systems by which desirable behavior can be promoted, as well as the methodologies, tools, data, and information resources necessary to define and support such behavior. One example might be the development of legal structures that promote environmentally appropriate behavior; another might be the development and diffusion throughout the global economy of environmentally preferable technologies. Examples of such policies might be environmentally sensitive government procurement regulations, military specifications, and military standards; the removal of environmentally and economically inefficient subsidies for virgin (as opposed to recycled) materials; and the removal of energy, transport, agricultural, fishery, and forestry subsidies, which distort production in those sectors in environmentally inappropriate ways.

Unlike the first three levels, the fourth level, application to practice, is concerned primarily with implementation. While the specific activities undertaken will be different for different firms, different consumers, different economic sectors, and different elements of the public, it will for all of them represent the level of immediate action, based on industrial ecology principles as currently understood and translated into policy. While still nascent, such implementation efforts represent important experimentation activities, and there is a dialog between this level, which plays with industrial ecology principles and theories, and the more theoretical higher levels of the industrial ecology framework.

INDUSTRIAL ECOLOGY

Recognizing that industrial ecology itself is a part of a broader framework makes it easier to understand this still nascent field of study. One of the earlier definitions can be found in the first textbook on industrial ecology:

Industrial ecology is the means by which humanity can deliberately and rationally approach and maintain a desirable carrying capacity, given continued economic, cultural, and technological evolution. The concept requires that an industrial system be viewed not in isolation from its surrounding systems, but in concert with them. It is a systems view in which one seeks to optimize the total materials cycle from virgin material, to finished material, to component, to product, to obsolete product, and to ultimate disposal. Factors to be optimized include resources, energy, and capital.[8]

The words "deliberately" and "rationally" make an important point about the motivation for studying industrial ecology: the desire to develop the technological and scientific basis for a considered path toward global sustainability, in contrast to unplanned, precipitous, and potentially quite costly and disastrous alternatives. "Desirable" indicates that, in view of the potential for different technologies, cultures, and forms of economic organization, a number of sustainable states are possible; it thus becomes an ethical imperative to choose among them, and to then act so as to approach the desired state.

The electronics industry, especially in the United States, has been a leader in developing industrial ecology and exploring its implementation through Design for Environment (DFE) practices. Accordingly, it is instructive to quote the definition provided in the White Paper on Sustainable Development and Industrial Ecology issued by the Institute of Electrical and Electronic Engineers:

Industrial ecology is the objective, multi-disciplinary study of industrial and economic systems and their linkages with fundamental natural systems. It incorporates, among other things, research involving energy supply and use, new materials, new technologies and technological systems, basic sciences, economics, law, management, and social sciences. Although still in the development stage, it provides the theoretical scientific basis upon which understanding, and reasoned improvement, of current practices can be based. Oversimplifying somewhat, it can be thought of as "the science of sustainability." It is important to emphasize that industrial ecology is an objective field of study based on existing scientific and technological disciplines, not a form of industrial policy or planning system.[9]

This definition makes the important point that industrial ecology strives to be objective, not normative. While industrial ecology studies may be undertaken for subjective reasons—to develop more environmentally efficient technology systems, or identify opportunities to conserve resources, for example—the studies themselves, like any scientific effort, should be transparent, replicable, and falsifiable. Thus, where cultural, political, or psychological issues arise in an industrial ecology study, they are evaluated as objective dimensions of the problem. Additionally, it is important to note that, while for a number of reasons industrial ecology studies to date have tended to focus on manufacturing and manufactured products, the scope of the field is broader, including all elements of economic activity, such as mining, agriculture, forestry, manufacturing, and consumer behavior. Both demand-side (consumer) and supply-side (producer) aspects of economic behavior, and the consequent impacts on natural systems at all temporal and spatial scales, are included. Equally important, industrial ecology includes not only economic activity in advanced economies but also subsistence human activity at the fringes of formal economic systems. After all, many critical impacts, both in terms of human health and in terms of ecosystem and biodiversity maintenance, are generated by activities at that level.

The focus on manufacturing to date reflects both the traditional focus of environmental science and regulation and the history of the evolution of the field. The term "industrial ecology" seems to have been first used in 1970, in the title of a journal with a small circulation and a short life (apparently only one issue was ever printed). Judging by the subject matter, it was intended simply to reflect the fact that industrial activities impacted nature. In 1972, however, Japan's Ministry of Industry and International Trade began considering such a metaphor as suggesting a model for structuring the Japanese industrial system; however, that effort was dropped with the coming of the energy crisis of 1973. In 1989, Robert Frosch and Nicholas Gallapoulos revived the term as an analogy, suggesting that industrial systems could be more efficient if their material flows were modeled after natural ecosystems.[10]

In parallel, several important studies of the interrelationship between technology and industrial activities and environmental impacts were being published. The National Academy of Engineering was particularly active in this area; its noteworthy early efforts include *Technology and Environment*[11] and *Energy: Production, Consumption, and Consequences*.[12] Another important contribution was *The Earth as Transformed by Human Action,*[13] a comprehensive and still valuable source. Although these earlier publications did not use the term "industrial ecology," they did not take the prevalent reductionist approach, but attempted to view the activities from a more comprehensive, systematic basis—that is, they began to take a true industrial ecology approach to the issues.

Meanwhile, the complex, systems-based nature of regional and global environmental perturbations—particularly global climate change, ozone depletion, and loss of habitat and biodiversity—and the growing inability of existing environmental policies, based on reductionist approaches, to address such issues adequately, became increasingly obvious. Additionally, industry was increasingly affected by rapidly rising costs of environmental compliance and cleanup, which were perceived as highly inefficient both environmentally and economically. These linked pressures generated a strong need for a new paradigm, a new way of thinking about these issues. Concomitantly, the 1987 report from the World Commission on Environment and Development, *Our Common Future,* had begun the dialog on the concept of sustainable development, a goal which clearly pushed beyond the existing environmental regulation approach.[14]

Responding in large part to these pressures, industry, particularly the chemical and electronics sectors in the United States, began to develop practices based on industrial ecology principles. In electronics, for example, such initial methodologies were generally developed as Design for Environment techniques, to indicate that they were intended to be a module of existing concurrent engineering practices in that industry, which are based on a "Design for X" approach, where X is any desirable characteristic of the product being designed (manufacturability, safety, testability, etc.). Thus, in 1993 the American Electronics Association published a collected set of White Papers under the title *The Hows and Whys of Design for the Environment: A Primer for Members of the American Electronics Association.*[15] More broadly, the Office of Technology Assessment of the US Congress issued *Green Products by Design,* which reinforced the focus on products, rather than specific materials or localized impacts.[16]

Despite the OTA report, the product-oriented focus has been generally ignored in the United States. It has, however, become a central tenet of European technology and environment policies, particularly in such Northern European countries as the Netherlands, Germany, and Sweden. Examples include the work on Integrated Substance Chain Management published by the Netherlands Department of Housing, Physical Planning and Environment in 1991 and the 1992 report Extended Producer Responsibility as a Strategy to Promote Cleaner Products from Lund University in Sweden. Thus, even though the term "industrial ecology" was not initially used in Europe, much related activity has occurred there, particularly in the Netherlands, in Germany, and in the Scandinavian countries. The Netherlands Department of Housing, Physical Planning and Environment, for example, initiated a unique effort to define a sustainable society in one generation, publishing a National Environmental Policy Plan in 1989 and a National Environmental Policy Plan Plus in 1990. European private industry also contributed significantly to the dialog on development and the environment; of particular note is *Changing Course,* by Stephan Schmidheiny and the Business Council for Sustainable Development (MIT Press, 1992).

The effect of these mutually reinforcing trends has been to evolve industrial ecology from just an interesting metaphor into a nascent field of study. Though this process has not yet been completed, and the term is still subject to misuse, a number of new publications have begun to formalize the field. Among the more notable are *The Greening of Industrial Ecosystems* (National Academy Press, 1994) and *Industrial Ecology: US-Japan Perspectives* (National Academy Press, 1994); *Industrial Ecology and Global Change* (Cambridge University Press, 1994); *Industrial Ecology,*[24] the first engineering textbook in the field (Prentice-Hall, 1995); *The Industrial Ecology of the Automobile* (Prentice-Hall, 1997); and *Industrial Ecology: Policy Framework and Implementation* (Prentice-Hall, 1999). A new scientific journal, the *Journal of Industrial Ecology,* based at the Yale University School of Forestry and Environmental Studies, began publication in 1997.

In parallel, there have been several efforts to begin to apply the models and understanding of biological ecology to industrial ecology questions: B. R. Allenby and W. E. Cooper, "Understanding Industrial Ecology from a Biological Systems Perspective," *Total Quality Environmental Management,* spring 1994, 343–354; T. E. Graedel, "On the Concept of Industrial Ecology," *Annual Review of Energy and the Environment* (1996); C. Schulze, ed., *Engineering Within Ecological Constraints* (National Academy Press, 1996). Such efforts, however, must always be undertaken cautiously and with care, as the intuitive appeal of the analogy can disguise the considerable differences between the two types of systems. Unlike economic systems, for example, biological communities do not contain agents with some foresight, and thus they lack the same element of contingency and free will. It is not that economic systems are the same as biological ecologies; rather, the point is that much work on complex systems has been done in the context of biological ecology, and some of this learning is applicable to economic and industrial systems, despite their differences.

The analogy between natural and human systems does help, however, to express important elements of industrial ecology. Thus, the concept of industrial ecology can be illustrated by considering three different models (figure 2).

A Type I system is linear. Virgin materials enter the system, are used only once, and are then disposed of as waste (that is, as materials for which there

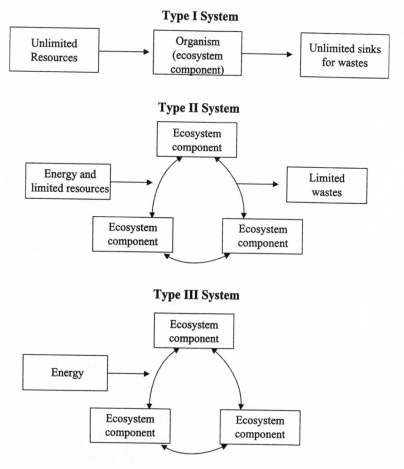

FIGURE 2
Types of industrial ecology systems.

is no further use within the system). Materials that may have no further use within the generating process (or firm) but could be reused within the system as a whole may be considered "residuals." A Type I flow of particular concern in most economies involves dispersive use of toxics—that is, uses of toxic materials that inevitably result in their dispersal into the environment. Lead in gasoline is a classic example of this type, as is cadmium in fertilizers containing mined phosphate.

A more complex Type II system arises as scarcity (or growing populations) makes a Type I system inadequate. Feedback and internal cycling loops develop, in large part through the evolution of new organisms—or, in the case of an economy, new technologies, products, or services. Accordingly, flows of material into, and waste out of, the system diminish on a per-unit basis (e.g., per weight of biomass supported, or per dollar of economic activity). Internal reuse of materials can become quite significant. In an economy, while the internal material use within the system may remain high, the velocity of materials through the system is reduced. Material management systems, either planned or spontaneous, become more prevalent. A common example of a Type II system in many economies today is the life cycle of most cars. The post-consumer car is first stripped of useful sub-assemblies, which are then reconditioned if necessary and recycled as used parts. The remaining hulk is shredded, and the steel (some 75 percent by weight of the automobile) is recovered for recycling. This is an internal material loop. The remaining plastic, glass, and miscellaneous materials and liquids, known as "fluff" or "ASR" (automobile shredder residue), are then landfilled. Landfills, which would be both environmentally and economically more efficient if conceptualized and designed as residual storage facilities as opposed to heterogeneous waste disposal sites, currently are not part of material cycling streams. Accordingly, the fluff stream offers an opportunity to shift from a Type I to a Type II system, from treating ASR as a waste to managing it as a residual.

A Type III system is one in which full cyclicity has been achieved. In many cases a Type III system should be a goal, but not always. For example, local energy recovery from plastics might well be preferable to shipping lightweight plastics long distances to reformulating facilities, because of the environmental impacts of the transportation required in the latter case. Materials recycling must also be considered broadly, in conjunction with related technologies and natural systems. Thus, for example, production of energy from biomass feedstocks should be evaluated in the context of the global carbon, nitrogen, and hydrologic cycles, and the environmental implications of all sectors involved in the process (e.g., agriculture or forestry; agricultural chemical production, distribution and application; biomass transportation and processing; energy production and transmission; and energy end use efficiency). Such industrial ecology approaches form the science and technology base for the practice of Earth Systems Engineering.[17]

With current technologies and economic incentives, achieving a total Type III system for many materials and technological systems will be extremely difficult because of the dissipative uses of materials discussed above. Tires, for example, wear under normal circumstances, and major categories such as food and many personal care products (e.g., soaps and shampoos) are inherently dissipative. In such instances, minimization of dissipative uses, especially any dissipative uses of toxics or materials that may be toxics, may be the best that can be hoped for.

In all such instances, the systemic principle underlying industrial ecology highlights the point that there is no inherent requirement for the materials management function in any of these recycling systems to include only one firm or one sector, although in some cases such "tighter" loops may be more environmentally efficient. In principle, rather, the goal should be a self-contained global economy, as illustrated in figure 3, with materials efficiency optimized over the larger system. Thus, for example, the concept of geographically co-located industrial material cycling systems, such as eco-parks, is interesting, but is only one possibility among many. To focus too intently on any one potential subsystem runs the risk of overlooking the need to improve system performance as a whole. Note also that these systems are energetically open, as are natural systems: the constraint here is that energy use by the system as a whole must be sustainable over some period of time.

This simple discussion illustrates several critical aspects of industrial ecology. Most important, industrial ecology is profoundly systems-oriented. This does not mean that every industrial ecology activity should include all possible impacts on any relevant system; it does, however, require a sensitivity to the systems aspect of industrial ecology research. Thus, for example, where boundaries are drawn around particular activities or subsystems, they should be justifiable and reasonable in the context of the relevant system, and the interfaces between the subsystem and the external environ-

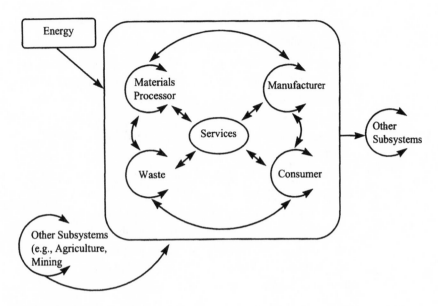

FIGURE 3
A Type III self-contained global economy.

ment should reflect the integrated nature of the system as a whole. Of particular concern is the need to ensure that "emergent behavior"—that is, behavior at a certain level of the system which cannot be predicted simply by summing up the activities at lower levels of the system—is not inadvertently overlooked.

An interesting hypothesis is that biological systems, as they increase in complexity, are able to support more biomass.[18] It is certainly possible that, in broad aggregate, such a relationship between complexity and population levels holds for human systems too. A hunter-gatherer society is simpler than an urban, industrialized society; however, all else being equal, it can support far fewer people per unit of land than the latter. This, of course, leads to the interesting hypothesis that appropriate deliberate increases in the complexity of the existing human economy, perhaps through increased information density, can increase the human population the globe can support, at least within certain (not well understood) constraints.

It is also important to recognize that systems of any complexity are composed of a number of subsystems, or cycles, functioning on widely differing temporal and spatial scales, a behavior that greatly complicates analysis and understanding of the system. This can be particularly difficult because many important natural systems and cycles, such as the carbon cycle, operate on time scales (decades or even centuries) that are considerably beyond the perceptual and psychological time horizons of people and many institutions. Much of the political and social concern involving nuclear power, for example, arises from the fact that some residuals are radioactive for centuries—and thus, must be managed for centuries—and social and technological systems are not routinely designed to incorporate such long-term activity.

Finally, it is important to remember that industrial ecology is an integrative, not a reductionist, field. It focuses on a comprehensive, holistic understanding of systems rather than the reductionist approach of developing more and more knowledge about increasingly specific subsystems, and, as table 1 illustrates, cuts across a number of disciplines in doing so. The Western scientific paradigm is, of course, highly reductionist, so it is not surprising that this facet of industrial ecology has a number of implications. For example, some scientists tend to view any approach which is not reductionist as lacking in scientific rigor or quality. More practically, the Western scientific edifice, from funding institutions to peer review systems to the organization of academic institutions, is not structured to support integrative disciplines, posing a substantial and very real barrier to the development of such fields. These barriers are especially strong between social science disciplines, which study human systems, and the natural sciences (with engineering somewhat in the middle), which makes collaboration in the field of industrial ecology even more difficult.

PRINCIPLES OF INDUSTRIAL ECOLOGY

Although industrial ecology is a nascent field, a number of relevant principles have already been developed. Several of the most significant of these are listed below. Though these are phrased in terms of a manufactured article, it is relatively easy to develop analogous principles for other types of products or services, such as food, fiber, infrastructure, and the built environment generally. Services tend to be more complicated: although they use material platforms, much of their value derives from their economic and cultural context and is thus more difficult to evaluate.

1. Products, processes, services, and operations should be designed to produce residuals but not waste.

2. Every process, product, facility, constructed infrastructure, and techno-logical system should be planned to the extent possible to be easily adapted to foreseeable environmentally preferable innovations. For example, build-ings should be designed to be capable of supporting photovoltaic systems by leaving an unshaded south-facing rooftop, even if such a system is not installed during the initial construction.

3. Every molecule that enters a specific manufacturing process should leave that process as part of a salable product. Thus, for example, processes should be designed so that process chemicals can be resold for other uses as a resid-ual material, rather than having to be disposed of as waste.

4. Every erg of energy used in manufacture should produce a desired material transformation.

5. Industries should make minimum use of materials and energy in prod-ucts, processes, services, and operations. The materials used should be the least toxic ones suitable for the purpose, all else equal.

6. Unfortunately, the cases where all else is equal are usually trivial; it is far more common to have difficult tradeoffs. For example, in many cases involving manufacture of complex objects, experience indicates that there is a tradeoff between energy consumption and toxicity of materials.

7. Industries should get most of the needed materials through recycling streams (theirs or those of others) rather than through the extraction of raw materials, even in the case of common materials.

8. Every process and every product should be designed to preserve the embedded utility of the materials used. An efficient way to accomplish this goal is by designing modular equipment and by remanufacturing.

9. Every product should be designed so that it can be used to create other useful products at the end of its current life; at the least, end-of-life mate-rials recovery should be facilitated by initial design.

10. Every industrial landholding, facility, or infrastructure system or com-ponent should be developed, constructed, or modified with attention to maintaining or improving local habitats and species diversity and to mini-mizing impacts on local or regional resources.

11. Close interactions should be developed with materials suppliers, customers, and representatives of other industries, with the aim of developing coopera-tive ways of minimizing packaging and of recycling and reusing materials.

These principles cannot be implemented unless environmental concerns are an integral part of the initial design activities for virtually every technoloical system. This may occur at a firm level; for example, process and product concurrent engineering activities in the manufacturing sectors should include all relevant environmental considerations. Similarly, engineers and architects should not design buildings or infrastructure components without relevant input regarding the environmental implications of, for example, their material choices, siting decisions, and mechanical and electrical system requirements. But it also may require sectoral integration. For example, in agriculture and forestry, crop selection and practices should be appropriate to local conditions, requiring minimal input of scarce or environmentally problematic resources (such as water or pesticides) and supporting long-term maintenance and enhancement of the pedosphere. Tools such as integrated pest management (IPM) will require coordination of the farmer, the manufacturer of the chemicals involved, and the IPM service provider (which may or may not be the chemical manufacturer). On a social level, governments may be the only party capable of evaluating new technological systems that span across several sectors, such as hydrogen-powered fuel-cell automobiles.

Robert Socolow of Princeton University has expressed the essence of industrial ecology in six principles:

1. Industrial ecology focuses on long-term, rather than short-term or ad hoc approaches, which tend to characterize current practice. This requirement, reflecting the systems approach of industrial ecology, encourages the understanding of anthropocentric disruption to fundamental life-supporting systems and cycles rather than just responding to the obvious localized perturbations.

2. Industrial ecology focuses on concerns which are of regional and global scope, and are persistent and difficult to manage. In this, it augments, but does not replace, current environmental activity.

3. Industrial ecology focuses on cases where the scale of human activity overwhelms the existing dynamics of natural systems. The human contribution may be a significant percentage of the stocks and flows of the natural system, as in the case of many heavy metals, or it may simply affect critical dynamics, as with the carbon cycle. In either case, the system no longer functions as it did before human activity grew to current levels.

4. Industrial ecology attempts to understand and protect the resiliency of natural and human systems, while identifying and minimizing impacts on

more vulnerable systems. Complex systems may be very resilient, but they may also have thresholds beyond which they shift to different states rapidly and unpredictably. Maintaining the performance of both human and natural systems within desirable limits is an important goal supported by industrial ecology.

5. Industrial ecology uses such systems techniques as mass-flow analysis to understand economic and environmental systems, and the linkages among them.[19]

6. Industrial ecology as an objective field of study views private firms as central to mitigating environmental impact, and seeks to understand how their behavior might become more environmentally appropriate. In this, it avoids the pattern of blame and ideology which has become rather prevalent in the environmental policy debates.

APPLYING THE PRINCIPLES: EARTH SYSTEMS ENGINEERING

Currently, industrial ecology principles and learnings are being integrated into methodologies such as Life Cycle Assessment, Design for Environment, and Integrated Pest Management. These are sector-specific, firm-specific, and sometimes product-specific applications. But perhaps a better idea of the power of the principles can be obtained by considering their application in Earth Systems Engineering, which I have defined as "the study and practice of engineering human technology systems and related elements of natural systems in such a way as to provide the required functionality while facilitating the active management of the dynamics of, and minimizing the risk and scale of unplanned or undesirable perturbations in, strongly coupled human and fundamental natural systems."[20] The raison d'etre of Earth Systems Engineering is the recognition that, primarily as a result of the Industrial Revolution, the scale and the scope of human impacts on natural systems are now such that their dynamics are increasingly dominated not by life processes as a whole but by the activity of one species—ours.[21] Many of the resulting perturbations, such as human impacts on the nitrogen, carbon, hydrologic, and heavy metal cycles, are unanticipated, problematic, and highly systemic. Engineering and managing these impacts on a going-forward basis will, inevitably, rely on industrial ecology studies and methodologies to provide critical elements of the required science and technology base. This is particularly difficult because the highly coupled nature of these hybrid human/natural systems makes their degree of complexity quite challenging.

It is not that the co-dependence of natural and human systems is new learning. Familiar examples include the numerous impacts of humans on biota, through direct predation and indirectly through the introduction of new species to indigenous habitats, which has been going on for centuries.[22] In fact, the distinction between a "human" system and a "natural" one is itself somewhat artificial in many cases. Economic systems are generally considered human, and estuarine systems natural, for example, although each is inevitably a complex mixture of both. What is different is the striking discontinuity between the relatively minor and localized impacts that predominated before the Industrial Revolution and the global, systemic impacts of human activity that now characterize the interrelationships between human systems and fundamental biological, physical, and chemical systems.[23] In general, these effects have been unintended, arising as the sum of human activities, grown to scales unprecedented for any species in the history of the globe. Myriads of economic and engineering decisions, evaluated and taken as if independent, are in reality tightly coupled to each other and to underlying natural systems. Each action in this process may, indeed, be planned, but the comprehensive, systemic impacts, which are just becoming apparent, are neither planned nor foreseen.

global climate change   Consider one specific example, global climate change. Perturbations of the carbon cycle arise from virtually all aspects of human economic behavior, from subsistence farming to manufacturing to use of fossil fuels to supply energy. An industrial ecology view of these perturbations reveals a system that is highly complex but which may be managed if the principles discussed above are kept in mind. Indeed, numerous options for "geoengineering" responses to global climate change forcing have already been identified,[24] including direct ocean disposal of CO_2, ocean fertilization with phosphate, ocean fertilization with iron, reforestation, solar shields increasing the planet's albedo, stratospheric SO_2 creation, injection of inert dust into the stratosphere, and injection of SO_2 into the troposphere. Such options can be evaluated in light of the principles of industrial ecology as well as on the basis of historic experience with complex engineering projects (especially those involving systems engineering).[25] As the following examples show, this background can be used to review the desirability and implications of proposed activities for managing the carbon cycle.

A DAM ACROSS THE STRAIT OF GIBRALTAR

The possibility that global climate change could result in abrupt shifts in global ocean circulation patterns is generally recognized.[26] More specifically, R. J. Johnson has hypothesized that, as a result of human reduction of freshwater flow into the Mediterranean Sea, increased salinity in that water body could lead to a higher volume of outflow from the Mediterranean Basin at Gibraltar.[27] This, in turn, could modify high-latitude oceanic-atmospheric circulation patterns in such a way as to cause extensive glaciation in Canada and cooling in northern Europe. As mitigation, Johnson proposes a partial dam across the Strait of Gibraltar, which would limit the outflow and reverse the climate deterioration. The dam would be a huge construction project designed to limit the Mediterranean flow through the Strait to some 20 percent of what it is now; it would likely take decades to construct and require about 1.27 cubic kilometers of material. The proposal has been strongly criticized and defended.[28]

Several points about this proposal are of interest. The first is that the shift in salinity in the Mediterranean is primarily a result of human diversion of water from the Nile: some 90 percent of the Nile's 2,700-square-meter-per-second flow is now diverted for agriculture or lost through evaporation. Significantly, much of this is due to the Aswan High Dam, so the (potential, and unanticipated) effects of shifts in the salinity of the Mediterranean—such as increased Canadian glaciation—are attributable, at least in part, to a single hydrologic construction project. The implications of failing to take a systems view of major engineering projects when they are first proposed, as industrial ecology would require, are apparent. This case also illustrates that Earth Systems Engineering is not something completely new—although understanding such projects in a systemic, rather than ad hoc, way has yet to occur.

The need for new ways to evaluate such systems—the use of industrial ecology principles, among others—is apparent from the scale and nature of the perturbation and impacts arising from such complex coupled human/natural systems. In this case, for example, the potential causal connections run from a historical pattern of increasing human water use around the Mediterranean, with a single large construction project as a possible triggering event, leading by a series of perturbations of natural oceanic and atmospheric circulation patterns to severe and disruptive climate changes in highly populated areas, with concomitant economic and social costs. As is frequently the case with environmental perturbations, the economic and environmental costs and benefits are differentially realized across classes of people, countries, economic sectors, and geographic regions.[29] Egypt benefits from the Aswan High Dam; Europe and Canada may face adjustment costs. How can institutional, scientific, ethical, and technological systems be evolved to manage such situations? And, when the system is considered as a whole, where is (are) the best place(s) to intervene to mitigate forcing functions, system response, and/or costs?

IRON-ENRICHED PHYTOPLANKTON BLOOMS

This option builds on the basic ecological principle that most communities are limited in their growth by a single factor. In this instance, growth of oceanic phytoplankton populations is limited by iron, so artificially introducing iron into oceanic systems would result in phytoplankton blooms which would absorb carbon, then die. At least some of the resulting carbonaceous material would sink to the bottom of the ocean, thereby depositing the carbon in a long-term sedimentary sink. This phenomenon has indeed been verified in tests on the open ocean.[30] These experiments also demonstrated, however, that the iron tended to precipitate out of solution rapidly, potentially reducing the effectiveness of this mechanism for capturing carbon.[31] In addition, the makeup of the planktonic community changed as a result of iron fertilization,[32] as did emissions of dimethyl sulfide (DMS) production by phytoplankton, which could lead to a cooling effect on the climate. However, even if this work is at a preliminary stage, it begins to create the knowledge base that will be needed if this option is to become part of a human-engineered carbon cycle. Of course, the gaps in the necessary knowledge are also apparent even at this early stage. However, unlike the proposal for a dam across the Mediterranean, this option can be explored at small scales, and, if justified, could be implemented incrementally, perhaps in locations where disruption of existing communities is minimal.

FOSSIL-FUEL SYSTEMS AS A BASIS FOR CARBON-CYCLE MANAGEMENT

The last case study is also conceptually simple. Carbon dioxide resulting from the combustion of fossil fuel in power plants is captured and then sequestered for centuries in the ocean, in deep aquifers, in geologic formations, or in other reasonably long-term sinks. Technologies to capture the CO_2 emissions and inject them into various sinks exist, and, especially if carbon capture is implemented at the initial design stage, rather than retrofitted, such systems appear to be technologically and economically feasible.[33] This technology is already in use: in the first instance of carbon sequestration for environmental reasons, Norway's state-owned petroleum company, Statoil, is sequestering the carbon dioxide content of the gas it is extracting from the North Sea Sleipner gas field back into an aquifer about 1,000 meters below the sea bed. (The CO_2 content of the gas is about 9 percent.) Statoil finds this economically preferable to paying the $55-per-ton CO_2 tax that would apply if the gas were simply vented.[34] This is only a proof of concept, however; issues of environmental impacts, of technological and economic feasibility, and of liability remain to be resolved.

When combined with the possibility of a hydrogen economy, such carbon sequestration raises the possibility of being able to exploit reserves of fossil fuel without a substantial increase in CO_2 emissions; thus, global climate change forces us, essentially, to decarbonize fossil-fuel consumption.[35] But an industrial ecology approach allows one to explore a more visionary, and perhaps far more important, possibility: the implementation of such carbon sequestration/hydrogen technologies as the basis of a deliberately engineered system of governance for the human carbon cycle (figure 4). Here, the global set of fossil-fuel plants are designed as a system to be tunable to help produce over time the desired atmospheric concentration of carbon dioxide, given other variables (e.g., impacts on vegetation, desired degree of global climate change, lag times of various components of the systems involved, changes in insolation, other carbon dioxide emissions, concentrations of other greenhouse gases, use of other mitigation technologies). The control functions of such a system are twofold: the ratio of biomass and municipal waste to fossil fuel input into the system, and the ratio of CO_2 emitted to CO_2

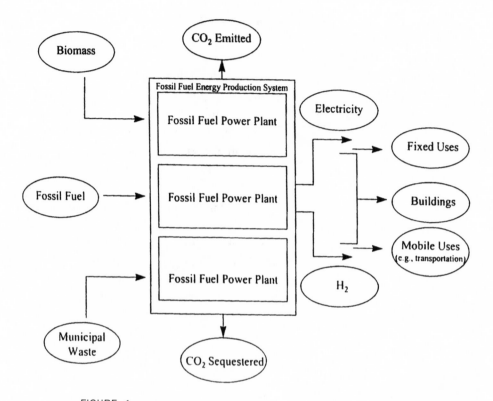

FIGURE 4

A governance system for the carbon cycle. Control functions are as follows. Input: B + αMW/Fossil fuel. Output: CO_2 emitted/CO_2 sequestered. Target metric: CO_2 concentration in atmosphere.

sequestered. By manipulating these ratios, the system is moved toward the target atmospheric concentration of CO_2.

All three of these examples pertain to elements of the carbon cycle, but, of course, that cycle is itself tightly coupled to other systems and cycles (figure 5); the implications for these coupled systems of any techniques used to modulate the carbon cycle must be part of a systems assessment. Industrial ecology assessments at this high level, and, indeed, Earth Systems Engineering as a practice, do not replace, but augment, existing engineering activities: even if a fossil-fuel power plant is a part of a system of carbon cycle stabilization, it and its components must still be engineered.

The science and technology base, the institutional capacity, and the ethical infrastructure needed to support such an assumption of human respon-

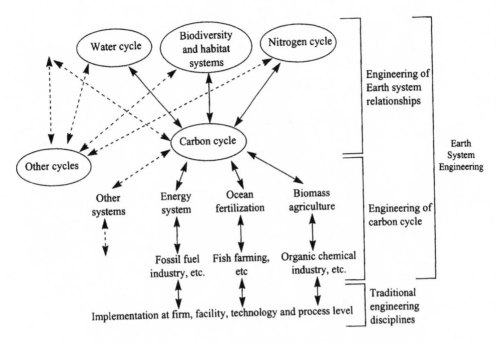

FIGURE 5
An Earth Systems Engineering schematic for the carbon cycle.

sibility for the functioning of a fundamental Earth system such as the carbon cycle do not yet exist. To urge immediate implementation of such an Earth Systems Engineering program in its totality in the immediate future would be irresponsible. On the other hand, it is also important to recognize the need to begin developing the industrial ecology base and associated framework for Earth Systems Engineering, which means that individual projects—such as building experimental coal-fired plants that capture and sequester their CO_2—should be begun immediately to facilitate learning and data acquisition. Moreover, it is necessary to support the R&D, the public dialog, and the institutional development that are needed to support the development of a capacity for Earth Systems Engineering. It bears repeating that the question is not whether humanity is manipulating the carbon cycle. The science clearly demonstrates that that question has already been resolved de facto. The only question remaining is how responsibly the species will perform that function—and industrial ecology, with its systems approach and integrative perspective, will be an important part of the answer.

ETHICAL AND INSTITUTIONAL INFRASTRUCTURE

Even though the focus of this discussion is the role of industrial ecology in developing Earth Systems Engineering capabilities, it is important to devote some time to ethical and institutional issues. In part, this reflects the importance of avoiding the trap of too much technocratic hubris: of thinking that, because technologies that can mitigate human impacts on the carbon cycle have been developed, the perturbation is therefore under control. In fact, while science and technology can inform progress toward that end, they are not by themselves sufficient—nor, for that matter, are current institutions and ethical systems.

To begin with, there is at least in some senses a significant difference between the concepts of stewardship, which is often adopted as the appropriate ethical positioning for achievement of sustainability, and active management of complex human-natural systems. Stewardship tends toward a passive approach: live on Earth and manage human activity in such a way that impacts on natural systems are minimized. The concept has connotations derived from a pre-industrial, pastoral mental model, and it retains the implication that humanity is set aside from (if not over) other natural systems. Earth Systems Engineering, however, requires that human institutions not only accept moral responsibility for human-natural systems but also assume an active role in the management of most global systems. For example, a stewardship approach to global climate change might suggest reducing global emissions of anthropogenic carbon dioxide and other greenhouse gas emissions to "safe" levels. An Earth Systems Engineering approach, however, would imply developing an institutional ability to develop, deploy, and monitor technology so as to deliberately modulate the carbon cycle within specified bounds (e.g., to maintain appropriate atmospheric concentrations of carbon dioxide and other greenhouse gases) so as to achieve and maintain a desired global state. This approach presupposes that it is no longer adequate simply to reduce the human impact; rather, explicit human management of relevant global systems, including both natural and human components, is required.

Even a brief consideration of the religious and ethical issues raised by these two approaches shows how different they really are. It is one thing to say "I will take care of the world," another to assert "I am actively responsible for the world and everything in it, and I will decide what lives and dies." The first approach is at the core of much of today's environmental activity, and is supported by a number of increasingly accepted principles, such as that the polluter should pay for cleaning up its pollution. The ethics of stewardship, which inherently assumes that the natural and human spheres are different (that is, there has to be a "natural" sphere for the "human" sphere to be a steward of), are inherent in the overhead approach, and are in that sense not radical (even if not fully implemented). To a great extent, the stewardship approach is also compatible with many existing theological systems, in principle if not in practice, especially because, by accepting the duality of the natural versus human world, such an approach does not displace the role of deities.

An Earth Systems Engineering approach based on industrial ecology is a profoundly different matter, in that human institutions are required to assert the right—indeed, the obligation—to manage global systems, and, concomitantly, to make tradeoffs that heretofore have widely been regarded as transcendent. Consider just briefly the question "How many people are you willing to kill to save a species?" or, conversely, "How many species are you willing to drive to extinction to save one human life?" Few people are willing to answer such questions. In fact, most would view them as profoundly offensive, in large part because they pose what many people regard as two moral absolutes—the value of an innocent human life, and the value of a species—against each other, and, in doing so, make them relative rather than absolute. Choosing either answer reduces the other to a conditional status. And yet, of course, such tradeoffs are occurring every day, in ways that are invisible to individuals and to social and cultural institutions.

This is not surprising. Just as scientific, technological, and policy institutions have not yet adjusted to an Earth increasingly engineered by humans, neither have ethical or religious institutions. In a world of plentiful resources, which the Industrial Revolution essentially created for the human species, it was virtually by definition possible to have adequate resources for both "humans" and "nature."[36] Moral systems which were evolved under such conditions would have no need to develop the ability to answer difficult questions about tradeoffs between, e.g., human lives and species extinctions. In a world of plenty, both could be absolutes. It is only in a world defined by human activity, however, that such absolutes no longer pertain.

Industrial ecology and Earth Systems Engineering also raise another fundamental question that is moral, not technical, in nature. Most simply, this may be stated as follows: "To what end are humans engineering—or should humans engineer—the Earth?" Though industrial ecology may be able to help understand what is possible within technological, population, and economic constraints, institutions and values must determine the answer to that question. What *could* be is a question, at least in part, for industrial ecology; what *should* be is one for religion and society. The need for serious theological, ethical, and moral dialog in this area is emphasized by the problem that choice at the level of the individual may not be relevant. In fact, the choice of path might be made by institutions at a relatively high hierarchical level in the system, or determined by systems dynamics arising from the interactions of science and technology; natural, institutional, and cultural systems; and national states in ways which we are not yet capable of modeling or understanding.[37] Thus, for example, no individual seems to be choosing the "barricade" sustainable world mentioned earlier, yet many trends (e.g., increasing inequality within and among countries) seem to be heading in that direction.

This, then, is a significant barrier to any substantial implementation of Earth Systems Engineering. The moral structure that is required if such activities are to become widespread is at a very primitive stage of development. The social acceptance of the assumption of such power by any existing institution, private or public, is likely to be minimal for a long time—and justifiably so, until the necessary competencies are developed. (As figure 1 illustrates, all three components—institutional, ethical, and knowledge base—need to evolve to support progress toward sustainability.) Moreover, it is not apparent where leadership in this dimension will come from. This does not mean that individual projects cannot go forward, particularly where they can be done in line with the guiding principles discussed below and where the perturbation to be addressed is immediate, difficult to reverse, and extensive in time and space. But it is cautionary: technologists sometimes assume technical solutions to problems that have substantial ethical and cultural dimensions, and such solutions often fail.[38] Particularly where, as here, the "ethical infrastructure" is weak, special attention to the non-technical aspects of such projects is crucial.

PRINCIPLES OF EARTH SYSTEMS ENGINEERING

Just as it was possible to develop principles of industrial ecology applicable to manufacturing activity (and other bounded economic activities), it is possible to develop principles of Earth Systems Engineering that form an operational basis for industrial ecology at this scale. These reflect not only the study of industrial ecology to date but also previous experience with large systems engineering projects (for example, large civil engineering activities such as hydrologic projects and the Boston Central Artery/Tunnel project, and complex technological systems such as the space shuttle program in the United States, global air transport control and safety programs, and nuclear power systems in many countries). These principles are both common-sense and daunting: they can be thought of as means by which the highly complex systems studies by industrial ecology can then be managed. They are clearly a work in progress.

1. Earth Systems Engineering must explicitly accept high levels of uncertainty and lack of knowledge as endogenous to the engineering function, rather than thinking of engineering as an effort to create a system certain. The traditional engineering mental model of centralized control of knowable systems is dysfunctional when applied to the complex, unpredictable (and quite possibly chaotic) systems involved in these cases, which involve coupled biological, physical, and traditional engineered systems, with control and feedback mechanisms widely distributed along many temporal and spatial scales. Rather than attempting to dominate a system—as is, for example, the case when a building or a chemical manufacturing complex is constructed—the Earth Systems Engineer will have to see herself or himself as an integral component of the system itself, closely coupled with its evolution and subject to many of its dynamics on an immediate feedback basis.
2. Whenever possible, engineered changes should be incremental and reversible, rather than fundamental and irreversible. Thus, for example, fertilizing oceanic planktonic populations with iron should begin (if at all) with small target plots, heavily monitored to determine whether the effects are as predicted, and what the unanticipated effects are. Conversely, a major problem with projects such as the proposed dam across the mouth of the Mediterranean, is that it is a large, single, relatively irreversible, engineered intervention in a complex physical system (ocean circulation). There is no room for the continuous learning and feedback that incremental engineering interventions support. In all cases, scale-up should allow for the fact that, especially in complex systems, discontinuities and emergent characteristics are the rule, not the exception, as scales change.
3. Continual learning at the personal and institutional level must be built into the Earth Systems Engineering process. This learning process is messy

and highly multi-disciplinary, and, accordingly, difficult to maintain even in the best of circumstances. It is also problematic because the learning will probably have to occur at an institutional, rather than personal level, because of the complexity of the systems involved and the inability of any single person, no matter how qualified, to understand them in their entirety.[39]

4. An important goal of Earth Systems Engineering projects should be to support the evolution of resiliency, not just redundancy, in the system. The two are different: a redundant system may have a backup mechanism for a particular subsystem, but still be subject to difficult-to-predict catastrophic failure; a resilient system will resist degradation, and, when it must, will degrade gracefully, even under unanticipated assaults.

5. Analogously, it is preferable to design (or encourage the evolution of) inherently safe systems, rather than engineered safe systems. An inherently safe system, when it fails, fails in a non-catastrophic way; an engineered safe system is designed to reduce the risk of catastrophic failure, but there is still a finite probability that such a failure may occur.

6. There must be adequate resources available to support both the project and associated research activities. Financial pressures can be particularly insidious with complex engineering technologies even today.[40] Earth system engineering projects are likely to be at least as complex as existing technological systems, and last over longer time periods than the usual budgetary cycles, meaning that they may be particularly prone to financial fluctuations.

7. If any Earth Systems Engineering project is to achieve public acceptance and social legitimacy, it must at all stages be characterized by an inclusive dialog among all stakeholders. Not all will agree, for a number of reasons, but to be successful, a project requires broad public support. The most obvious example of a complex technology system where this principle was not followed is, of course, civilian nuclear power. In the United States, for example, the secrecy and technological hubris which grew out of the nuclear weapons program and the nuclear navy meant that both the regulators and the experts were culturally adverse to open communication and dialog, with eventual results for the industry that were both predictable and disastrous.[41]

8. It is not enough to evolve the scientific and technological base for Earth Systems Engineering activities which, broadly speaking, falls under the field of industrial ecology. As the discussion above indicates, there must be parallel evolution along at least two other fundamental paths: ethical/religious and institutional. Industrial ecologists must remain sensitive to these dimen-

sions of their work even while striving for objectivity in their research activities.

CONCLUSION

The nascent field of industrial ecology is one component of an integrated effort to understand and approach a sustainable global system. Although the outlines of such a system are currently undefined, it is possible even now to develop principles, tools, and methodologies that will support more environmentally and economically efficient practices. Moreover, as the example of Earth Systems Engineering demonstrates, industrial ecology, with its systems-based, objective approach, is crucial to developing the base of knowledge that will be required to design and manage a world increasingly dominated by the activities of our species.

NOTES

1. B. R. Allenby, "Clueless," *Environmental Forum* 14 (1997), no. 5, 35–37; D. Rejeski, "An incomplete picture," *Environmental Forum* 14 (1997), no. 5, 26–34.

2. Netherlands Department of Housing, Physical Planning and Environment, Recyclable vs. Disposable: A Comparison of the Environmental Impact of Polystyrene, Paper/Cardboard and Porcelain Crockery, 1992.

3. Franklin Associates, Resource and Environmental Profile Analysis of a Manufactured Apparel Product: Woman's Knit Polyester Blouse: Final Report, 1993. The "consumer use life-cycle segment" accounted for 82 percent of the total energy requirements and 66 percent of the solid waste volume.

4. B. R. Allenby, *Industrial Ecology: Policy Framework and Implementation* (Prentice-Hall, 1999).

5. Netherlands Department of Housing, Physical Planning and Environment, National Environmental Policy Plan: To Choose or to Lose, 1989; Netherlands Department of Housing, Physical Planning and Environment, National Environmental Policy Plan Plus, 1990.

6. T. E. Graedel and B. R. Allenby, *Industrial Ecology* (Prentice-Hall, 1995).

7. J. E. Cohen, *How Many People Can the Earth Support?* (Norton, 1995); Allenby, *Industrial Ecology.*

8. Graedel and Allenby, *Industrial Ecology.*

9. Environment, Health and Safety Committee, Institute of Electrical and Electronics Engineers, White Paper on Sustainable Development and Industrial Ecology, 1994.

10. R. A. Frosch and N. E. Gallopoulos, "Strategies for Manufacturing," *Scientific American* 261 (1989), no. 3, 144–153.

11. J. H. Ausubel and J. E. Sladovich, eds., *Technology and Environment* (National Academy Press, 1989).

12. J. L. Helm, ed., *Energy: Production, Consumption and Consequences* (National Academy Press, 1990).

13. B. L. Turner II, W. C. Clark, R. W. Kates, J. F. Richards, J. T. Mathews, and W. B. Meyers, *The Earth as Transformed by Human Action* (Cambridge University Press, 1990).

14. World Commission on Environment and Development, *Our Common Future* (Oxford University Press, 1987).

15. American Electronics Association, The Hows and Whys of Design for the Environment, 1993.

16. Office of Technology Assessment, *Green Products by Design* (Government Printing Office, 1992).

17. B. R. Allenby, "Earth Systems Engineering: The Role of Industrial Ecology in an Engineered World," *Journal of Industrial Ecology* 2 (1999), no. 3, 73–93.

18. The work of Tilman et al. with regard to the positive relationship between biological productivity and diversity in grasslands is suggestive in this regard. See D. Tilman, D. Wedlin, and J. Knops, "Productivity and Sustainability Influenced by Biodiversity in Grassland Ecosystems," *Nature* 379 (1996), 718–720. (See also P. Kareiva, "Diversity and Sustainability on the Prairie," ibid. 673, 674.)

19. R. Socolow, C. Andrews, F. Berkhout, and V. Thomas, eds., *Industrial Ecology and Global Change* (Cambridge University Press, 1994). See also R. U. Ayers and U. E. Simonis, eds., *Industrial Metabolism* (United Nations University Press, 1994).

20. B. R. Allenby, "Earth Systems Engineering: The Role of Industrial Ecology in an Engineered World, *Journal of Industrial Ecology* 2 (1999), no. 3, 73–93.

21. See *Scientific American* 261 (1989), no. 3, 46–174; B. L. Turner II, W. C. Clark, R. W. Kates, J. F. Richards, J. T. Mathews, and W. B. Meyers, *The Earth as Transformed by Human Action* (Cambridge University Press, 1990); R. Gallagher and B. Carpenter, "Human-Dominated Ecosystems: Introduction," *Science* 277 (1997), 485; D. Mallakoff, "Extinction on the High Seas," *Science* 277 (1997), 486–488.

22. D. Jablonski, "Extinctions: A Paleontological Perspective," *Science* 253 (1991), 754–757.

23. See R. Ehrlich and E. O. Wilson, "Biodiversity Studies: Science and Policy," *Science* 253 (1991), 758–762; B. L. Turner II, W. C. Clark, R. W. Kates, J. F. Richards, J. T. Mathews, and W. B. Meyers, *The Earth as Transformed by Human Action* (Cambridge University Press, 1990); special report on human-dominated ecosystems, *Science* 277 (1997), 485–525.

24. D. W. Keith and H. Dowlatabadi, "A Serious Look at Geoengineering," *Eos* 73 (1992), no. 27, 289, 292, 293.

25. On the Atlas missile program, the Internet, and the Boston Central Artery/Tunnel project, see T. Hughes, *Rescuing Prometheus* (Pantheon, 1998).

26. W. S. Broecker, "Thermohaline Circulation, the Achilles' Heel of Our Climate System: Will Manmade CO_2 Upset the Current Balance?" *Science* 278 (1997), 1582–1588.

27. R. J. Johnson, "Climate Control Requires a Dam at the Strait of Gibraltar," *Eos* 78 (1997), no. 27, 277, 280, 281.

28. J. Marotzke and A. Adcroft, "Comment on 'Climate Control Requires a Dam at the Strait of Gibraltar,'" *Eos* 78 (1997), no. 45, 507; R. J. Johnson, "Reply," *Eos* 78 (1997), no. 45, 507.

29. Allenby, *Industrial Ecology.*

30. See M. J. Behrenfeld, A. J. Bale, Z. S. Kolber, J. Aiken, and G. Falkowski, "Confirmation of Iron Limitation of Phytoplankton Photosynthesis in the Equatorial Pacific Ocean," *Nature* 383 (1996), 508–511; D. J. Cooper, A. J. Watson, and D. Nightingale, "Large Decrease in Ocean-Surface CO_2 Fugacity in Response to *in situ* Fertilization," *Nature* 383 (1996), 511–513.

31. R. A. Kerr, "Iron Fertilization: A Tonic, but No Cure for the Greenhouse," *Science* 263 (1994), 1089, 1090.

32. K. H. Coale, K. S. Johnson, S. E. Fitzwater, R. M. Gordon, S. Tanner, F. Chavez, L. Ferioli, C. Sakamoto, Rogers, F. Millero, Steinbert, Nightengale, D. Cooper, W. Cochlan, M. R. Landry, J. Constantinou, G. Rollwagen, A. Tranvina, and R. Kudela, "A Massive Phytoplankton Bloom Induced by an Ecosystem-Scale Iron Fertilization Experiment in the Equatorial Pacific Ocean," *Nature* 383 (1996), 495–501.

33. R. C. Socolow, ed., Fuel Decarbonization and Carbon Sequestration (report 302, Princeton University Center for Energy and Environmental Studies, 1997).

34. B. Hileman, "Fossil Fuels in a Greenhouse World," *Chemical and Engineering News* 75 (1997), no. 33, 34–37.

35. Socolow, ed., Fuel Decarbonization and Carbon Sequestration.

36. Allenby, *Industrial Ecology.*

37. Ibid.

38. Consider the example of nuclear power. See R. Pool, *Beyond Engineering: How Society Shapes Technology* (Oxford University Press, 1997).

39. Allenby, *Industrial Ecology.*

40. Pool (*Beyond Engineering*) cites the Bhopal Union Carbide chemical plant and the *Challenger* incident as examples where pressures generated in part as a result of chronic underfunding resulted in catastrophic failure of such systems.

41. Pool, *Beyond Engineering.*

PORTRAIT OF INNOVATION: ROBERT H. SOCOLOW

MARTHA DAVIDSON

> My object in living is to unite
> My avocation and my vocation
> As my two eyes make one in sight
>
> —Robert Frost, "Two Tramps in Mud Time"

Robert Socolow recites these lines from a favorite poem. The verses, which he first encountered in high school, reverberate in his life. Socolow, a physicist, is a former director of Princeton University's Center for Energy and Environmental Studies, the present editor of the *Annual Review of Energy and the Environment,* and a pioneer of energy efficiency research. Though drawn to science from an early age, he has also had a lifelong love of the humanities, especially the arts and languages, and a keen interest in other cultures. These diverse perspectives have led him to pose challenging questions about the environment and to foster multi-disciplinary efforts to answer them.

Born in New York in 1937, Socolow, the eldest of three children, acquired strong ethical values from both home and school. His mother, a remedial reading specialist, cultivated in him an appreciation of music and museums. His father, an attorney who wrote an early text on radio broadcasting law, provided a role model for community service through his work with Jewish organizations. The family's involvement with Reconstructionism, an emerging modernizing movement in Judaism, was an important influence. Socolow attended Fieldston, a high school of the Society for Ethical Culture. "Between Reconstructionism and Ethical Culture," he says, "I got a double dose of liberal values: public service, internationalism, anti-prejudice, pro-science, anti-sectarian, pro-rationality."

It was school, too, that nourished his interest in science. Although there were no scientists in his family, Socolow had inspiring science teachers at

Fieldston and felt the aura of the renowned physicist J. Robert Oppenheimer, who had attended the same school about 30 years earlier. Socolow, like Oppenheimer, chose Harvard for college.

Entering Harvard in 1955, Socolow intended to major in chemistry. Parental encouragement and his own inclinations ("I had a strong belief that I should learn everything. . . . I thought I could try to learn all the ideas taught at Harvard!") led him to select a broad range of courses along with science classes. Among the most memorable were a survey of the fine arts and a poetry class taught by Archibald MacLeish. Socolow also took a remarkable course in the Russian language taught outside the university. Already conversant in French, which he had studied in high school and during a summer spent with a French family, Socolow became comfortable with Russian as well.

In 1957, Socolow was invited to work as a summer student at the Brookhaven National Laboratory, where groundbreaking work in physics was being done. "I lived on site, breathed the excitement of physics. I changed my major from chemistry to physics upon returning to Harvard [and] I decided I wanted to participate in the discovery of the laws of fundamental particles. . . . It was a fascination."

Graduating summa cum laude in 1959 with a B.A. in physics, Socolow was awarded Harvard's Sheldon Travel Fellowship. It enabled him to travel for a full year through Russia, Asia, and Africa, returning via the Middle East. "My agenda was to be a sponge," he says. The journey left him with thousands of lasting impressions, particularly regarding the effects of colonialism and the strength of nationalism at that time. While traveling, Socolow read a book in which Albert Schweitzer described his decision to spend his twenties pursuing music and philosophy and to delay answering the call he felt to devote himself to social and medical problems. "I needed permission from myself to stay in physics, and here it was," Socolow recalls. He returned to Harvard in 1960 for graduate studies, and he completed a Ph.D. thesis in theoretical high-energy physics under a young professor, Sidney Coleman.

In August of 1961, after a summer spent working on arms control issues at RAND in California, Socolow was an aide at a Pugwash Conference held in Stowe, Vermont. Pugwash Conferences, inspired by a 1955 manifesto of Bertrand Russell and Albert Einstein and named for the site in Nova Scotia where they were first held, bring together scientists to discuss controversial issues of global importance, such as nuclear disarmament.

Although serving officially as a driver, Socolow acted also as an unofficial interpreter for some of the Russian scientists at that meeting, including Nikolai Bogoliubov and Igor Tamm.

Back at Harvard, Socolow decided that he had to reject a career in arms control. "One had to become as knowledgeable about weapons as those who loved them," he concluded. "I so disliked weapons that I couldn't force myself to learn about them. Arms control couldn't be my field."

While in graduate school, Socolow met Elizabeth Sussman, a Vassar undergraduate. They were married in 1962. Elizabeth became a graduate student in English at Harvard, where she received a Ph.D. in 1967. In 1964, Socolow received his Ph.D. and accepted a postdoctoral fellowship from the National Science Foundation for study at Berkeley and at the European Center for Nuclear Research (CERN) in Geneva. At a meeting in Budapest during his time at CERN he developed a strong bond with a North Vietnamese physicist. The two of them hoped to contribute to a resolution of the growing conflict in Vietnam, but their efforts were unsuccessful.

Socolow returned to the United States in 1966 as an assistant professor at Yale University. There he fulfilled an aspiration to teach quantum mechanics, and he also continued his antiwar efforts, though with a sense of futility. He openly supported draft resistance and organized, along with three other faculty members, Yale's 1969 "Day of Reflection," a symposium on scientists and war work. To represent the views of researchers who worked with the Department of Defense, Socolow invited Marvin Goldberger of Princeton University. That contact changed the course of Socolow's career.

Socolow had planned to spend the summer of 1969 in California, working at the Stanford Linear Accelerator. Goldberger told Socolow and a Yale colleague, John Harte, about a special summer study run by the National Academy of Sciences at Stanford that was to examine institutions for the management of the environment, using a proposed jetport in the Everglades as a case study. Socolow decided to stay away from the Accelerator for four weeks to participate in the Everglades study as a volunteer.

The National Academy study argued against the construction of the jetport. It cautioned developers about the importance of water conservation in the Florida interior to the development of the state's west coast. Not long after the study was completed, plans for the jetport were abandoned, and the federal and state governments created the Big Cypress National

Preserve, a huge interior water conservation area, to help both the Everglades and the development of Florida's west coast. "It was a heady beginning to a career," Socolow says. "I had found a way to combine my social concern and my science, and I didn't look back." Along with that discovery, there was another major development in his life that year: the birth of his first child, David. His second child, Seth, was born 2 years later.

Socolow's personal awakening to the environment that summer of 1969 coincided with a larger burst of national awareness of the Earth's fragility. We had seen the first photos of our planet taken from space. Environmental issues commanded the public's attention. President Nixon created the Council on Environmental Quality and the Environmental Protection Agency.

At Yale, Socolow's perspective had shifted. "Look at this environmental science—why didn't I learn it when I was learning physics? Why aren't there examples that convey environmental reality in the physics textbooks?" he wondered. "And, so, why not provide a supplementary text and try to bring environmental problem solving into introductory science courses?" With John Harte, he proposed the volume that was published 2 years later as *Patient Earth*.

The double meaning of the title was intentional: the Earth is patient with us, who mistreat it; and it is a patient, deserving of our care. The book was a compilation of case studies representative of environmental conflicts or collisions of values that its authors thought were likely to recur for many years. Topics included urban blight, population control, resource management, conservation, the ecological impact of the military, and alternative uses of land. The book explored philosophical and moral aspects of environmental issues as well as scientific ones, and it promoted social activism by individuals and citizen groups as well as action by legislatures and courts to remedy the problems described.

In 1969–70, in addition to editing *Patient Earth,* Socolow learned about the environmental research conducted by faculty in other Yale departments—biologists, geologists, economists, and professors in the School of Forestry. For the academic year 1970–71, Socolow had a Yale University Fellowship, a kind of sabbatical. He pondered his next career move and considered offers of research positions at science policy centers newly created at Harvard and Cornell. Ultimately, Marvin Goldberger, the Princeton professor who first awakened Socolow's interest in the environment through the Everglades study, persuaded him to join Princeton's faculty, as an associate professor in the Department of Mechanical and Aerospace Engineering (MAE). Socolow's principal responsibility would be to articulate and lead the research program of a new Center for Energy and Environmental Studies (CEES), which the University was forming in the School of Engineering and Applied Science. In addition to Goldberger, four senior Princeton professors would guide the effort: the physicist George Reynolds, the electric propulsion expert Robert Jahn, the combustion expert Irvin Glassman, and the economist William Bowen.

At Yale, senior professors had raised the funds to sponsor Socolow's research. At Princeton, proposal writing was Socolow's responsibility, and he began reading successful proposals from various departments. One from the School of Architecture described a sociological study of a New Jersey planned-unit housing development called Twin Rivers. "In an 'aha!' moment," Socolow remembers, "I thought, 'What if I study the same community, with energy flow questions in mind, to understand what determines the energy used in the most common kinds of housing?'" Richard Grot, another member of MAE, quickly responded to the idea, pointing out that the replicated units of the development provided a research opportunity.

Over the next 7 years, Socolow, Grot, David Harrje (a rocket engineer), some colleagues in statistics and psychology, and several graduate students made extensive studies of the units at Twin Rivers. They monitored energy use and experimented with ways to reduce it by modifying the building shell. They showed that savings in annual heating of up to 75 percent were possible. The team published its findings in a book titled *Saving Energy in the Home: Princeton's Experiments at Twin Rivers*. Their discoveries about common construction practices and the importance of small details for energy efficiency stimulated the practice of retrofitting, which became widespread by the 1980s.

Socolow's decision to study middle-class housing was deliberate, he says: "I'm interested in the environmental impact of the way we live. That's dominated by middle class consumption. . . . Decisions that determine middle-income housing are replicated millions of times."

At Princeton, in the wake of the 1973 oil crisis, Socolow ran the American Physical Society 1974 summer study on energy use. The meeting helped legitimate the study of energy efficiency by physicists. Socolow, along with Marc Ross at Michigan, Arthur Rosenfeld at Berkeley, Robert Williams then at Michigan and later a colleague at CEES, and quite a few other physicists, questioned two widely held assumptions among scientists: that society can achieve well-being only through ever-greater expenditures of energy, and that physicists should work only on problems of energy supply, not of energy use. The Twin Rivers project, carried out by physicists and engineers, was a model of a new application of physics research.

In his first years at CEES, parallel with his work on energy efficiency, Socolow launched a second multi-disciplinary study, this one focusing on the proposed construction of Tocks Island Dam on the Delaware River, just above the Delaware Water Gap, between New Jersey and Pennsylvania. Tocks Island Dam was slated to be the largest dam in the Northeast. The study explored the analytical methods of ecologists, hydrologists, energy analysts, and economists, calling into question the applicability of the assumptions used in each field. Early work on the resulting book, *Boundaries of Analysis: An Inquiry into the Tocks Island Dam Controversy,* co-edited with colleagues Harold Feiveson and Frank Sinden, may have influenced the decision of Governor William Cahill of New Jersey to question the dam in his role as a member of the Delaware River Basin Commission. Cahill's concerns changed the political balance, and the project was scuttled a few years later. Today, a stretch of the Delaware is a part of the National Wild and Scenic River System. Socolow observes: "It is not much of an oversimplification to say that in the United States, until the Tocks Island dam controversy, all dams of that type that had been proposed were built; after Tocks Island, all similar dam proposals were rejected before construction."

Socolow, succeeding Reynolds and Glassman, served as director of CEES from 1979 to 1997. Among his colleagues were Robert Williams, an influential analyst of energy technology and policy, and Frank von Hippel, a specialist in nuclear energy and arms control and a leader of Russian–US arms control collaborations. Socolow saw his job as twofold: to connect CEES with the rest of the university through formal teaching and supervision of work by undergraduates and graduate students, and to "infect the disciplines" at Princeton, to nudge the academic enterprise to take the environmental challenge seriously.

Socolow believes CEES has had an impact on a lot of individual careers of people with straight science backgrounds, typically physics backgrounds: "I describe our place as a roundhouse. They come in, oriented in one direction . . . and we help them turn about thirty degrees, and they leave in a different direction, . . . still using physics, but in a different way." Socolow advises students to get a firm grounding in physics or another discipline before getting involved in multi-disciplinary policy work. He feels strongly that traditional scientific training is an irreplaceable foundation for innovative work on environmental issues.

In 1983, Socolow participated in a Pugwash Conference in Venice. His motive was to meet the physicist Evgenie Velikhov, a senior officer of the Soviet Academy of Sciences, who was seeking collaborations with American scientists. Socolow was concerned about the demonization of the USSR during the early Reagan era and saw science as an arena to build communication and to address common aims. In Moscow in 1984 at Velikhov's invitation, Socolow worked with Russian counterparts to launch a collaboration between Soviet and US scientists in the general area of energy efficiency, ultimately involving the National Academy of Sciences in the United States as well. The collaboration continued for a full decade.

In 1984, at the 25th-year reunion of his Harvard class, Socolow met someone else of importance in his life. His first marriage had ended 2 years earlier, and at the reunion he was introduced to Jane Ries Pitt, widow of a former classmate and herself a Harvard alumna. They married in 1986, and Socolow became stepfather to her two children, Jennifer and Eric. Jane Pitt Socolow, a physician and a professor at Columbia University, directs a research program in perinatal and pediatric HIV infection. She shares her husband's commitment to use science to solve society's problems.

One of the societal issues that concerns Socolow is the relationship between technologies used in countries of the Earth's northern hemisphere and those used in less industrialized countries of the southern hemisphere. Thanks especially to the regular visits of Amulya Reddy (from Bangalore, India) and Jose Goldemberg (from Sao Paulo, Brazil), CEES conducts much original research on technologies that support the industrialization of developing countries in environmentally responsive ways. In *Perspectives in Energy* (January 1991), Socolow wrote: "To solve the problems of the South, there is no reason to confine attention to those technologies and policies that have worked in the North. Indeed, one of the great stimuli to innovation over the next decade will be to confront the problems of the South as new problems, and to devise original solutions for them."

Socolow, Williams, and their colleagues and students have been exploring a number of technologies for transportation and electricity tailored to the needs of developing countries. Working against a widely held assumption that technologies should be deployed in such countries only after they have been fully tested in industrialized societies, Socolow and his colleagues are advocating the deployment of certain technologies first in developing areas of the world.

Socolow has been the editor of the *Annual Review of Energy and the Environment* since 1993. With the help of an editorial committee, he selects themes and solicits articles that bring research in the natural and social sciences and in technology to the attention of a wide community of scientists, engineers, and policy analysts in the academy, government, in industry, and in non-governmental organizations. Socolow stepped down as director of CEES in 1997, when he took a sabbatical that included travel in China and India. He teaches innovative courses in environmental science, technology, and policy based at the MAE Department and at Princeton's Woodrow Wilson School of Public and International Affairs.

Socolow's most recent interest is industrial ecology. Industrial ecology encompasses two lines of analysis. The first, introduced by Robert Frosch and Nicholas Gallopoulos and developed by Braden Allenby, Thomas Graedel, and Robert Laudise at AT&T, is concerned with material flows within industry. It takes natural ecology as a model, particularly in looking at waste products. In nature, one organism's wastes become another organism's food. These researchers seek opportunities for the waste products of one industry to become the raw materials of another. The second line of thinking, led by Robert Ayers at Carnegie Mellon University, traces the flow of materials—arsenic, for example, or lead—through both the natural and industrial environments, on a regional or a global scale.

Socolow became involved with industrial ecology when he was asked to head a 1992 workshop on industrial ecology and global change at Snowmass, Colorado. "Industrial ecology gave my career a second wind," he says. "With industrial ecology, I'm able to return to the resource and environmental issues and themes that first brought me into environmental work and that motivated *Patient Earth.*" Socolow has been using the industrial ecology approach to address the fate of three elements: carbon, lead, and nitrogen.

The carbon problem that interests Socolow is called carbon sequestration. It is a way to slow down global warming, by reducing the rate of increase of the atmospheric concentration of carbon dioxide. Socolow explains that the version of carbon sequestration that interests him "involves continuing to use fossil fuels, but preventing most of the carbon in these fuels from reaching the atmosphere." He continues: "For example, after capturing the carbon at a power plant as carbon dioxide, one might send the carbon dioxide back below ground into deep saline aquifers capable of retaining the carbon dioxide for thousands of years with very little leakage back into the environment. Carbon sequestration is essentially a challenge to the conventional thinking, which holds that the fossil fuel industries cannot be part of the solution to the greenhouse problem." Socolow helped promote this field of research by running a workshop on carbon sequestration in Washington in 1997. Early indications from research at CEES on technical approaches to the sequestration of the carbon in fossil fuels suggest that these technologies offer one of the least costly approaches to mitigating the greenhouse problem. From this perspective, hydrogen is the transportation fuel of the future: hydrogen is most of what is left chemically when carbon is extracted from a fossil fuel, and hydrogen fuel becomes harmless water when its energy is used.

Socolow's work on lead has focused on the lead battery. He and his CEES collaborator Valerie Thomas concluded that the lead battery could become one of the first examples of a hazardous product managed in an environmentally acceptable fashion. Industrial ecology makes the distinction between dispersive and recyclable uses of materials: because lead used as a gasoline additive cannot be recovered, it is a dispersive use, while the use of lead in a battery is a recyclable use. Accordingly, Socolow found himself confronting a new question: What should be the criteria for deciding when recycling is environmentally acceptable? These criteria include, Thomas and Socolow decided, nearly 100 percent collection of used batteries; environmentally clean battery dismantlement, secondary lead refining, and battery reassembly; low worker exposures at each step; and exports of used batteries only to places where equivalent stringent environmental standards are in effect.

Socolow has also published a number of papers on nitrogen. Fertilizer production and other human activities have more than doubled the Earth's natural rate of nitrogen fixation, contributing to ecosystem imbalances, air pollution, ozone depletion, and greenhouse effects. Socolow's paper "Nitrogen Management and the Future of Food: Lessons from the Management of Energy and Carbon," published in the *Proceedings of the National Academy of Sciences*, suggests how productive approaches to carbon management, such as focusing on end-use efficiency, encouraging markets in pollution rights, and conducting targeted research and development, can be applied to the emerging challenge of nitrogen management, at scales ranging from the cornfield to the entire globe.

Socolow's work increasingly focuses on ethical aspects of environmentalism. In a 1996 talk at the Yale Institute for Social and Policy Studies, he argued for a moral and reverent response to the Earth's vulnerability:

Wherein is the moral imperative to enhance those portions of the scientific enterprise likely to illuminate critical environmental issues? It arises from our obligation to preserve the capacity of future generations to enjoy experiences that they value as much as we enjoy what we value. . . . Each generation must provide the next generation with new capabilities in order to compensate for bequeathing to the next generation a natural environment more degraded than the one it inherited. Where geology threatens to impoverish, the intergenerational accounts must be balanced by scientific understanding, new instruments and devices, and more subtle and effective policies.

Socolow remains optimistic about the Earth's future. "The problem," he says, "is that the Earth is small, compared to the exuberance of the human species. We will have a very challenging time adjusting to the fact that the cumulative effect of the many wonderful things so many of us want to do on this planet is a changed planet. But I believe people will negotiate their way through the environmental challenge. I believe in democracy more than I believe in technocracy. My optimism originates in a conviction that people have a lot of common sense."

CONCLUSION: THE NEW ENVIRONMENTALISM

RODERICK NASH, WITH MARTHA DAVIDSON

What we have been engaged in, in this book and in the program that inspired it, is the creation of a new definition of environmentalism. In the past, we have defined the term in various ways, and each definition has, in turn, affected our relationship with nature. Late in the nineteenth century and early in the twentieth, there was a vision of environmentalism as efficiency. It was the era of big dams and of managed forests with straight lines of trees. Environmentalism, in that sense, was really just a more efficient form of exploiting natural resources. That concept was followed in mid-century by pastoralism—environmentalism as a revival of the Jeffersonian ideal of a society of gentleman farmers, though now transformed to suburbia. And that was superseded, in the 1960s and the 1970s, by environmentalism as primitivism: the back-to-the-wilderness and back-to-the-land movements. By that definition, environmentalism was a call to turn our backs on technology, to reclaim our human heritage as hunters and gatherers living in a more direct relationship with the natural world.

But in these essays, a new definition seems to be taking shape: environmentalism as a paradigm change. Environmentalism is being linked with the human capacity to imagine, to invent, to innovate, and to shape the future. We are realizing, as we enter a new millennium, that we can have a huge role in shaping a new culture, and that environmentalism should be an integral part of that culture.

The cusp of the millennium is a rare vantage point for looking both backward and forward. Let us look a thousand years into the past and a thousand years into the future. We know from the study of history and archaeology what was happening in the first century of the last millennium, around the year 1000. The Crusades were being launched to oust Muslims from the Holy Land. In Africa, the kingdom of Ghana reached its apogee and was overtaken by Islamic forces. The Normans invaded and conquered Britain, the Vikings reached the coast of North America, and in China trade was flourishing under the Northern Sung dynasty. In Meso-America, the Maya and Toltec civilizations were building cities. These migrations of populations and contacts of cultures brought exchanges of biota. The quarrying of stone and the building of temples and fortifications changed the landscape. Agricultural improvements led to population growth and to the clearing of forests. Human impact on the environment continued in the centuries that followed. A hundred years ago, people in this country were talking about the end of the frontier, and now there are books with ominous titles such as Bill McKibben's *The End of Nature.*

So let us now think about the year 3000. What will this planet be like in a thousand years? I don't have an answer to that question, but I offer you a dream of what could be. This Smithsonian publication seems an appropriate place to talk about dreams of the future, because it was not far from the Smithsonian, on the National Mall, that another man once spoke of such a dream. Martin Luther King Jr. didn't know exactly how to get to the place he dreamed of, but he envisioned people coming together in an ethical community. He saw black children and white children joining hands as sisters and brothers. Now I suggest that we think also about the children of salmon, the children of elk, the children of blue-green algae. Perhaps it is time to widen the circle, to define society as an ecological community. Just as a social contract flowed from John Locke's seventeenth-century concept of person-to-person relationships, with authority deriving solely from the consent of the governed, so an ecological contract could flow out of our sense of being part of a much larger community, a community in which we must behave with consideration for all forms of life.

The dream I offer is of "island civilization." It may not be the best of all possible futures, it may not be the sort of civilization that market forces are propelling us toward, but I believe it is the vision that ethics compels us to choose. Island civilization, quite simply, is a reversal of the direction in which we are now going. Today we have islands of land that are wilderness, surrounded by sprawling mega-cities. Michael Robinson refers to bioparks as "islands of nature," but I am thinking of wilderness on a much larger scale. If you look at photographs of Earth at night, these patches of uninhabited land appear as islands of darkness in seas of light, the light produced by human settlements. I am concerned that the islands are small and that the areas of light are becoming extensive.

In island civilization, that relationship is reversed. I envision civilization as occupying only about two percent of the land mass, with the rest of the land allowed to be wild and uncontrolled. As Thoreau said in *Walden*, the trick is to secure all the advantages of civilization without suffering any of the disadvantages. Most of us, particularly in the United States, have grown up believing that growth and development are good. Growth and development have in fact become synonymous with progress. I suggest that the new paradigm of environmentalism sees growth as potentially not good. It is not that progress was wrong; it is simply that progress went on too long. Most things in nature have limits, and things that grow far beyond normal size—a hand the size of a chair, for example—are considered abnormal, unfortunate, freakish. How about our civilization, then? How about our population? As we look toward the next thousand years, when do we begin to say that we need to make big corrections to population growth and material development—decisions that will be against our short-term interests?

To move toward an island civilization requires setting goals. The first one, I suggest, is to reduce the world's population from 6 billion to 1.5 billion by the year 3000. I choose this figure because Paul Ehrlich and others have concluded that 1.5 billion people could live sustainably, justly, with other forms of life on this planet. They could enjoy a high quality of life, realizing their full human potential. The second goal is the implosion of the civilized environment. By implosion, I mean a concentration of the human population in small areas, rather than the explosion and sprawl that characterize our cities today. We must scale back to habitats that take up less space—1.5 billion people living in perhaps 500 concentrated habitats. To do this, we may have to eliminate much of the infrastructure of our present civilization, including roads, parkways, power lines, and dams. Perhaps physical transportation will be instantaneous or unnecessary. Perhaps these habitats will be suspended in the air, or will be built underground, or partly under the oceans. Such notions may seem preposterous today, but then the prospect of someone walking on the moon seemed just as preposterous 100 years ago. Remember, I am speaking of a very long-term perspective, a thousand years in the future. I am offering a dream of what could be, what I believe ethically should be, though I don't know exactly how we can achieve it. I do know that we need a vision of the future that sees as its end not growth, not a higher Dow Jones average, not ever greater development, but rather a technologically gentle and kind island civilization. We will need engineers and other innovators, like the contributors to this book, to guide us and to devise the technologies that will make it possible.

We need to redefine progress away from growth, away from what Paolo Soleri calls the "pursuit of improved wrongness," and toward sustainability and justice for all Earth's creatures. Soleri tells us that if we don't like cities we should make bigger ones, not build "sub-exurbias." Following his reasoning—that we should not attempt to solve the problems of cities by bombarding the landscape with single-family houses—I say "If you don't like technology, make it more sophisticated and lessen its impact on nature." I do not mean that we should return to pastoralism or primitivism; I mean that we should conserve nature by developing higher but more benign technologies. And that, in effect, is what the contributors to this book are exploring: how to lessen our impact on the planet, how to live without a huge flow of resources to meet our needs, how to leave a smaller footprint.

This vision of a more efficient, lower-impact technology is a common denominator of many of the essays in this book. This new environmentalism encompasses architecture and planning, water systems and public health, and the use of innovative building materials. It demands new approaches to energy and transportation. It aims to preserve biodiversity. Ultimately, its realization depends on economics, education, and government based in an ethical concept of community not limited to humans but including the whole biosphere.

Richard White's discussion of hybridity reminds us that human technology and the natural, non-human environment are so intertwined as to be inseparable. There can be no turning back to a pristine natural paradise, nor can we ever fully control or predict the outcome of our best-laid plans. Yet White sets us the task of imagining new and better worlds.

The historians Robert Kargon and Arthur Molella remind us of utopian efforts of the 1930s in their stories of two planned techno-cities. Both Norris and Salzgitter reflected the social, environmental, economic, and political values of their time, and both drew on the knowledge, insights, and skills of respected professionals. And both were failures owing to "contingency," that chain of historical and ecological interconnections that White cautions us can never be fully controlled or predicted.

Timothy Davis gives us another historical example of urban planning in his essay on the evolution of American parkways. Here was a nineteenth-century attempt to re-create the natural landscape in the city while serving needs created by the introduction of the automobile. This yearning for the natural landscape is reiterated by Michael Robinson in his discussion of bioparks, which he sees as necessary both for maintaining biodiversity and for teaching the bioliteracy that is essential for an educated, informed populace in the twenty-first century. Jon Coe sees in the design of zoo exhibits a way to engender in people a respect for wildlife and an appreciation of biodiversity.

Erick Valle and his colleagues in the New Urbanism movement are breaking with traditions of "traditional urbanism" or "cultural urbanism." They draw on features that have evolved historically in successful communities—accessibility to shopping and services, so that cars are not necessary; economic diversity in neighborhoods; a reflection of local social patterns; and use of regional materials and vernacular architectural styles rather than imported ones. Their integration of the natural, the built, and the cultural environment is another manifestation of the new environmentalism.

Fundamental to the health and survival of any community are functional, affordable water supply and sewage systems. Martin Melosi has given us insights into the connections between belief systems, technologies, and health in England and the United States from the seventeenth century through the nineteenth. He has also revealed the extent to which our current water systems are based on outmoded and flawed science. Ashok Gadgil, in contrast, offers a state-of-the-art yet simple and inexpensive solution to the problem of water disinfection. His invention, now marketed under the name UVWaterworks, is an energy-efficient, compact, lightweight device that can rapidly disinfect unpressurized water, such as water that is carried from wells or rivers. For five cents of electricity per day, it can produce potable water for a village of up to 2,000 people. This is an example of the more benign technological solutions that I envision for the future. Gadgil's impetus to develop the apparatus was his compassionate response to the plight of cholera victims in one of the poorest areas of the world. It is this sort of ethical motivation that I also see as part of island civilization.

Kathryn Henderson and Marley Porter both have contributed essays about straw-bale architecture. Here again is a reversal of the twentieth-century trend toward ever more complex technologies. The return to a low-cost, natural material and the revival of a 100-year-old construction method prove not to be just a nostalgic evocation of a romanticized way of life, but actually provide high levels of energy efficiency as well as design options that other materials do not. Part of the appeal of straw-bale construction is the experience of community and connection one finds in working with others—a kind of direct connection that has nearly disappeared in our cell-phone society, but that might be regained in human communities of the future.

David Hertz has a different approach to materials in his ingenious recycling of consumer castoffs (nuts and bolts, vinyl records, golf tees) into products for architectural surface applications. This kind of recycling is one small aspect of the complex system of industrial ecology discussed by Christine Rosen and Braden Allenby. Amory Lovins, in his essay on end-use/least-cost planning, hypercars, and natural capitalism, touches on other issues that are integral to industrial ecology, such as the minimal use of materials and energy.

In the energy sector, Subhendu Guha gives us another fine example of higher but more benign technology. His invention of flexible solar shingles holds the promise of a low-cost, low-impact source of electricity that does not depend on a municipal power grid. And Thomas Lovejoy tells us that some corporations, such as BP and Shell, have redefined themselves as energy companies rather than fossil-fuel companies and are instituting environmentally sensitive practices.

It is in the area of industrial ecology that the paradigm shift of the new environmentalism is most apparent. Whereas environmental management of industry has to this point focused largely on the end products and impacts of manufacturing, industrial ecology moves to a higher view that encompasses the complex interrelationships of technology, culture, economics, and natural systems. It has led to such innovative programs as the waste exchanges among Danish companies that Rosen describes. But it goes beyond management of waste products to grapple with fundamental questions of institutions and values. The role of government—including politicians, bureaucrats, and the electorate—in regulating technology is brought to our attention by Rudi Volti in his case study of emission-control policy in California. There is always a high level of uncertainty in planning and implementing projects and policies, as Allenby reminds us. Allenby speaks of the need for continual learning at the personal and institutional levels and emphasizes the need for theological, ethical, and moral discussions on these issues. "What *could* be is a question, at least in part, for industrial society," he tells us. "What *should* be is one for religion and society."

In thinking about Earth, I am reminded of an encounter I once had with Bill Anders, an astronaut who was in the first manned spacecraft to orbit the moon. He told me that as he emerged from the dark side of the moon and saw Earth again, suspended in space, he held his thumb out in front of his eye and blotted out the entire planet. Now we are all familiar with that famous photo of the blue planet, and we can easily do the same thing with that image. But Anders had the experience, in effect, of making the whole planet, the real Earth, disappear in an instant. He said it made him, so totally dependent on the systems of his small spacecraft at that moment, profoundly understand how dependent we are on the systems of Earth. So I say, although it has become a cliché, we must care for this spaceship and all of its systems. We must save as much biodiversity as we can and give those systems a chance to survive and function. We seem to be accelerating into a murky future, and I am trying here to provide a few beacons to guide our technology in the next century and millennium. All of the areas examined in this book are among the great frontiers to be explored.

As a professor of history, I am hopeful that the younger generations may be developing a greater concern for Earth's systems. Some of my students once placed an ad in the personals column of a local newspaper's classified section:

Temperate but endangered planet, enjoys weather, photosynthesis, evolution, continental drift. Seeks caring relationship with intelligent life form.

Maybe that intelligent life form could be us, the sapient primate. Maybe our capacity to think, to invent, to imagine, to innovate, could truly be an asset rather than the liability it has become without ethical restraints. Maybe we could create a higher and gentler technology. Maybe we could prove to be that intelligent life form, in a caring relationship with a finite planet.

CONTRIBUTORS

BRADEN ALLENBY
vice president, Environment, Health and Safety, AT&T

JOYCE BEDI
historian, Lemelson Center for the Study of Invention and Innovation, National Museum of American History, Behring Center, Smithsonian Institution

JON C. COE
principal, CLRdesign, Inc.

MARTHA DAVIDSON
researcher and writer

DEVRA DAVIS
visiting professor, Heinz School for Public Policy and Management, Carnegie Mellon University

TIMOTHY DAVIS
lead historian, Park Historic Structures and Cultural Landscapes Program, National Park Service

ASHOK GADGIL
senior staff scientist, Environmental Energy Technologies Division, Lawrence Berkeley National Laboratory

SUBHENDU GUHA
president, United Solar Systems Corporation

KATHRYN HENDERSON
associate professor of sociology, Texas A&M University

DAVID HERTZ
AIA architect; president, Syndesis, Inc.

ROBERT KARGON
professor of history of science, Johns Hopkins University

ERIC LEMELSON
Lemelson Foundation

THOMAS LOVEJOY
president, H. John Heinz III Center for Science, Economics and the
Environment

AMORY LOVINS
co-founder, chief executive officer (research), and treasurer, Rocky Mountain Institute

MARTIN MELOSI
Distinguished University Professor of History, University of Houston

ARTHUR MOLELLA
director, Lemelson Center for the Study of Invention and Innovation,
National Museum of American History, Smithsonian Institution

RODERICK NASH
professor emeritus of history, University of California, Santa Barbara

MARLEY PORTER
architect and builder, Living Architecture and Construction Management,
Inc.

STEPHEN J. PYNE
professor, Biology and Society Program, Arizona State University

HARRY RAND
senior curator, Division of Cultural History, National Museum of American History, Smithsonian Institution

MICHAEL H. ROBINSON
former director, National Zoological Park, Smithsonian Institution

CHRISTINE MEISNER ROSEN
associate professor of business and public policy, University of California, Berkeley

ROBERT SOCOLOW
professor of mechanical and aerospace engineering, Princeton University

PAOLO SOLERI
co-founder, Cosanti Foundation

ERICK VALLE
architect; partner, Correa Valle Valle, Inc.

RUDI VOLTI
professor of sociology, Pitzer College

RICHARD WHITE
professor of history, Stanford University

INDEX